Computational Physics

Springer-Verlag Berlin Heidelberg GmbH

Karl Heinz Hoffmann
Michael Schreiber (Eds.)

Computational Physics

Selected Methods
Simple Exercises
Serious Applications

With 145 Figures, 15 Tables
and a 3.5″ MS-DOS Diskette

 Springer

Professor Dr. Karl Heinz Hoffmann
Professor Dr. Michael Schreiber
Institut für Physik
Technische Universität Chemnitz-Zwickau
D-09107 Chemnitz
Germany

Additional material to this book can be downloaded from http://extras.springer.com

Library of Congress Cataloging-in-Publication Data.

Hoffmann, K. H. (Karl Heinz), 1953– . Computational physics : selected methods, simple exercises, serious applications / K. H. Hoffmann, M. Schreiber. p. cm. Includes bibliographical references and index.

1. Physics–Problems, exercises, etc. – Methodology. 2. Numerical calculations. 3. Mathematical Physics. I. Schreiber, Michael, 1954– . II. Title.
QC32.H67 1996 530'.01'13–dc20 96-2830

ISBN 978-3-642-85240-4   ISBN 978-3-642-85238-1 (eBook)
DOI 10.1007/978-3-642-85238-1

© Springer-Verlag Berlin Heidelberg 1996
Softcover reprint of the hardcover 1st edition 1996

Typesetting: Camera-ready copy from the editors using a Springer TEX macro package
Cover design: Erich Kirchner, Heidelberg

SPIN 10525808   56/3144 – 5 4 3 2 1 0 – Printed on acid-free paper

Preface

Computational physics is the field in physics that has experienced probably the most rapid growth in the last decade. With the advent of computers, a new way of studying the properties of physical models became available. One no longer has to make approximations in the analytical solutions of models to obtain closed forms, and interesting but intractable terms no longer have to be omitted from models right from the beginning of the modeling phase. Now, by employing methods of computational physics, complicated equations can be solved numerically, simulations allow the solution of hitherto untractable problems, and visualization techniques reveal the beauty of complex as well as simple models. Many new and exciting results have been obtained by numerical calculations and simulations of old and new models.

This book presents samples of many of the facets that constitute computational physics. Our aim is to cover a broad spectrum of topics, and we want to present a mixture ranging from simple introductory material including simple exercises to reports of serious applications. This is not meant to be an introductory textbook on computational physics, nor is it a proceedings volume of a research conference. This book instead provides the reader with an overview of computational physics, its basic methods, and its many areas of application. Our coauthors lead the reader into new and "hot" topics of research, but the presentation does not require any specific knowledge of the topics and methods. We hope that a reader who has gone through the book can appreciate the wealth of computational physics and is motivated to proceed with further reading.

The topics covered in this book cover a wide spectrum, with a coarse division into "Monte Carlo" type and "molecular dynamics" type chapters. We start with discussing random numbers and their generation on computers. Then these random numbers are used in a variety of applications, which center around "Monte Carlo methods". In these applications the focus is first on classical systems in physics, chemistry, biology, material science, and optimization. Then quantum-mechanical problems are investigated by Monte Carlo procedures. On our way we also encounter quantum chaos and fractal concepts, which are of increasing importance nowadays. The transition from "Monte Carlo" to "molecular dynamics" occurs in the chapter on hybrid methods, which combine elements of both. Then "molecular dynamics" methods are presented, with fluids and solids covered. A chapter on finite-element methods follows, and the two final chapters present principles of parallel computers and associated programming models.

As usual in physics, only active interaction with the matter at hand provides deep insight, and thus we include a diskette that contains sample programs and demonstrations to support the interaction of the reader with the text. The sample programs and demonstrations are selected to provide a glimpse of current research activities, even though the limitations of the available hardware and/or the limited patience of some readers might require a reduction in the dimensionality or size of the application. Also some exercises are included to further foster an active use of this book.

The material in this book is born out of lectures the authors gave at a Heraeus Summer School on computational physics at the Technical University in Chemnitz. The aim of the summer school was the same as the aim of this book: to give a sampler of the field. Due to the gracious funding by the Dr. Wilhelm Heinrich Heraeus and Else Heraeus Foundation the editors (see figure) were able to present two weeks of intense lecturing and "learning by doing" to more than 80 students. We would like to use this opportunity to thank the Heraeus Foundation for making the summer school and this book possible.

But most important we like to thank our coauthors for their contributions to this volume (as well as for their lectures at the summer school). We very much appreciate their willingness to contribute even under the severe limitations that their everyday teaching and research activities (and administrative duties) put on their time. And finally we thank Jörg Arndt, Peter Blaudeck, Andre Fachat, Göran Hanke, Karin Kumm, Sven Schubert, and Peter Späht for their technical help and Springer-Verlag for making this volume a reality.

Chemnitz, December 1995

Karl Heinz Hoffmann and *Michael Schreiber*

With this original answer to the question "How to measure the height of the building of the Institut für Physik in Chemnitz with a computer and a stop watch only?" the editors give a peculiar interpretation of the topic "Physics with a computer".

Contents

---

* Software included on the accompanying diskette.

Quantum Dynamics in Nanoscale Devices*
Hans De Raedt

Quantum Chaos
Hans Jürgen Korsch and Henning Wiescher

Numerical Simulation in Quantum Field Theory*
Ulli Wolff

Modeling and a Simulation Method for Molecular Systems

Constraints in Molecular Dynamics, Nonequilibrium Processes in Fluids via Computer Simulations

Molecular-Dynamic Simulations
of Structure Formation in Complex Materials
Thomas Frauenheim, Dirk Porezag, Thomas Köhler, and Frank Weich . . 294

Finite Element Methods for the Stokes Equation
Jochen Reichenbach and Nuri Aksel 329

Random Number Generation*

Dietrich Stauffer

Institut für Theoretische Physik, Universität zu Köln, D-50923 Köln, Germany
e-mail: **stauffer@thp.uni-koeln.de**

Abstract. The sad situation of random number generation is reviewed: there are no good random numbers. But life has to go on anyhow, and thus we explain how to produce reasonable random numbers efficiently, emphasizing multiplication with 16807 and the Kirkpatrick–Stoll R250 generator.

1 Introduction

Molecular Dynamics and Monte Carlo are the two standard simulation methods of the last decades. Monte Carlo simulations use random numbers to produce random fluctuations. Today, they are no longer made at the roulette tables in Monaco, but on computers. In the good old days, people printed tables of random numbers from which the user could read them off. This, of course, is somewhat tedious when simulating a square lattice of size one million times one million, today's world record [1]. About a decade ago, computer chips became available which produced random numbers through the thermal noise of the electrons, about one number per microsecond. This is not fast enough for many quality applications. Besides, for testing purposes we would like to have *reproducible* random numbers: when we have made a program more efficient without changing the results, we want to run it again and indeed get exactly the same results, and not just roughly the same, within the statistical errors of the Monte Carlo simulations. Moreover, when we switch from one computer to another, we would like again to get the same results: portability is important. Thus special chips using thermal noise are not suitable for this purpose.

Also, the random numbers should be produced quickly since Monte Carlo simulations consume lots of time and we never have enough of it. Thus we need efficient methods, and on many computers it is very slow to call a function or subroutine to produce one random number. Thus a good random number generator should be:
- (1) random
- (2) reproducible
- (3) portable
- (4) efficient

Using the built-in random number generator of your computer can make your program inefficient and nonportable. (Seymour Cray knew what he was doing:

---

* Software included on the accompanying diskette.

his random number generators for the good old CDC series or modern Crays were efficient.) Besides, the user then does not understand what is going on.

Thus we now review why the above criteria are difficult to fulfill and what to do about it, by programming your own random numbers.

2   The Miracle Number 16807

Linear congruential random number generators multiply the last random integer by some big factor, add another integer to it, treat the sum modulo some power of two, and normalize this integer to the interval between zero and unity. This all sounds very complicated, sometimes is presented in this complicated fashion in the literature, and may cause you to give up programming your own random numbers. Thus simply forget these complications and look at the following Fortran or Basic statement, which works for most 32-bit machines:

```
IBM = IBM*16807
```

(fans of Pascal and C should end this line with a semicolon; and enemies of International Bussiness Machines may use a different variable name). If you start with an odd integer for IBM, e.g., through IBM = 2*ISEED-1, then this single program line should give you, again and again, integers IBM distributed randomly between -2^{31} and 2^{31}. Just try it out. Why does it work ?

If you multiply two ten-digit integers, the result will be an integer with about twenty digits and is difficult to obtain by paper and pencil. You may estimate, however, the leading digit correctly without too much effort. On a computer, you may not be able to store more then ten digits for each integer. Then most computers simply throw away, without any error message, the somewhat predictable leading digits and keep only the ten least significant digits. Of course, computers work with binary digits (bits) and not with decimal ones, and with four-byte integers (32 bits) all leading bits beyond the least significant 32 bits are thrown away if the product of two integers has more than 32 bits. In terms of decimal numbers restricted to be at most 999, this would mean that the product of 123 and 899 is not 110577 but merely 577. It is clear that these least significant digits are difficult to predict, that means for a user they look pretty random.

In your youth you have learned that a*b equals b*a, and that the product of two positive numbers is again positive. In linear algebra or quantum mechanics you found out that the first statement was a lie, and now you realize the same for the second statement: IBM*16807 may be negative even when IBM was positive. The reason is that the first (most significant) bit of an integer indicates the sign. Thus before the leading bits of the product were thrown away, the product was positive; but then only the last 32 bits were kept, and the leftmost (most significant) bit may be zero (positive 32-bit number) or one (negative 32-bit number). So, plus times plus is minus, in about half the cases.

Some ancient DEC computers may not have liked this overflow above the 32-bit limit, but otherwise I am not aware of computers where the above Fortran statement causes trouble. Thus we have not only an efficient one-line random number generator, but also a portable one.

If for some reason you want only positive random numbers IBM, then you have to add 2^{31} to them if they are negative. This number 2^{31} is too large to handle for the 32-bit computer, but $2^{31} - 1 = 2147483647$ is fine. Thus try

    IF(IBM.LT.0) IBM = IBM + 2147483647 + 1

and it works if the computer is too stupid to find out that you really want to add 2^{31}.

If you want to normalize this number to the interval between 0 and 1, you multiply a positive random integer by $2^{-31} = 4.656612 \times 10^{-10}$. If they are both positive and negative, use

    Z = 0.5+2.328306 E-10 * IBM

to get a random number z between 0 and 1. Of course, this normalization from an integer to a real number costs a lot of computer time; you could do it faster by learning how a floating point number is stored in your computer and then constructing one via bit operations treating the random integer as a bit string for the mantissa.

However, in most cases this normalization is not needed, and you may stay within integer arithmetic. For example, some command GOTO 1 should be executed with probability p. Normally this is done with

    IF(Z.LE.P) GOTO 1

requiring a random number between 0 and 1. This normalization is avoided by

    IF(IBM.LE.IP) GOTO 1

provided you have defined once (and not millions of times, i.e., for each random number) the variable IP = 2147483648.0*(2.0*P-1.0)

    IF(P.GT.0.999) IP=2147483647

    IF(P.LT.0.001) IP=-2147483648

which varies between -2^{31} and 2^{31}. Now the computer runs faster. (The last two "if" statements are precautions, seldomly needed, in case rounding errors cause trouble in the conversion to integers if $p = 0$ or $= 1$.)

The number $16807 = 7^5$ is not entirely arbitrary; historically earlier was 65539, and 65549 has also been used. So you may mix them, using in most of your program lines multiplication with 16807, but sometimes also 65539. Do not try to produce different samples just by changing the multiplicator from 16807 to 16809, then 16811, and so on. Also, your IBM numbers must always be odd integers; to be safe I start with an integer ISEED and then state once IBM = 2*ISEED-1, as mentioned already above.

If you simulate at zero temperature (see Sect. 4), then the probabilities are 0, 1/2, and 1 only. With integer random numbers IBM varying between -2^{31} and $+2^{31}$ the conditions and Boltzmann integers then have to be formulated exactly as stated above (not IBM .LT. IP for example), to avoid a spin flipping when it should not flip. With floating point random numbers you need double precision real*8. This detail may be important for the fraction of frozen spins in spin glasses [7].

3   Bit Strings of Kirkpatrick–Stoll

The principle of Tausworth shift generators has been around for a long time but physicists started to use it mainly after Kirkpatrick and Stoll made it popular in a physics journal [2]. For many years it was regarded as superior to multiplication with 16807; this is no longer true but at least it offers a completely different alternative. It requires bit-manipulation functions which had not yet been standardized before Fortran 90 (they are part of the C language standard) and which were initially demanded by the Pentagon for signal analysis.

Imagine you have two 32-bit integers M and N. Then the exclusive-or operation puts a bit equal to one if and only if the two corresponding bits in M and N are different; otherwise exclusive-or puts this bit to zero. Thus this bit-by-bit exclusive-or IEOR(M,N) (Cray called it M.xor.N but unfortunately this did not become the standard) treats 32 bits in parallel and does for each of these bits what a logical operation would do for one bit only with logical (Boolean) variables. Obviously such bit-handling operations can be used in lots of problems where the essential information consists of independent bits, such as in Ising models or cellular automata where it is called multispin coding [3]. Fortran manuals usually hide these tricks in an appendix on the functions which the compiler has stored.

Imagine you have an array of 250 integers N consisting of completely random bits. Then the next integer N(251) is produced via N(251) = IEOR(N(1), N(148)) and generally

N(K) = IEOR(N(K-250), N(K-103))

where again 250 and 103 are magic numbers which should not be changed. An alternative choice is the simple subtraction

N(K) = N(K-250) - N(K-103)

but this is less widespread than the exclusive-or method, also called R250.

To work with it we first need 250 random integers. It is not recommended to take the results of IBM*16807 directly as such integers since the last bits are not random enough; for example the least significant bit is always one since IBM is always odd. Instead we set a bit in N equal to one if and only if the result of IBM*16807 is negative. Thus our 32-bit integers N are initialized through

```
DO K = 1, 250
ICI=0
DO I=1,32
ICI=ISHFT(ICI,1)
IBM=IBM*16807
IF(IBM.LT.0) ICI=ICI+1
ENDDO
N(K)=ICI
ENDDO
```

Here again ISHFT is a bit-manipulation function shifting the first argument by one bit to the left. Instead of ICI=ICI+1, one could also have used a bit-by-bit or-function ICI=IOR(ICI,1); on most compilers integer and bitstring operations can be mixed. (Cray vector computers may even have separate hardware for addition and bit handling, and then run faster if one line contains a mixture of bit and arithmetic operations since now the various hardware parts work in parallel.)

Had we used R250 as described the above we would need a huge array N(K) of random integers. To avoid this explosion of memory demand it is practical to recycle the indices k; that means to start counting again with $k = 1, 2, 3$ when possible. So we could treat k, $k - 103$, $k - 250$ modulo 250 but we would waste time. It is usually faster to treat these three numbers modulo 256 since 256 is a power of two. Modulo 256 thus means " take the last eight bits" and is realized by a bit-by-bit AND operation with 255:

```
N(IAND(255, K)) = IEOR(N(IAND(255, K-250)),N(IAND(255, K-103)))
```

Now k can run from one to "infinity" while the indices of the array N vary only between 0 and 255, saving memory. Note, however, that we require

DIMENSION N(0:255)

instead of DIMENSION N(256) at the beginning of the program.

4   A Modern Example

Let us now discuss a modern example, besides the typical random walk, percolation, and Ising models. In 1994 Derrida et al [4] found that nontrivial and at that time unexplained exponents govern the relaxation of the one-dimensional Ising model into equilibrium, if initially all spins are randomly oriented and if the absolute temperature is set to zero in a Glauber (heat bath) simulation: "spinodal decomposition at $T = 0$". The number of spins which have never flipped decreases to zero asymptotically proportional to $(\text{time})^{-\theta}$ in one, two and three dimensions. Even though the one-dimensional Ising model was found by Ernst Ising in 1925 not to have a phase transition at finite temperatures, here at $T = 0$ it has an unexpected exponent θ which was estimated within months by better and better simulations as being close to 3/8 until this result was proven analytically also by Derrida et al. This is one of the rare cases where computer simulations indicated a new effect which thereafter could be explained theoretically. (The behavior in higher dimensions is not yet explained.)

So we fill a one-dimensional array IS randomly with +1 and −1. Then we go again and again regularly through the lattice and leave every spin as it is if its two neighbors on the chain have the same value as the "spin" IS(I). If both neighbors have the orientation opposite to the center spin, this center spin is always flipped into the orientation of its neighbors. If the two neighbors differ, the center spin does not know whom to follow and flips with probability 1/2. If we imagine our variables IS to be ferromagnetic spins, then we never flip if this would increase the energy, always if this decreases the energy, and with probability 1/2 if the energy remains unchanged.

Thats all; a simple program is listed here.

```
c        one-dimensional spinodal decomposition in T=0 Ising model
c        (Derrida et al)
         parameter(L=10000, Lp1=L+1)
         dimension is(0:Lp1),never(0:Lp1)
         byte is,never
         data max/1000/,iseed/1/
         print *, L,max,iseed
         ibm=2*iseed-1
         do 1 i=0,Lp1
           never(i)=1
           ibm=ibm*16807
           is(i)=1
c          random initialization: half up and half down
 1       if(ibm.lt.0) is(i)=-1
         do 2 itime=1,max
           do 3 i=1,L
             ien=is(i)*(is(i-1)+is(i+1))
             if(ien.gt.0) goto 3
             ibm=ibm*16807
             if(ien.lt.0.or.ibm.lt.0) then
c            flip spin if this lowers E, or
c            with probability 1/2 if E=const
               is(i)=-is(i)
               never(i)=0
             endif
 3         continue
           m=0
           n=0
           do 4 i=1,L
            m=m+is(i)
c          computer magnetization m and fraction of
c          never flipped spins
 4          n=n+never(i)
 2       print *, itime,n,m
         stop
         end
```

5   Problems

A famous and particularly simple example to prove that random numbers are
not always random enough is the exercise of emptying a cube. Let is(i,j,k)
be a $L \times L \times L$ array initially filled with one's. During each iteration, we select
randomly L^3 times first an i, then a j, and finally a k as coordinates, and

empty the corresponding site, i.e., we set is(i,j,k) = 0 if it was still occupied before. Theoretically, the number of still occupied sites should decay as $\exp(-t)$ in t iterations. Using multiplication with 65539, we indeed find good results for $L = 10$, but already at $L = 20$ serious deviations are found after a few interations: a quarter of the sites are not reached. This effect is seen within a few seconds at most and is not a slight deviation observed only in high-quality computations.

We can avoid this problem by replacing 65539 with 16807 (in at least one of the three appearances), or by using one-dimensional storage: instead of three indices running each from 1 to L, we use one running from 1 to L^3. This trick is generally recommended since it speeds up the simulation, particularly on vector computers for small L. In our case, the correlations causing our difficulties no longer disturb us since each site now requires only one random number and no longer three.

Until recently, multiplication with 16807 was regarded as simpler but less reliable than Kirkpatrick–Stoll. In 1992, a group at University of Georgia (USA) [5] found errors they called "dramatic" in simulations of the two-dimensional Ising model, using Kirkpatrick–Stoll, whereas things worked well with *16807. Even the New York Times reported on it at length. With all due respect to the world's best newspaper, I cannot find these deviations by less than one percent in the energy very dramatic. However, further investigations by other groups confirmed that R250 leads to difficulties which are avoided with *16807. There are correlations between random numbers which are 250 iterations apart. If one uses larger numbers than the above pair (103,250) then these correlations show up at the corresponding longer intervals, and thus in general are less disturbing. No difficulties of this type are yet known to me for Ziff's random number generator [6] which combines four random integers (from an array of nearly 10 000) via exclusive-ors to produce the next random integer.

However, *16807 may have even more dramatic errors, ignored by the New York Times. Using the above-mentioned one-dimensional storage to simulate a five-dimensional Ising model right at its critical point, with initially all spins up, I found the magnetization to deviate after only 20 iterations by a factor of two from the correct value if the linear lattice size is $L = 32$. For $L = 31$ or $L = 33$ the effect vanishes. (Also in three dimensions at $L = 128$ I had problems.) Apparently there are some correlations between random numbers separated by a power of two in the number of random number generations.

Thus the sad truth is that there is no good and simple random number generator. What is good for one problem may be bad for another one. The best way is not to rely on some mathematical tests which a generator is said to have passed successfully, but to test it for your particular application. Also, use widespread generators like *16807 and R250, and not some new one claimed by somebody to be excellent: it may have some hidden errors which have not yet been found simply because it is new and not widespread.

So I stopped worrying and started to love *16807, relying on the famous last words of the movie "Some like it hot."[1] I avoid powers of two for L, prefer prime numbers for L, vary the lattice size to check for suspicious deviations and use one-dimensional storage. Moreover, outside the inner loop I may produce one extra time a new random number and throw it away, to destroy periodicities; some people even throw away most of their random numbers and use only a small fraction of them. Repetitions of random numbers can also be avoided if outside the innermost loop some intermediate result of the simulation is used to modify the random numbers, e.g., by IBM = IBM*(2*M+1) when M is the magnetization. (This helped M. Siegert for large 2D Ising models. After about 10^9 iterations, *16807 alone will produce exactly the same sequence of random numbers.)

Then, to check my results, I may replace a few of the 16807 by 65539, or use R250 instead. If nothing changes within the statistical errors then I am satisfied; otherwise I should worry and try the Ziff complication. In other words, its not an exact science.

6   Summary

You may feel disappointed because of these unreliable foundations of Monte Carlo methods. However, in most cases, published results which turned out to be wrong were erroneous not because of random numbers or programming errors but because of systematic errors due to finite relaxation times in finite lattices. And the fact that people today argue about the sixth digit in the value $J/k_B T_c = 0.22165\ldots$ of the critical point in the 3D Ising model suggests that inspite of all these difficutlies, high-precision Monte Carlo studies are possible.

References

[1] A. Linke, D.W. Heermann, and P. Altevogt, Comp. Phys. Comm. **90**, 66 (1995); Physica A, in press
[2] S. Kirkpatrick and E.P. Stoll, J. Comput. Phys. **40**, 517 (1981)
[3] D. Stauffer, J. Phys. A **24**, 909 (1991)
[4] B. Derrida, V. Hakim, and V. Pasquier, Phys. Rev. Lett. **75**, 751 (1995); D. Stauffer, Physics World, Nov. 1995, p. 26
[5] A.M. Ferrenberg, D.P. Landau, and Y.J. Wong, Phys. Rev. Lett. **69**, 3382 (1992)
[6] R.M. Ziff, Phys. Rev. Lett. **69**, 2670 (1992), footnote 15
[7] U. Gropengiesser, Physica A **220**, 239 (1995); N. Jan, priv. comm.

---

[1] Note to editors: "Nobody is perfect."

A Few Exercises with Random Numbers

Peter Blaudeck

Institut für Physik, Technische Universität, D-09107 Chemnitz, Germany
e-mail: blaudeck@physik.tu-chemnitz.de

1. Generate sequences of real random numbers:
   (a) following the algorithm

$$k_{n+1} = (ck_n)(\mathrm{mod}M) \quad ,$$
$$r_n = \frac{k_n}{M} \quad ,$$

   with the parameters

$$M = 2^7, \qquad c = 5, \qquad k_0 = 1 \quad ;$$

   (b) using the "16807 generator".

2. Using any kind of graphics software at hand, draw points with coordinates
   (r from exercise 1)

$$x = r_{2i-1} \quad ,$$
$$y = r_{2i}, \qquad i = 1, 2, 3, \dots \quad ,$$

   into a cartesian coordinate system. What can be concluded from the distribution of the points according to the correlation of consecutive numbers and
   the quality of the generators?

3. Verify that also the "16807 generator" produces correlations by magnifying
   cut-outs of the graphics in exercise 2, for example the square $0 \le x \le \frac{1}{1000}$,
   $0 \le y \le \frac{1}{1000}$.

Monte Carlo Simulations of Spin Systems[*]

Wolfhard Janke

Institut für Physik, Johannes Gutenberg-Universität, D-55099 Mainz, Germany
e-mail: janke@miro.physik.uni-mainz.de

Abstract. This chapter gives a brief introduction to Monte Carlo simulations of classical $O(n)$ spin systems such as the Ising ($n = 1$), XY ($n = 2$), and Heisenberg ($n = 3$) models. In the first part I discuss some aspects of the use of Monte Carlo algorithms to generate the raw data. Here special emphasis is placed on nonlocal cluster update algorithms which proved to be most efficient for this class of models. The second part is devoted to the data analysis at a continuous phase transition. For the example of the three-dimensional Heisenberg model it is shown how precise estimates of the transition temperature and the critical exponents can be extracted from the raw data. I conclude with a brief overview of recent results from similar high-precision studies of the Ising and XY models.

1 Introduction

The statistical mechanics of complex physical systems pose many hard problems which are difficult to solve by analytical approaches. Numerical simulation techniques will therefore be indispensable tools on our way to a better understanding of systems like (spin) glasses, disordered magnets, or proteins, to mention only a few classical problems. Quantum statistical problems in condensed matter or the broad field of elementary particles and quantum gravity in high-energy physics would fill many other volumes such as this.

The numerical tools can roughly be divided into molecular dynamics (MD) and Monte Carlo (MC) simulations. With the ongoing advances in computer technology both approaches are expected to gain even more importance than they already have today. In the past few years the predictive power of the MC approach in particular was considerably enhanced by the development of greatly improved simulation techniques. Not all of them are already well-enough understood to be applicable to really complex physical systems. But as a first step it is gratifying to see that at least for relatively simple spin systems orders of magnitude of computing time can be saved by these refinements. The purpose of this lecture is to give a brief overview on what is feasible today.

From a theoretical view spin systems are also of current interest since on the one hand they provide the possibility of comparing completely different approaches such as field theory, series expansions, and simulations, and on the other hand they are the ideal testing ground for conceptual considerations such as universality or finite-size scaling. And last but not least they have found a revival in slightly disguised form in quantum gravity and conformal field theory,

---

[*] Software included on the accompanying diskette.

where they serve as idealized "matter" fields on Feynman graphs or fluctuating manifolds.

The rest of this chapter is organized as follows. In the next section I first recall the definition of $O(n)$ spin models and the definition of standard observables such as the specific heat and the susceptibility. Then some properties of phase transitions are summarized and the critical exponents are defined. In Sect. 3, Monte Carlo methods are described, and Sect. 4 is devoted to an overview of reweighting techniques. In Sect. 5, applications to the three-dimensional classical Heisenberg model are discussed, and in Sect. 6, I conclude with a few comments on similar simulations of the Ising and XY models.

2   Spin Models and Phase Transitions

2.1   Models and Observables

In the following we shall confine ourselves to $O(n)$ symmetric spin models whose partition function is defined as

$$Z_n(\beta) = \sum_{\{\sigma_i\}} \exp(-\beta H_n) \ , \tag{1}$$

with

$$H_n = -J \sum_{\langle ij \rangle} \sigma_i \cdot \sigma_j; \quad \sigma_i = (\sigma_i^{(1)}, \sigma_i^{(2)}, \ldots, \sigma_i^{(n)}); \quad |\sigma_i| = 1 \ . \tag{2}$$

Here $\beta = 1/k_B T$ is the inverse temperature, the spins σ_i live on the sites i of a D-dimensional cubic lattice of volume $V = L^D$, and the symbol $\langle ij \rangle$ indicates that the lattice sum runs over all 2D nearest-neighbor pairs. We always assume periodic boundary conditions.

Standard observables are the internal energy per site, $e = E/V$, with $E = -\mathrm{d} \ln Z_n / \mathrm{d}\beta \equiv \langle H_n \rangle$, and the specific heat,

$$C/k_B = \frac{\mathrm{d}e}{\mathrm{d}(k_B T)} = \beta^2 \left(\langle H_n^2 \rangle - \langle H_n \rangle^2 \right) /V \ . \tag{3}$$

On finite lattices the magnetization and susceptibility are usually defined as

$$m = M/V = \langle |\sigma_{\mathrm{av}}| \rangle; \quad \sigma_{\mathrm{av}} = \sum_i \sigma_i /V \ , \tag{4}$$

$$\chi = \beta V \left(\langle \sigma_{\mathrm{av}}^2 \rangle - \langle |\sigma_{\mathrm{av}}| \rangle^2 \right) \ . \tag{5}$$

In the high-temperature phase one often employs the fact that the magnetization vanishes in the infinite volume limit and defines

$$\chi' = \beta V \langle \sigma_{\mathrm{av}}^2 \rangle \ . \tag{6}$$

Similarly, the spin-spin correlation function can then be taken as

$$G(\mathbf{x}_i - \mathbf{x}_j) = \langle \sigma_i \cdot \sigma_j \rangle \ . \tag{7}$$

At large distances, $G(\mathbf{x})$ decays exponentially and the correlation length ξ can be defined as

$$\xi = -\lim_{|\mathbf{x}|\to\infty} |\mathbf{x}|/\ln G(\mathbf{x}) \ . \tag{8}$$

For $n = 1$, the partition function $Z_n(\beta)$ describes the standard Ising model where the spins can take only the two discrete values, $\sigma_i \equiv \sigma_i^{(1)} = \pm 1$. For $n = 2, 3, \ldots$ the spins vary continuously on the n-dimensional unit sphere. Particularly thoroughly studied cases are $n = 2$ (XY model) and $n = 3$ (Heisenberg model). The limit $n \longrightarrow \infty$ is known to be equivalent to the spherical model of Berlin and Kac [1]. In three dimensions (3D) the model exhibits for all n a continuous phase transition from an ordered low-temperature phase to a disordered high-temperature phase. The associated critical exponents are generic for the so-called $O(n)$ universality classes. In two dimensions (2D) the situation is little more complex. For the 2D Ising model the exact solution by Onsager and later Yang predicts a continuous order-disorder phase transition similar to 3D. For all $n \geq 2$, however, the spin degrees of freedom are continuous and, as a consequence of the Mermin–Wagner–Hohenberg theorem, the magnetization vanishes for all temperatures. The 2D XY model nevertheless displays a very peculiar (infinite order) Kosterlitz–Thouless transition [2, 3, 4]. Due to the $O(2)$ symmetry this model admits point like topological defects (vortices) which are tightly bound to pairs at low temperatures. With increasing temperature isolated vortices are entropically favored, and the transition is usually pictured as the point where vortex pairs start to dissociate: for a review see, e.g., Kleinert [5]. For the 2D Heisenberg model and all other 2D $O(n)$ models with $n > 3$, on the other hand, it is commonly believed that there is no phase transition at finite temperature.[1]

For later reference we also recall another generalization of the Ising model, the q-state Potts model [7] with Hamiltonian

$$H_{\text{Potts}} = -J\sum_{\langle ij\rangle} \delta_{\sigma_i\sigma_j}; \quad \sigma_i \in 1,\ldots,q \ . \tag{9}$$

This generalization has in 3D for all $q \geq 3$ a first-order transition and in 2D it is exactly known to exhibit a second-order transition for $q \leq 4$ and a first-order transition for all $q \geq 5$ [8, 9].

2.2 Phase Transitions

In limiting cases such as low and high temperatures (or fields, pressure, etc.) the physical degrees of freedom usually decouple and the statistical mechanics of even complex systems become quite manageable. Much more interesting is the region in between these extremes where strong cooperation effects may cause phase transitions, e.g., from an ordered phase at low temperatures to a disordered phase at high temperatures. The prediction of properties of this most difficult region

---

[1] For an alternative view see, however, [6] and references to earlier work therein.

of a phase diagram as accurately as possible is the most challenging objective of all statistical mechanics approaches, including numerical simulations.

The theory of phase transitions is a very broad subject described comprehensively in many textbooks. Here we shall be content with a rough classification into *first-order* and *second-order* (or, more generally, continuous) phase transitions, and a summary of those properties that are most relevant for numerical simulations. Some characteristic properties of first- and second-order phase transitions are sketched in Fig. 1.

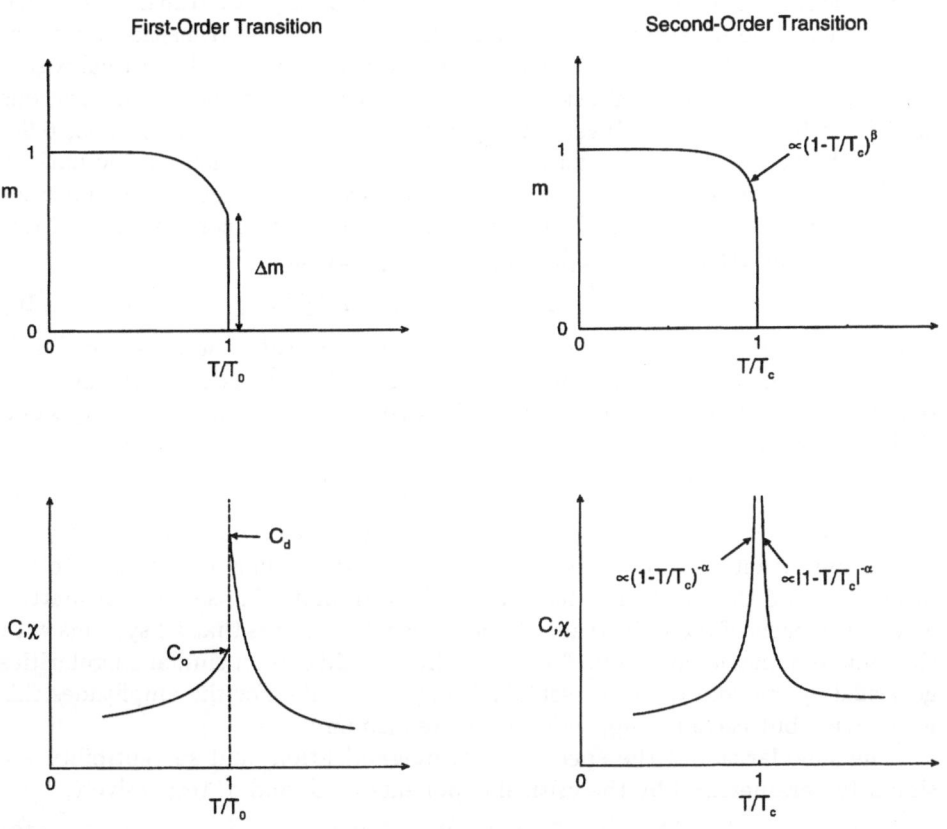

Fig. 1. The characteristic behavior of the magnetization, m, specific heat, C, and susceptibility, χ, at first- and second-order phase transitions

Most phase transitions in nature are of first order [10, 11, 12, 13]. The best-known example is the field-driven transition in magnets at temperatures below the Curie point, and the paradigm of a temperature-driven first-order transition experienced every day is ordinary melting [14, 15]. In general, first-order phase transitions are characterized by *discontinuities* of the order parameter (the jump

Δm of the magnetization m in Fig. 1), or the energy (the latent heat Δe), or both. This reflects the fact that, at the transition temperature T_0, two (or more) phases can coexist. In the example of a magnet at low temperatures the coexisting phases are the phases with positive and negative magnetization, whereas at the melting transition they are the solid (ordered) and liquid (disordered) phases. The correlation length in the coexisting pure phases is finite.[2] Consequently the specific heat, C, and the susceptibility, χ, also do not diverge in the pure phases. Mathematically there are, however, superimposed delta function-like singularities associated with the jumps of e and m.

In this chapter we will mainly consider second-order phase transitions, which are characterized by a *divergent* correlation length ξ at the transition temperature $T_c \equiv 1/\beta_c$. The growth of correlations as one approaches the critical region from high temperatures is illustrated in Fig. 2, where six typical configurations of the 2D Ising model at inverse temperatures $\beta/\beta_c = 0.50$, 0.70, 0.85, 0.90, 0.95, and 0.98 are shown. Because for an infinite correlation length fluctuations on all length scales are equally important, one expects power-law singularities in thermodynamic functions. The leading singularity of the correlation length is usually parametrized in the high-temperature phase as

$$\xi = \xi_{0+}|1 - T/T_c|^{-\nu} + \dots \quad (T \geq T_c) , \tag{10}$$

where the ... indicate subleading corrections (analytical as well as confluent). This defines the critical exponent ν and the critical amplitude ξ_{0+} on the high-temperature side of the transition. In the low-temperature phase one expects a similar behavior,

$$\xi = \xi_{0-}(1 - T/T_c)^{-\nu} + \dots \quad (T \leq T_c) , \tag{11}$$

with the same critical exponent ν but a different critical amplitude $\xi_{0-} \neq \xi_{0+}$.

An important feature of second-order phase transitions is that due to the divergence of ξ the short-distance details of the Hamiltonian should not matter. This is the basis of the universality hypothesis which states that all systems with the same symmetries and same dimensionality should exhibit similar singularities governed by one and the same set of critical exponents. For the amplitudes this is not true, but certain amplitude ratios are also universal.

The singularities of the specific heat, magnetization, and susceptibility are similarly parametrized by the critical exponents α, β, and γ, respectively,

$$C = C_{\text{reg}} + C_0|1 - T/T_c|^{-\alpha} + \dots , \tag{12}$$

$$m = m_0(1 - T/T_c)^{\beta} + \dots , \tag{13}$$

$$\chi = \chi_0|1 - T/T_c|^{-\gamma} + \dots , \tag{14}$$

where C_{reg} is a regular background term, and the amplitudes are again different on the two sides of the transition, cf. Fig. 1.

---

[2] For the 2D q-state Potts model with $q \geq 5$, where many exact results are known, this is illustrated by the recent simulations of Janke and Kappler [16, 17, 18, 19, 20]; for details see [21].

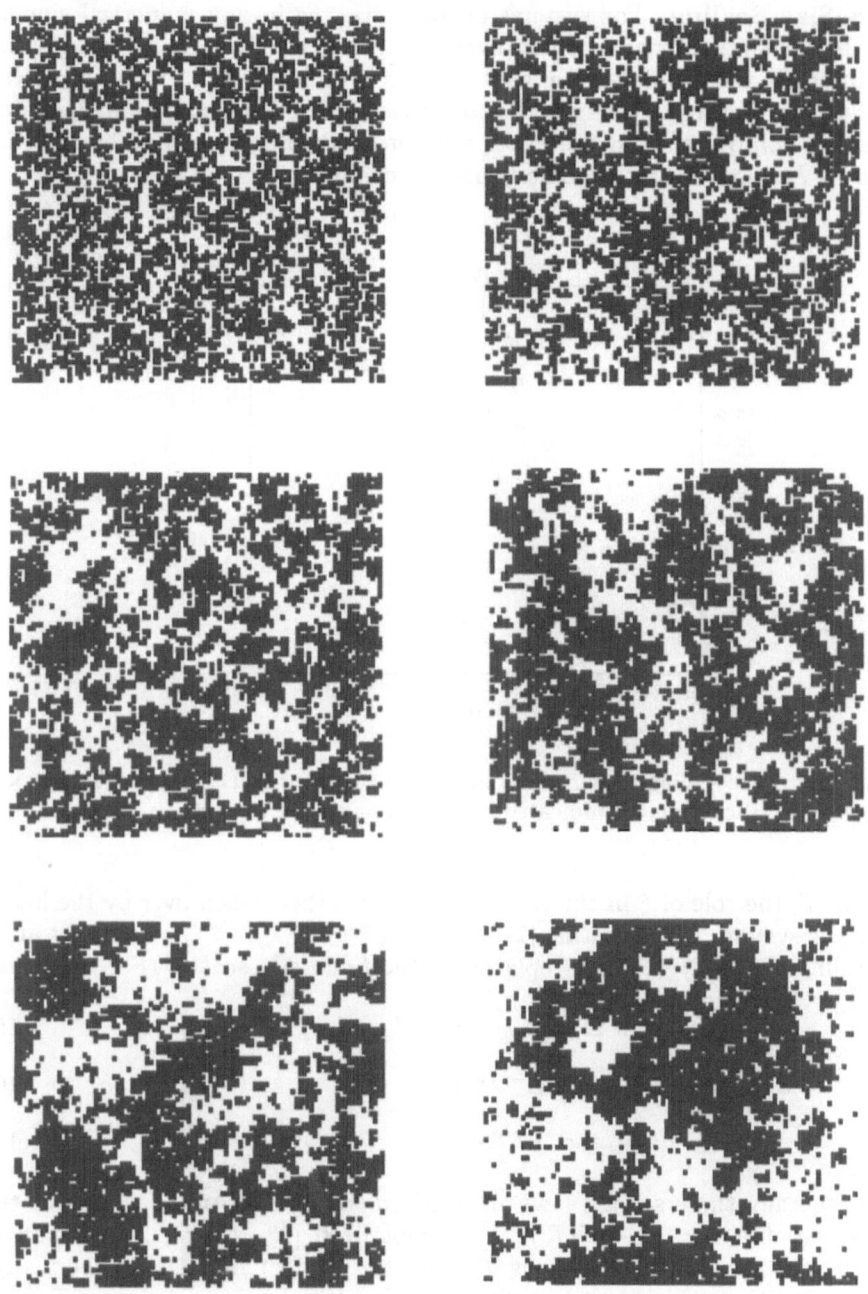

Fig. 2. The growth of correlations from high temperatures (*upper left*) to the critical region (*lower right*), characterized by large spatial correlations. Shown are actual 2D Ising configurations for a 100×100 lattice at $\beta/\beta_c = 0.50$, 0.70, 0.85, 0.90, 0.95, and 0.98

Finite-Size Scaling. For systems of finite size, as in any numerical simulation, the correlation length cannot diverge, and also the divergencies in all other quantities are then rounded and shifted. This is illustrated in Fig. 3, where the specific heat of the 2D Ising model on various $L \times L$ lattices is shown. The curves are computed from the exact solution of Ferdinand and Fisher [22] for any $L_x \times L_y$ lattice with periodic boundary conditions.

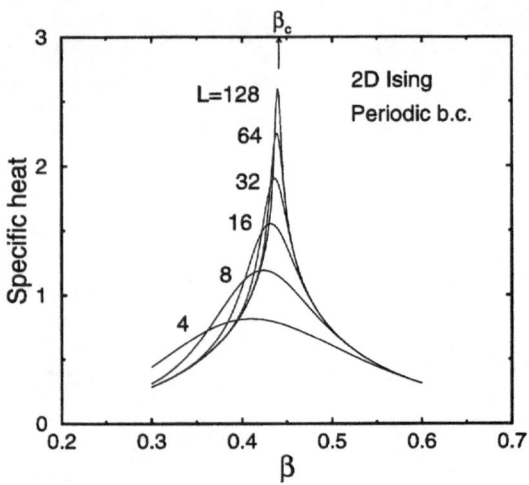

Fig. 3. Finite-size scaling behavior of the specific heat of the 2D Ising model on $L \times L$ lattices. The critical point is indicated by the *arrow* on the top axis

Near T_c the role of ξ in the scaling formulas is then taken over by the linear size of the system, L. By writing $|1 - T/T_c| \propto \xi^{-1/\nu} \longrightarrow L^{-1/\nu}$, we see that at T_c the scaling laws (12)–(14) are replaced by the *finite-size scaling* (FSS) ansätze,

$$C = C_{\mathrm{reg}} + aL^{\alpha/\nu} + \dots \ , \tag{15}$$

$$m \propto L^{-\beta/\nu} + \dots \ , \tag{16}$$

$$\chi \propto L^{\gamma/\nu} + \dots \ . \tag{17}$$

More generally these scaling laws are valid in the vicinity of T_c as long as the scaling variable $x = (1 - T/T_c)L^{1/\nu}$ is kept fixed [23, 24, 25, 26]. In particular this is true for the locations T_{max} of the (finite) maxima of thermodynamic quantities, which are expected to scale with the system size as $T_{\mathrm{max}} = T_c(1 - x_{\mathrm{max}}L^{-1/\nu} + \dots)$. In this more general formulation the scaling law for, for example, the susceptibility reads $\chi(T, L) = L^{\gamma/\nu} f(x)$. By plotting $\chi(T, L)/L^{\gamma/\nu}$ vs the scaling variable x, one thus expects that the data for different T and L would fall onto a kind of master curve. While this is a nice way to demonstrate the scaling properties qualitatively, it is not particularly suited for quantitative analyses.

Since the goals of most simulation studies of spin systems are high-precision estimates of the critical temperature and the critical exponents, one therefore prefers fits either to the "thermodynamic" scaling laws (12)–(14) or to the FSS ansätze (15)–(17).

Similar considerations for first-order phase transitions also show that the delta-function-like singularities originating from the phase coexistence are smeared out for finite systems [27, 28, 29, 30, 31]. In finite systems they are replaced by narrow peaks whose height (width) grows proportional to the volume (1/volume) [32, 33, 34, 35, 36, 37, 38].

3   The Monte Carlo Method

Let us now discuss how the expectation values in (3)–(7) can be computed numerically. A direct summation of the partition function is impossible, since even for the Ising model with only two possible states per site the number of terms would be enormous: $2^{2500} \approx 10^{753}$ for a 50×50 lattice! Also a naive random sampling of the spin configurations does not work. Here the problem is that the relevant region in the high-dimensional phase space is relatively narrow and hence too rarely hit by random sampling. The solution to this problem has been known for a long time. One has to use the *importance sampling* technique [39] which is designed to draw configurations according to their Boltzmann weight,

$$P^{\text{eq}}[\{\sigma_i\}] \propto \exp\left(-\beta H[\{\sigma_i\}]\right) \ . \tag{18}$$

In more mathematical terms one sets up a Markov chain,

$$\dots \xrightarrow{W} \{\sigma_i\} \xrightarrow{W} \{\sigma'_i\} \xrightarrow{W} \{\sigma''_i\} \xrightarrow{W} \dots \ ,$$

with a transition operator W satisfying the conditions

(a) $W(\{\sigma_i\} \longrightarrow \{\sigma'_i\}) > 0$ for all $\{\sigma_i\}, \{\sigma'_i\}$, $\tag{19}$

(b) $\sum_{\{\sigma'_i\}} W(\{\sigma_i\} \longrightarrow \{\sigma'_i\}) = 1$ for all $\{\sigma_i\}$, $\tag{20}$

(c) $\sum_{\{\sigma_i\}} W(\{\sigma_i\} \longrightarrow \{\sigma'_i\}) P^{\text{eq}}[\{\sigma_i\}] = P^{\text{eq}}[\{\sigma'_i\}]$ for all $\{\sigma'_i\}$. $\tag{21}$

From (21) we see that P^{eq} is a fixed point of W. A somewhat simpler sufficient condition is detailed balance,

$$P^{\text{eq}}[\{\sigma_i\}]W(\{\sigma_i\} \longrightarrow \{\sigma'_i\}) = P^{\text{eq}}[\{\sigma'_i\}]W(\{\sigma'_i\} \longrightarrow \{\sigma_i\}) \ . \tag{22}$$

After an initial equilibration time, expectation values can then be estimated as an arithmetic mean over the Markov chain, e.g.,

$$\langle H \rangle = \sum_{\{\sigma_i\}} H[\{\sigma_i\}]P^{\text{eq}}[\{\sigma_i\}] \approx \frac{1}{N} \sum_{j=1}^{N} H[\{\sigma_i\}]_j \ . \tag{23}$$

A more detailed exposition of the basic concepts underlying any Monte Carlo algorithm can be found in many textbooks and reviews [23, 40, 41, 42].

3.1   Estimators and Autocorrelation Times

In principle there is no limitation on the choice of observables. The expectation value $\langle \mathcal{O} \rangle$ of any observable $\mathcal{O}$ can be estimated in a MC simulation as a simple arithmetic mean over the Markov chain, $\bar{\mathcal{O}} = \frac{1}{N} \sum_{j=1}^{N} \mathcal{O}_j$, where $\mathcal{O}_j \equiv \mathcal{O}[\{\sigma_i\}]_j$ is the measurement after the jth iteration. Conceptually it is important to distinguish between the expectation value $\langle \mathcal{O} \rangle$, which is an ordinary number, and the estimator $\bar{\mathcal{O}}$, which is a *random* number fluctuating around the theoretically expected value. Of course, in practice one does not probe the fluctuations of the estimator directly (which would require repeating the whole MC simulation many times), but rather estimates its variance $\sigma_{\bar{\mathcal{O}}}^2 = \langle [\bar{\mathcal{O}} - \langle \bar{\mathcal{O}} \rangle]^2 \rangle = \langle \bar{\mathcal{O}}^2 \rangle - \langle \bar{\mathcal{O}} \rangle^2$ from the distribution of $\mathcal{O}_j$. If the N measurements $\mathcal{O}_j$ were all uncorrelated then the relation would simply be $\sigma_{\bar{\mathcal{O}}}^2 = \sigma_{\mathcal{O}_j}^2 / N$, with $\sigma_{\mathcal{O}_j}^2 = \langle \mathcal{O}_j^2 \rangle - \langle \mathcal{O}_j \rangle^2$. For correlated measurements one obtains after some algebra

$$\sigma_{\bar{\mathcal{O}}}^2 = \frac{\sigma_{\mathcal{O}_k}^2}{N} 2\tau_{\bar{\mathcal{O}},\text{int}} \ , \tag{24}$$

where the *integrated* autocorrelation time,

$$\tau_{\bar{\mathcal{O}},\text{int}} = \frac{1}{2} + \sum_{j=1}^{N} A(j) \left(1 - \frac{j}{N} \right) \ , \tag{25}$$

turns out to be a sum ("integral") over the the autocorrelation function,

$$A(j) = \frac{\langle \mathcal{O}_i \mathcal{O}_{i+j} \rangle - \langle \mathcal{O}_i \rangle \langle \mathcal{O}_i \rangle}{\langle \mathcal{O}_i^2 \rangle - \langle \mathcal{O}_i \rangle \langle \mathcal{O}_i \rangle} \ . \tag{26}$$

For large time separations the autocorrelation function decays exponentially,

$$A(j) \overset{j \to \infty}{\longrightarrow} a e^{-j/\tau_{\bar{\mathcal{O}},\text{exp}}} \ , \tag{27}$$

with a being a constant. This defines the *exponential* autocorrelation time $\tau_{\bar{\mathcal{O}},\text{exp}}$. Due to the exponential decay of $A(j)$, in any meaningful simulation with $N \gg \tau_{\bar{\mathcal{O}},\text{exp}}$, the correction term in parentheses in (25) can safely be neglected. Notice that only if $A(j)$ is a pure exponential do the two autocorrelation times, $\tau_{\bar{\mathcal{O}},\text{int}}$ and $\tau_{\bar{\mathcal{O}},\text{exp}}$, coincide (up to minor corrections for small $\tau_{\bar{\mathcal{O}},\text{int}}$) [43].

The important point is that for correlated measurements the statistical error $\epsilon_{\bar{\mathcal{O}}} \equiv \sqrt{\sigma_{\bar{\mathcal{O}}}^2}$ on the MC estimator $\bar{\mathcal{O}}$ is enhanced by a factor of $\sqrt{2\tau_{\bar{\mathcal{O}},\text{int}}}$. This can be rephrased by writing the statistical error similar to the uncorrelated case as $\epsilon_{\bar{\mathcal{O}}} = \sqrt{\sigma_{\mathcal{O}_j}^2 / N_{\text{eff}}}$, but now with an effective statistics parameter $N_{\text{eff}} = N/2\tau_{\bar{\mathcal{O}},\text{int}}$. This shows more clearly that only at every $2\tau_{\bar{\mathcal{O}},\text{int}}$ iteration are the measurements approximately uncorrelated and gives a better idea of the relevant effective size of the statistical sample. Since some quantities (e.g., the specific heat or susceptibility) can severely be underestimated if the effective statistics are too small [44], any serious simulation should therefore provide at least a rough order-of-magnitude estimate of autocorrelation times.

Unfortunately, it is very difficult to give reliable a priori estimates, and an accurate numerical analysis is often too time consuming (as a rough estimate it is about ten times harder to get precise information on dynamic quantities than on static quantities such as critical exponents). To get at least an idea of the orders of magnitude, it is useful to record the "running" autocorrelation time

$$\tau_{\mathcal{O},\text{int}}(k) = \frac{1}{2} + \sum_{j=1}^{k} A(j) \; , \tag{28}$$

which approaches $\tau_{\mathcal{O},\text{int}}$ for large k. Approximating the tail end of $A(j)$ by a single exponential as in (27), one derives [45]

$$\tau_{\mathcal{O},\text{int}}(k) = \tau_{\mathcal{O},\text{int}} - \frac{a}{e^{1/\tau_{\mathcal{O},\text{exp}}} - 1} e^{-k/\tau_{\mathcal{O},\text{exp}}} \; . \tag{29}$$

The latter expression may be used for a numerical estimate of both the exponential and integrated autocorrelation times.

To summarize this subsection, any realization of a Markov chain (i.e., MC update algorithm) is characterized by autocorrelation times which enter directly in the statistical errors of MC estimates. Since correlations always increase the statistical errors, it is a very important issue to develop MC update algorithms that keep autocorrelation times as small as possible. In the next subsection we first discuss the classical Metropolis algorithm as an example of an update algorithm that near criticality is plagued by huge temporal correlations. The discussion of cluster updates in the next subsection then demonstrates that there indeed exist clever ways of overcoming this critical slowing-down problem.

3.2 Metropolis Algorithm

In the standard Metropolis algorithm [46] the Markov chain is realized by *local* updates of single spins. If E and E' denote the energy before and after the spin flip, respectively, then the probability of accepting the proposed spin update is given by

$$W(\{\sigma_i\} \longrightarrow \{\sigma_i'\}) = \begin{cases} \exp\left[-\beta(E' - E)\right] & E' > E \\ 1 & E' \leq E \; . \end{cases} \tag{30}$$

If the energy is lowered, the spin flip is always accepted. But even if the energy is increased, the flip has to be accepted with a certain probability to ensure the proper treatment of entropic contributions. In thermal equilibrium the *free* energy is minimized and not the energy. Only at zero temperature ($\beta \longrightarrow \infty$) does this probability tend to zero and the MC algorithm degenerates to a minimization algorithm for the energy functional. With some additional refinements, this is the basis of the simulated annealing technique, also discussed in this volume, which is usually applied to hard optimization and minimization problems.

There are many ways of choosing the spins to be updated. The lattice sites may be picked at random or according to a random permutation, which can be updated every now and then. But also a simple fixed lexicographical order is permissible. Or one updates first all odd and then all even sites, which is the usual choice in vectorized codes. A so-called lattice sweep is completed on average[3] when an update was proposed for all spins.

The advantage of this simple algorithm is its flexibility which allows its application to a great variety of physical systems. The great disadvantage, however, is that this algorithm is plagued by large autocorrelation times, as most other *local* update algorithms (one exception is the overrelaxation method [47, 48, 49, 50]. Empirically one finds that the autocorrelation time grows proportional to the spatial correlation length,

$$\tau \propto \xi^z \, , \tag{31}$$

with a dynamical critical exponent $z \approx 2$. Heuristically this can be understood by assuming that local excitations diffuse through the system like a random walk. Since ξ diverges at criticality, the Metropolis algorithm thus severely suffers from *critical slowing down*. Of course, in finite systems ξ cannot diverge. Then ξ is replaced by the linear lattice size L, yielding $\tau \propto L^z$.

The problem of critical slowing down can be overcome by *nonlocal* update algorithms. In the past few years several different types of such algorithms have been proposed. Quite promising results were reported with Fourier acceleration [51] and multigrid techniques [52, 53, 54, 55, 56, 57]. A very nice pedagogical introduction to these techniques is given by Sokal [58, 59]. For the $O(n)$ spin models considered here, however, the best performance was achieved with cluster update algorithms which will be described in the next subsection in more detail.

3.3   Cluster Algorithms

As we shall see below, cluster update algorithms [60, 61] are much more powerful than the Metropolis algorithm. Unfortunately, however, they are less generally applicable. We therefore consider first only the Ising model, where the prescription for cluster update algorithms can easily be read off from the equivalent Fortuin–Kasteleyn representation [62, 63, 64, 65],

$$Z = \sum_{\{\sigma_i\}} \exp\left(\beta \sum_{\langle ij \rangle} \sigma_i \sigma_j\right) \tag{32}$$

$$= \sum_{\{\sigma_i\}} \prod_{\langle ij \rangle} e^{\beta} \left[(1-p) + p\delta_{\sigma_i \sigma_j}\right] \tag{33}$$

$$= \sum_{\{\sigma_i\}} \sum_{\{n_{ij}\}} \prod_{\langle ij \rangle} e^{\beta} \left[(1-p)\delta_{n_{ij},0} + p\delta_{\sigma_i \sigma_j}\delta_{n_{ij},1}\right] \, , \tag{34}$$

---

[3] This is only relevant in the case where the lattice sites are picked at random.

with

$$p = 1 - e^{-2\beta} \ . \tag{35}$$

Here the n_{ij} are bond variables which can take the values $n_{ij} = 0$ or 1, interpreted as "deleted" or "active" bonds. In the first line we used the trivial fact that the product $\sigma_i \sigma_j$ of two Ising spins can only take the two values ± 1, so that $\exp(\beta \sigma_i \sigma_j) = x + y \delta_{\sigma_i \sigma_j}$ can easily be solved for x and y. In the second line we made use of the "deep" identity $a + b = \sum_{n=0}^{1} (a\delta_{n,0} + b\delta_{n,1})$.

Swendsen–Wang Cluster. According to (34) a cluster update sweep then consists of alternating updates of the bond variables n_{ij} for given spins with updates of the spins σ_i for a given bond configuration. In practice one proceeds as follows:

1. Set $n_{ij} = 0$ if $\sigma_i \neq \sigma_j$, or assign values $n_{ij} = 1$ and 0 with probability p and $1 - p$, respectively, if $\sigma_i = \sigma_j$, cf. Fig. 4.
2. Identify clusters of spins that are connected by "active" bonds ($n_{ij} = 1$).
3. Draw a random value ± 1 independently for each cluster (including one-site clusters), which is then assigned to all spins in a cluster.

Technically the cluster identification part is the most complicated step, but there are by now quite a few efficient algorithms available which can even be used on parallel computers. Vectorization, on the other hand, is only partially possible.

Notice the difference between the just-defined *stochastic* clusters and *geometric* clusters whose boundaries are defined by drawing lines through bonds between unlike spins. In fact, since in the stochastic cluster definition bonds between like spins are also "deleted" with probability $p_0 = 1 - p = \exp(-2\beta)$, stochastic clusters are on average smaller than geometric clusters. Only at zero temperature ($\beta \longrightarrow \infty$) does p_0 approache zero and the two cluster definitions coincide. As described above, the cluster algorithm is referred to as Swendsen–

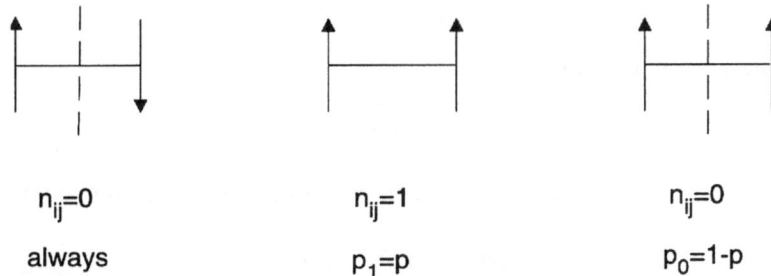

$n_{ij}=0$ $n_{ij}=1$ $n_{ij}=0$

always $p_1=p$ $p_0=1\text{-}p$

Fig. 4. Illustration of the bond variable update. The bond between unlike spins is always "deleted" as indicated by the *dashed line*. A bond between like spins is only "active" with probability $p = 1 - \exp(-2\beta)$. Only at zero temperature ($\beta \longrightarrow \infty$) do stochastic and geometric clusters coincide

Wang (SW) or multiple-cluster update [60]. The distinguishing point is that the *whole* lattice is decomposed into stochastic clusters whose spins are assigned a random value +1 or −1. In one sweep one thus attempts to update all spins of the lattice.

Wolff Cluster. Shortly after the original discovery of cluster algorithms, Wolff [61] proposed a somewhat simpler variant in which only a single cluster is flipped at a time. This variant is therefore sometimes also called a single-cluster algorithm. Here one chooses a lattice site at random, constructs only the cluster connected with this site, and then flips all spins of this cluster. A typical example is shown in Fig. 5. In principle, one could also here choose a value +1 or −1 at random, but then nothing at all would be changed if one hits the current value of the spins. Here a sweep consists of $V/\langle C \rangle$ single cluster steps, where $\langle C \rangle$ denotes the average cluster size. With this definition autocorrelation times are directly comparable with results from the Metropolis or Swendsen–Wang algorithm. Apart from being somewhat easier to program, Wolff's single-cluster variant is usually more efficient than the Swendsen–Wang multiple-cluster algorithm, especially in 3D. The reason is that with the single-cluster method on average larger clusters are flipped.

Embedded Cluster. While it is quite easy to generalize the derivation (32) – (35) to q-state Potts models, for the $O(n)$ spin models with $n \geq 2$ one needs a new strategy [61, 66, 67, 68]. Here the basic idea is to isolate Ising degrees of freedom by projecting $\boldsymbol{\sigma}_i$ onto a randomly chosen unit vector $\mathbf{r}$,

$$\boldsymbol{\sigma}_i = \boldsymbol{\sigma}_i^{\parallel} + \boldsymbol{\sigma}_i^{\perp} ; \quad \boldsymbol{\sigma}_i^{\parallel} = \epsilon |\boldsymbol{\sigma}_i \cdot \mathbf{r}| \mathbf{r}; \quad \epsilon = \mathrm{sign}(\boldsymbol{\sigma}_i \cdot \mathbf{r}) . \qquad (36)$$

If this is inserted into the original Hamiltonian one ends up with an effective Hamiltonian

$$H = - \sum_{\langle ij \rangle} J_{ij} \epsilon_i \epsilon_j + \mathrm{const} , \qquad (37)$$

with positive random couplings,

$$J_{ij} = J |\boldsymbol{\sigma}_i \cdot \mathbf{r}||\boldsymbol{\sigma}_j \cdot \mathbf{r}| \geq 0 , \qquad (38)$$

whose Ising degrees of freedom ϵ_i can be updated with a cluster algorithm as described above.

For $O(n)$ spin models the performance of both types of cluster algorithms is excellent. As is demonstrated in Table 1 and Fig. 6, critical slowing down is drastically reduced. We see that especially in three dimensions the Wolff cluster algorithm performs better than the Swendsen–Wang algorithm. Compared with the Metropolis algorithm, factors of up to 10 000 in CPU time have been saved in realistic simulations [69, 70].

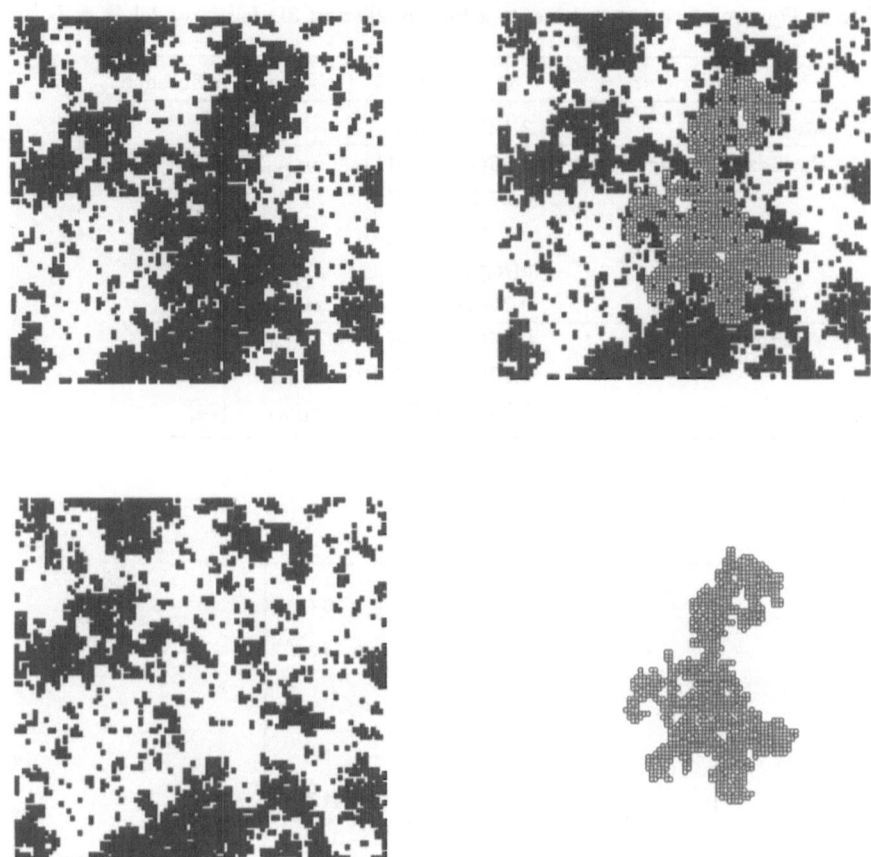

Fig. 5. Illustration of the Wolff cluster update. *Upper left* Initial configuration. *Upper right* The stochastic cluster is marked. *Lower left* Final configuration after flipping the spins in the cluster. *Lower right* The flipped cluster. The shown spin configuration is from an actual simulation of the 2D Ising model at $0.97 \times \beta_c$ on a 100×100 lattice

Improved Estimators. A further advantage of cluster algorithms is that they lead quite naturally to so-called improved estimators which are designed to further reduce the statistical errors. Suppose we want to measure the expectation value $\langle \mathcal{O} \rangle$ of an observable $\mathcal{O}$. Then any estimator $\hat{\mathcal{O}}$ satisfying $\langle \hat{\mathcal{O}} \rangle = \langle \mathcal{O} \rangle$ is permissible. This does not determine $\hat{\mathcal{O}}$ uniquely since there are infinitely many other possible choices, $\hat{\mathcal{O}}' = \hat{\mathcal{O}} + \hat{\mathcal{X}}$, where the added estimator $\hat{\mathcal{X}}$ has zero expectation, $\langle \hat{\mathcal{X}} \rangle = 0$. The variances of the estimators $\hat{\mathcal{O}}'$, however, can be quite different and are not necessarily related to any physical quantity (contrary to the standard mean-value estimator of the energy whose variance is proportional to the specific heat). It is exactly this freedom in the choice of $\hat{\mathcal{O}}$ which allows the construction of improved estimators.

Table 1. Dynamic critical exponents z for the 2D and 3D Ising model ($\tau \propto L^z$)

Algorithm	D=2	D=3	Observable	Reference
Metropolis	2.125	2.03		
Swendsen–Wang cluster	0.35(1)	0.75(1)	$z_{E,\exp}$	[60]
	0.27(2)	0.50(3)	$z_{E,\mathrm{int}}$	[71]
	0.20(2)	0.50(3)	$z_{\chi,\mathrm{int}}$	[71]
	0(log L)	–	$z_{M,\exp}$	[72]
	0.25(5)	–	$z_{M,\mathrm{rel}}$	[73]
Wolff cluster	0.26(2)	0.28(2)	$z_{E,\mathrm{int}}$	[71]
	0.13(2)	0.14(2)	$z_{\chi,\mathrm{int}}$	[71]
	0.25(5)	0.3(1)	$z_{E,\mathrm{rel}}$	[74]

3D XY model

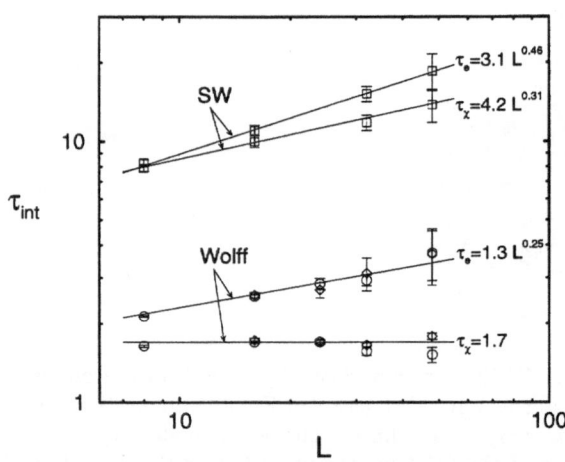

Fig. 6. Double logarithmic plot of the integrated autocorrelation times for the Swendsen Wang (SW) and Wolff algorithm of the 3D XY model near criticality. The *squares* ($\beta = 0.45421$) are taken from [75], the *circles* ($\beta = 0.4539$) and *diamonds* ($\beta = 0.4543$) from [76]

For the single-cluster algorithm an improved "cluster estimator" for the spin-spin correlation function in the high-temperature phase, $G(\mathbf{x}_i - \mathbf{x}_j) \equiv \langle \boldsymbol{\sigma}_i \cdot \boldsymbol{\sigma}_j \rangle$, is given by [67]

$$\hat{G}(\mathbf{x}_i - \mathbf{x}_j) = n \frac{V}{|C|} \mathbf{r} \cdot \boldsymbol{\sigma}_i \, \mathbf{r} \cdot \boldsymbol{\sigma}_j \Theta_C(\mathbf{x}_i) \Theta_C(\mathbf{x}_j) \; , \tag{39}$$

where $\mathbf{r}$ is the normal of the mirror plane used in the construction of the cluster of size $|C|$ and $\Theta_C(\mathbf{x})$ is its characteristic function (=1 if $\mathbf{x} \in C$ and 0 otherwise). For the Fourier transform, $\tilde{G}(\mathbf{k}) = \sum_{\mathbf{x}} G(\mathbf{x}) \exp(-i\mathbf{k} \cdot \mathbf{x})$, this implies the improved estimator

$$\hat{\tilde{G}}(\mathbf{k}) = \frac{n}{|C|} \left[\left(\sum_{i \in C} \mathbf{r} \cdot \sigma_i \cos \mathbf{k}\mathbf{x}_i \right)^2 + \left(\sum_{i \in C} \mathbf{r} \cdot \sigma_i \sin \mathbf{k}\mathbf{x}_i \right)^2 \right] , \qquad (40)$$

which, for $\mathbf{k} = 0$, reduces to an improved estimator for the susceptibility χ' in the high-temperature phase,

$$\hat{\tilde{G}}(0) = \hat{\chi}'/\beta = \frac{n}{|C|} \left(\sum_{i \in C} \mathbf{r} \cdot \sigma_i \right)^2 . \qquad (41)$$

For the Ising model ($n = 1$) this reduces to $\chi'/\beta = \langle |C| \rangle$, i.e., the improved estimator of the susceptibility is just the average cluster size of the single-cluster update algorithm. For the XY and Heisenberg model one finds empirically that in two as well as in three dimensions $\langle |C| \rangle \approx 0.81 \chi'/\beta$ for $n = 2$ [66, 76] and $\langle |C| \rangle \approx 0.75 \chi'/\beta$ for $n = 3$ [67, 77], respectively.

It should be noted that by means of the estimators (39)-(41) a significant reduction of variance should only be expected outside the FSS region where the average cluster size is small compared with the volume of the system.

3.4 Multicanonical Algorithms for First-Order Transitions

Let us finally make a few brief comments on numerical simulations of first-order phase transitions [13]. Since here the correlation lengths in the pure phases are finite, the numerical problems are completely different from those in the case of second-order phase transitions. Here the origin of numerical difficulties near the transition point can be traced back to the coexistence of two phases which, for finite systems, is reflected by a double-peak structure of the corresponding order parameter or energy distribution. The minimum between the two peaks is governed by mixed-phase configurations which are strongly suppressed by an additional Boltzmann factor $\propto \exp(-2\sigma L^{D-1})$. Here σ denotes the interface tension at the phase boundaries, L^{D-1} is the cross section of the (cubic) system of size $V = L^D$, and the factor 2 takes into account the usually employed periodic boundary conditions [78, 79]. The problem of numerical simulations is to achieve equilibrium between the two phases. The system spends most of the time in the pure phases. Only very rarely does it "tunnel" through the exponentially suppressed mixed-phase region from one phase to the other. These rare tunneling events, however, are necessary to achieve equilibrium between the pure phases. The relevant time scale of equilibrium simulations is thus given by the inverse of the additional Boltzmann factor, i.e., the characteristic time τ grows exponentially with the system size, $\tau \propto \exp(2\sigma L^{D-1})$ [80]. Since for an accurate numerical study the simulation (and thus computing) time must be much larger than τ, this phenomenon has been termed the *exponential* or *super-critical slowing down* problem.

A surprisingly simple solution to this problem was discovered by Berg and Neuhaus [81]. In what they call "multicanonical" simulations one determines iteratively artificial weight factors which modify the original Hamiltonian in such a way that the order parameter or energy distribution is approximately flat between the two peaks of the canonical distribution. Since then the system has no longer to pass through an exponentially suppressed region one expects in multicanonical simulations a drastic reduction of the characteristic time scale τ. A simple random walk argument suggests a power-law behavior, $\tau \propto V^\alpha$, which has indeed been confirmed in numerical simulations of 2D Potts models [82, 83, 84].

The multicanonical technique is strictly speaking not an update algorithm but a reweighting procedure as discussed in detail in the next section. In principle, it can therefore be combined with any legitimate update algorithm. The earlier studies all employed the Metropolis or heat-bath algorithm. In more recent work it was shown that combinations with multigrid techniques [45, 85, 86] and cluster update algorithms [84, 87, 88, 21] are also feasible and can further reduce autocorrelation times.

4   Reweighting Techniques

Even though the physics underlying reweighting techniques is extremely simple and the basic idea has been known for a long time (see the list of references by Ferrenberg and Swendsen [89]), their power in practice has been realized only quite recently [89, 90]. The best performance is achieved *near* criticality, and in this sense reweighting techniques are complementary to improved estimators.

If we denote the number of states (spin configurations) that have the same energy by $\Omega(E)$, the partition function at the simulation point β_0 can always be written as[4]

$$Z(\beta_0) = \sum_E \Omega(E)e^{-\beta_0 E} \ . \tag{42}$$

This shows that the energy distribution $P_{\beta_0}(E)$ (normalized to unit area) is given by

$$P_{\beta_0}(E) = \Omega(E)e^{-\beta_0 E}/Z(\beta_0) \ . \tag{43}$$

It is then easy to see that, given $P_{\beta_0}(E)$, the energy distribution is actually known for any β,

$$P_\beta(E) = cP_{\beta_0}(E)e^{-(\beta-\beta_0)E} \ , \tag{44}$$

where c is a normalization constant [which in practice is trivially determined by enforcing the condition $\sum_E P_\beta(E) = 1$. Formally, one easily finds that

---

[4] For simplicity we consider here only models with *discrete* energies. If the energy varies continuously, sums have to be replaced by integrals, etc. Also lattice size dependencies are suppressed to keep the notation short.

$c = 1/\sum_E \mathcal{P}_{\beta_0}(E)e^{-(\beta-\beta_0)E} = Z(\beta_0)/Z(\beta)]$. Knowing $\mathcal{P}_\beta(E)$, we find that expectation values of the form $\langle f(E)\rangle$ are easy to compute,

$$\langle f(E)\rangle(\beta) = \sum_E f(E)\mathcal{P}_\beta(E) . \qquad (45)$$

Since the relative statistical errors increase in the wings of $P_{\beta_0}(E)$ one expects (44), (45) to give reliable results only for β near β_0. If β_0 is near criticality, the distribution is relatively broad and the method works best. In this case reliable estimates from (44) can be expected for β values in an interval around β_0 of width $\propto L^{-1/\nu}$, i.e., just in the FSS region. As a rule of thumb the reweighting range can be determined by the condition that the peak location of the reweighted distribution should not exceed the energy values at which the input distribution had decreased to one third of its maximum value [91]. In most applications this range is wide enough to locate from a single simulation, e.g., the specific-heat maximum by using any standard maximization routine.

This is illustrated in Figs. 7 and 8, again for simplicity for the 2D Ising model. In Fig. 7 the filled circle shows the result of a MC simulation at $\beta_c = \log(1 + \sqrt{2})/2 \approx 0.440686\ldots$, using the Swendsen Wang cluster algorithm with 5000 sweeps for equilibration and 50000 sweeps for measurements. The results of the reweighting procedure are shown as open circles (recall that the spacing between the circles can be made as small as desired, here it was chosen quite large for clarity of the plot) and compared with the exact curve [22]. We see that even with these rather modest statistics the whole specific-heat peak can be obtained with reasonable accuracy from a single simulation. But we also notice significant deviations in the tails of the peak. To understand the origin of the deviations it is useful to have a look at the energy histograms in Fig. 8. The curve labeled $\beta_0 = \beta_c$ is the histogram of the MC data at the simulation point, and the other two histograms at $\beta = 0.375$ and $\beta = 0.475$ are computed from this input histogram by reweighting. For comparison we have also included the histograms obtained from additional MC simulations (with the same statistics) at the two β values, indicated by the black dots. We see that the reweighted histogram at $\beta = 0.475$ looks smooth to the eye – and indeed agrees very well with the "direct" result of the additional MC simulation at this temperature. In Fig. 7 this is reflected by the still very good agreement of the numerical and the exact result. For $\beta = 0.375$, on the other hand, even visually one would not trust the reweighted histogram. While the tail on the right-hand side is still in reasonable agreement with the "direct" simulation, the left tail is obviously hopelessly wrong. By recalling that the reweighted histograms are computed by multiplying the input histogram with exponential factors, this is no surprise at all. For $-e \lesssim 1$ there are hardly any entries in the input histogram and hence the relative statistical errors ($\propto 1/\sqrt{\text{counts}}$) are huge. This is the source for the large deviations from the exact curve in Fig. 7 for $\beta \lesssim 0.4$.

The information stored in $P_{\beta_0}(E)$ is not yet sufficient to also calculate the magnetic moments $\langle m^k\rangle(\beta)$ as a function of β from a single simulation at β_0.

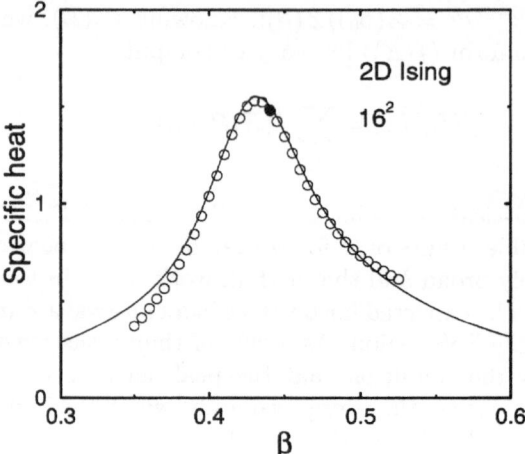

Fig. 7. Specific heat of the 2D Ising model computed by reweighting (○) from a single MC simulation at $\beta_0 = \beta_c$ (●). The *continuous line* shows for comparison the exact result of Ferdinand and Fisher [22]

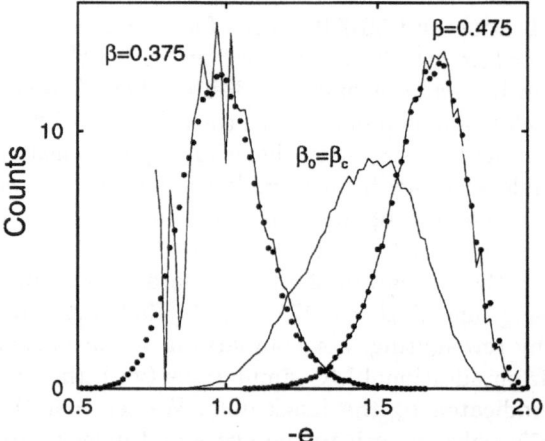

Fig. 8. The energy histogram at the simulation point $\beta_0 = \beta_c$, and reweighted to $\beta = 0.375$ and $\beta = 0.475$. The *black dots* show histograms obtained in additional simulations at these temperatures

Conceptually, the simplest way to proceed is to record the two-dimensional histogram $P_{\beta_0}(E, M)$, where $M = mV$ is the total magnetization. Because of disk space limitations one sometimes prefers to measure instead the "microcanonical averages"

$$\langle\langle m^k \rangle\rangle(E) = \sum_M P_{\beta_0}(E, M)m^k/P_{\beta_0}(E) \ , \tag{46}$$

where the relation $\sum_M P_{\beta_0}(E, M) = P_{\beta_0}(E)$ was used. In practice one simply accumulates the measurements of m^k in different slots or bins according to the energy of the configuration and normalizes at the end by the total number of hits of each energy bin. Clearly, once $\langle\!\langle m^k \rangle\!\rangle(E)$ is determined, this is a special case of $f(E)$ in (45), so that

$$\langle m^k \rangle(\beta) = \frac{\sum_E \langle\!\langle m^k \rangle\!\rangle(E) P_{\beta_0}(E) e^{-(\beta-\beta_0)E}}{\sum_E P_{\beta_0}(E) e^{-(\beta-\beta_0)E}} \; . \tag{47}$$

Similar to $\Omega(E)$ in (42), theoretically the microcanonical averages $\langle\!\langle m^k \rangle\!\rangle(E)$ also do not depend on the temperature at which the simulation is performed. Due to the limited statistics in the wings of $P_{\beta_0}(E)$, however, there is only a finite range around $E_0 \equiv \langle E \rangle(\beta_0)$ where one can expect reasonable results for $\langle\!\langle m^k \rangle\!\rangle(E)$. Outside of this range it simply can happen (and does happen) that there are no events to be averaged. This is illustrated in Fig. 9, where $\langle\!\langle m^2 \rangle\!\rangle(E)$ is plotted for the 3D Heisenberg model as obtained from three runs at different temperatures. We see that the function looks smooth only in the range where the statistics of the corresponding energy histogram is high enough. To take full advantage of the histogram reweighting technique it is therefore advisable to perform a few simulations at slightly different inverse temperatures β_i. Instead of spending all computer time in a single long run, it is usually more efficient to perform three or four shorter runs. To find the best solution, however, is a very difficult optimization problem, which depends on many details of the model under study! Now the question arises how to combine the data from different runs most efficiently. A very clear way is to compute for each simulation (at β_i) the β dependence of $O_i \equiv O^{(\beta_i)}(\beta)$ plus the associated statistical error ΔO_i, using jack-knife techniques [92, 93], say. A single optimal expression for $O \equiv O(\beta)$ is then obtained by combining the values O_i in such a way that the relative error $\Delta O/O$ is minimized [77],

$$O = \left(\frac{O_1}{(\Delta O_1)^2} + \frac{O_2}{(\Delta O_2)^2} + \frac{O_3}{(\Delta O_3)^2} \right) (\Delta O)^2 \;, \tag{48}$$

with

$$\frac{1}{(\Delta O)^2} = \frac{1}{(\Delta O_1)^2} + \frac{1}{(\Delta O_2)^2} + \frac{1}{(\Delta O_3)^2} \; . \tag{49}$$

A different procedure at the level of distribution functions was discussed by Ferrenberg and Swendsen [94]. For the specific heat the two methods were found [77] to give comparable results within the statistical errors. The optimization at the level of observables, however, is simpler to apply to quantities involving constant energy averages such as $\langle\!\langle m \rangle\!\rangle(E)$, and, more importantly, minimizes the error on each observable of interest separately.

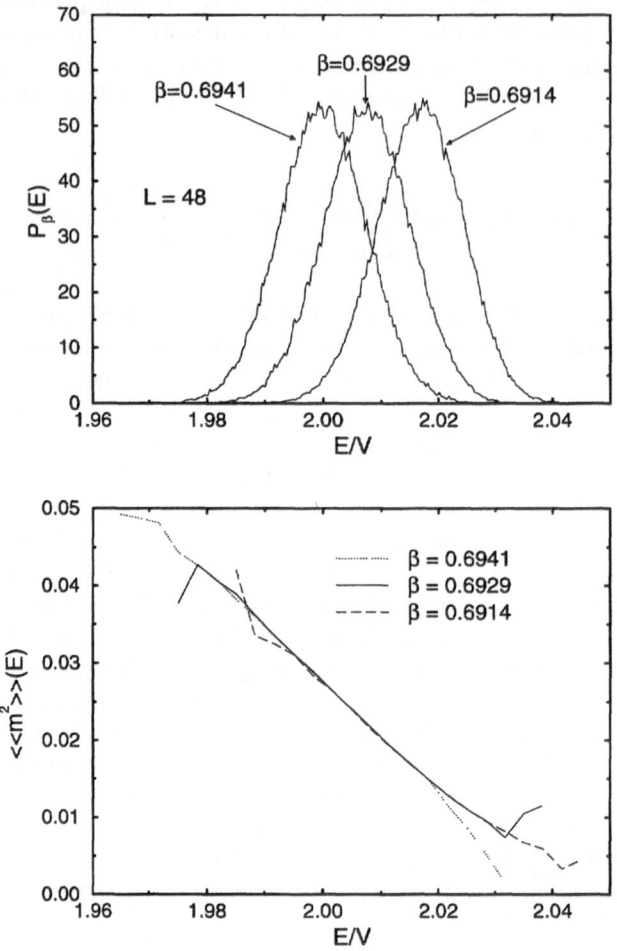

Fig. 9. Energy histograms and microcanonical magnetization squared of the 3D Heisenberg model at the three simulation points for $L = 48$

5   Applications to the 3D Heisenberg Model

Let us now turn to applications of the just-described techniques to the 3D classical Heisenberg model ($n = 3$), focussing on an accurate determination of the transition temperature and the critical exponents. To this end the cluster update algorithm proved to be a very important tool. Previous studies [95, 96, 97, 98] employing the Metropolis algorithm reported for the magnetization an exponential autocorrelation time of $\tau_{m,\mathrm{exp}} = aL^z$, with $a \approx 3.76$ and $z = 1.94(6)$. In simulations with the single-cluster algorithm we obtained for the susceptibility values of $\tau_{\chi,\mathrm{int}} \approx 1.5 - 2.0$ [77, 99, 100]. As for the 3D Ising and 3D XY model critical slowing down is thus almost completely eliminated. Compared to the Metropolis algorithms this implies for a 80^3 lattice an acceleration of the simulation by about four orders of magnitude.

In the paper by Holm and Janke [77] two sets of MC simulations are reported. The first set of data consists of 18 simulations for $T > T_c$ in the range $\beta = 0.650$ to $0.686 \approx 0.99\beta_c$, with typically about 10^5 (almost uncorrelated) measurements. The correlation length varies in this β range from $\xi \approx 3$ to $\xi \approx 12$ (see Fig. 11 below). Here the use of improved estimators was very useful and led to a further reduction of the statistical errors by a factor of about 2-3.

The second set of simulations was performed near criticality. For each lattice size typically three independent runs with more than 10^5 measurements each at different β values around β_c were combined using the optimized reweighting technique, which is the most important additional tool in the finite-size scaling region.

5.1   Simulations for $T > T_c$

Conceptually the easiest way to measure critical exponents are simulations in the high-temperature phase. In principle one simply has to fit the MC data for ξ, C, m, χ, ... with the expected power-law divergencies (10), (12)-(14) at criticality. For high-precision estimates, however, the procedure is far from being trivial. The problem is to locate the temperature range in which a simple power-law ansatz like (10) is valid. Clearly, since the omitted correction terms are positive powers of $T/T_c - 1$, at first sight, one would like to perform the simulations as close to T_c as possible. However, very close to T_c the correlation length gets very large, and on finite lattices one starts seeing finite-size corrections. The only way around these correction terms is to use large enough lattice sizes. In many models one finds empirically that the thermodynamic limit is approached when the linear lattice size satisfies $L \gtrsim (6-8)\xi$. But since the amplitude ξ_{0+} is non-universal, this estimate is not guaranteed to be always true. Therefore this question must be investigated very carefully for each model separately. With increasing temperature the correlation length decreases and finite-size corrections are no longer a problem, however then it is, for a different reason, again not clear if the simple power-law ansatz is valid. Very far away from T_c the lattice structure becomes important and the observables show a completely different behavior. In an intermediate range one sees the confluent and analytic correction terms which are very difficult to take into account in the fits. So in essence the problem is to locate a temperature window in which $1 \ll \xi \ll L$.

There are many ways to extract the correlation length ξ from the asymptotic decay of the spatial correlation function,

$$G(\mathbf{x}_i - \mathbf{x}_j) = \langle \boldsymbol{\sigma}_i \cdot \boldsymbol{\sigma}_j \rangle \propto \exp\left(-|\mathbf{x}_i - \mathbf{x}_j|/\xi\right) \ . \tag{50}$$

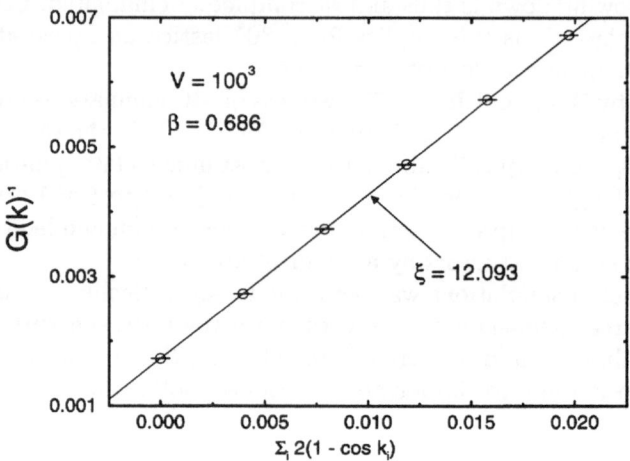

Fig. 10. Fit to the inverse of the Fourier transformed correlation function to compute the correlation length ξ

One way is measuring the Fourier transform, $\tilde{G}(\mathbf{k}) = \sum_{\mathbf{x}} G(\mathbf{x}) \exp(-i\mathbf{k} \cdot \mathbf{x})$, for a few long-wavelength modes and performing least-square fits to

$$\tilde{G}(\mathbf{k})^{-1} = c \left[\sum_{i=1}^{3} 2(1 - \cos k_i) + (1/\xi)^2 \right] \approx c \left[\mathbf{k}^2 + (1/\xi)^2 \right] , \qquad (51)$$

where c is a constant and $k_i = (2\pi/L)n_i$, $n_i = 1, \ldots, L$. Recall that for zero momentum the susceptibility is recovered, $\tilde{G}(\mathbf{k} = 0) = \chi'/\beta$. As an example Fig. 10 shows the data for the 3D Heisenberg model [$\mathbf{n} = (0,0,0)$, $(1,0,0)$, $(1,1,0)$, $(1,1,1)$, $(2,0,0)$, and $(2,1,0)$] at $\beta = 0.686 \approx 0.99\beta_c$.

By repeating this analysis for different temperatures one obtains the data shown in Fig. 11. The solid lines are fits according to

$$\xi(T) = \xi_{0+}(T/T_c - 1)^{-\nu} , \qquad (52)$$

with

$$\beta_c = 0.69281 \pm 0.00004 , \qquad (53)$$
$$\nu = 0.698 \pm 0.002 , \qquad (54)$$
$$\xi_{0+} = 0.484 \pm 0.002 , \qquad (55)$$

and a goodness-of-fit parameter $Q = 0.92$, and

$$\chi'(\beta)/\beta = \bar{\chi}'_0(1 - \beta/\beta_c)^{-\gamma} , \qquad (56)$$

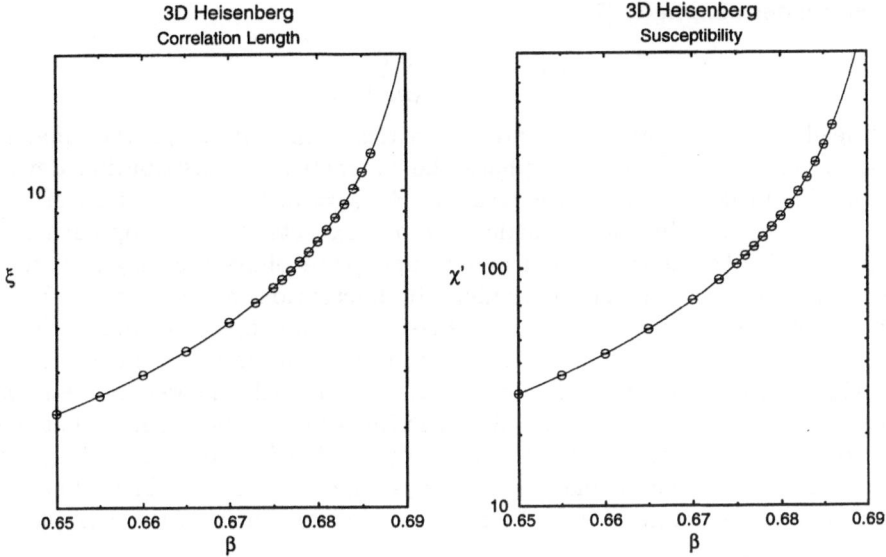

Fig. 11. Scaling behavior of the correlation length and susceptibility. The *solid lines* are fits to $\xi(T)$ and $\chi'(\beta)$ according to the asymptotic power laws

with

$$\beta_c = 0.69294 \pm 0.00003 \, , \tag{57}$$

$$\gamma = 1.391 \pm 0.003 \, , \tag{58}$$

$$\bar{\chi}'_0 = 0.955 \pm 0.006 \, , \tag{59}$$

and $Q = 0.93$. Notice that the correlation length in (52) is written as a function of T and the susceptibility in (56) as a function of β. The high quality of the fits a posteriori justifies this choice and indicates that in the chosen temperature range confluent as well as analytic correction terms are negligible. If we rewrite $T/T_c - 1 = 1 - \beta/\beta_c + (1 - \beta/\beta_c)^2 + \ldots$ and consider $\xi(\beta)$ instead of $\xi(T)$, we expect (and indeed confirmed) an analytical correction to asymptotic scaling. There is, however, so far no theoretical understanding of why for the particular choice of arguments the analytical correction terms should vanish.

5.2   Simulations near T_c

The second set of data consists of simulations near T_c on lattices of size up to 48^3 [77, 99]. In a later study focussing on topological defects, the maximal size could even be increased to 80^3 [100]. In the vicinity of T_c, finite-size corrections are dominant and one has to employ finite-size scaling (FSS) concepts to analyze the data. Usually one starts with an analysis of ratios of magnetization moments, e.g., the Binder parameter [78],

$$U_L(\beta) = 1 - \frac{1}{3} \frac{\langle m^4 \rangle}{\langle m^2 \rangle^2} \quad , \tag{60}$$

which leads to estimates of β_c and the critical exponent ν. In the spontaneously broken low-temperature phase the magnetization distribution develops a double-peak structure with peaks at $\pm m_0 \neq 0$. Since the width of the peaks decreases with increasing lattice size one expects that U_L approaches $\frac{2}{3}$ for all $T < T_c$. In the disordered high-temperature phase the magnetization vanishes and the moments are determined by fluctuations alone. In the infinite volume limit the fluctuations become Gaussian and a simple calculation yields $U_L \longrightarrow 2(n-1)/3n = 4/9$ for $n = 3$. Only at the transition point one expects a nontrivial limiting value which has been estimated by field theoretical methods [101] to be $U^* = 0.59684\ldots$ for $n = 3$. For finite systems, FSS predicts that the curves $U_L(\beta)$ for different L intersect around (β_c, U^*) with slopes $\propto L^{1/\nu}$, apart from confluent corrections explaining small systematic deviations. This allows an almost unbiased estimate of β_c, U^* and the critical exponent of the correlation length ν.

The data for the 3D Heisenberg model in Fig. 12 clearly confirm the theoretical expectations with a pronounced intersection point at $(\beta_c, U^*) = (0.6930(1), 0.6217(8))$. The final numbers are actually obtained from a slightly more elab-

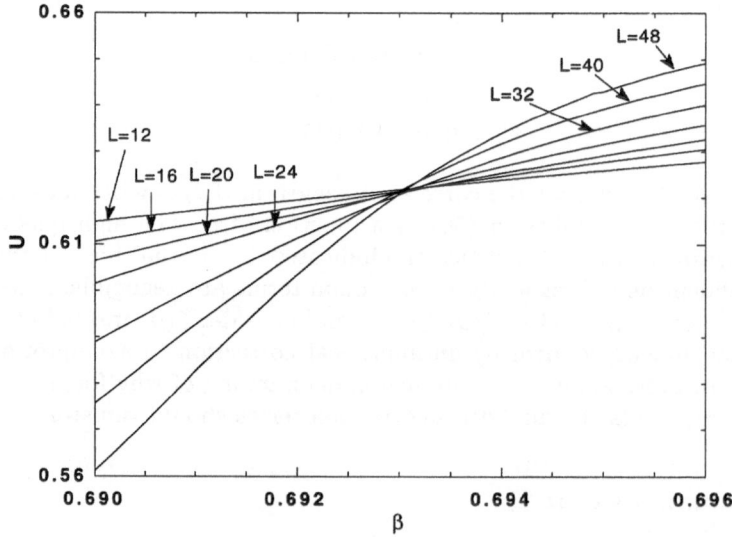

Fig. 12. The Binder parameter of the 3D Heisenberg model for various lattice sizes. The intersection points determine the inverse critical temperature $\beta_c = 0.6930(1)$

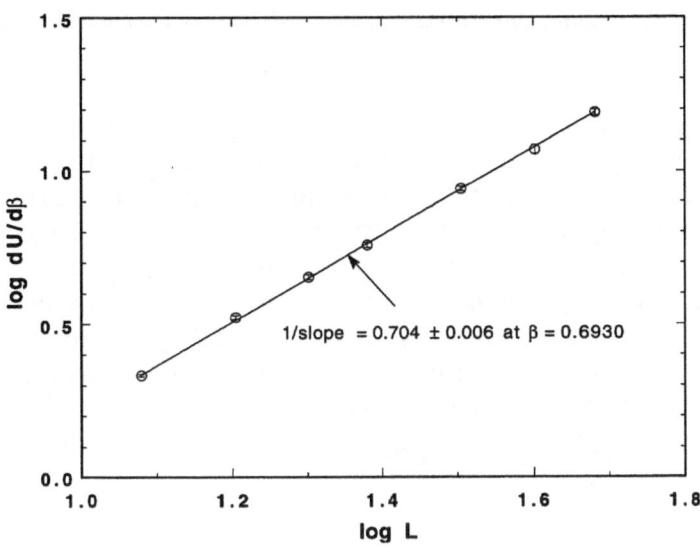

Fig. 13. FSS of the Binder parameter slopes at $\beta = 0.6930 \approx \beta_c$. The linear fit yields an estimate of the correlation length exponent, $\nu = 0.704(6)$

orate analysis taking into account also the confluent corrections to asymptotic FSS [77, 99].

Also the slopes $dU_L/d\beta$ at $\beta = 0.6930 \approx \beta_c$ in Fig. 13 show the expected behavior and a fit to the FSS prediction $dU_L/d\beta \propto L^{1/\nu}$ yields $\nu = 0.704(6)$, consistent with (54) and in very good agreement with estimates obtained by field theoretical methods or from series expansions, cf. Table 2.

In the analysis of the magnetization and susceptibility one proceeds similarly. The fit to the magnetization data reweighted to $\beta = 0.6930 \approx \beta_c$ shown in Fig. 14 yields $\beta/\nu = 0.514(1)$, and from the FSS of the susceptibility one reads off $\gamma/\nu = 1.9729(17)$. By multiplying the exponent ratios with the estimate of $\nu = 0.704(6)$, we finally arrive at the values for the critical exponents β and γ given in Table 2. The analysis of the specific heat is much more complicated since the critical exponent α is usually quite small and therefore the singularity in C not very pronounced. For the 3D Heisenberg model α is actually negative, so that we do not expect a divergence at all. By using the additional data on lattices up to 80^3 [100], we obtained from the fit $C = C_{\text{reg}} + C_0 L^{\alpha/\nu}$ shown in Fig. 15 an estimate of $\alpha/\nu = -0.225(80)$, resulting in $\alpha = -0.158(59)$. Due to the rather large statistical error this estimate is still consistent with the value obtained from hyperscaling, $\alpha = 2 - 3\nu = -0.112(18)$. Actually a much more precise result was obtained from the corresponding FSS fit to the energy at β_c. Using the ansatz $e = e_{\text{reg}} + e_0 L^{(\alpha-1)/\nu}$, we obtained $\alpha/\nu = -0.166(31)$, translating into $\alpha = -0.117(23)$. Obviously this value is in a much better agreement with the hyperscaling prediction.

Table 2. Critical coupling and critical exponents of the 3D classical Heisenberg ($n = 3$) model

Method	β_c	ν	γ	β	α	δ
g-expansion[a]	–	0.705(3)	1.386(4)	0.3645(25)	−0.115(9)	4.802(37)
ϵ-expansion[b]	–	0.710(7)	1.390(10)	0.368(4)	−0.130(21)	4.777(70)
MC[c]	0.6929(1)	0.706(9)	1.390(23)	0.364(7)	−0.118(27)	4.819(36)
MC[d]	0.6930(2)	0.73(4)	–	0.36(2)	–	–
MC[e]	0.6930(1)	0.704(6)	1.389(14)	0.362(4)	−0.112(18)	4.837(11)
MC[f]	0.693035(37)	0.7036(23)	1.3896(70)	0.3616(31)	−0.1108(69)	–
series[g]	0.6929(1)	0.712(10)	1.400(10)	0.363(10)	−0.136(30)	4.86(10)
series[h]	0.69302(7)	0.715(3)	1.403(6)	–	–	–

[a] [102, 103],   [b] [104],   [c] [98],   [d] [105],   [e] [77, 99],   [f] [106],   [g] [107],   [h] [108]

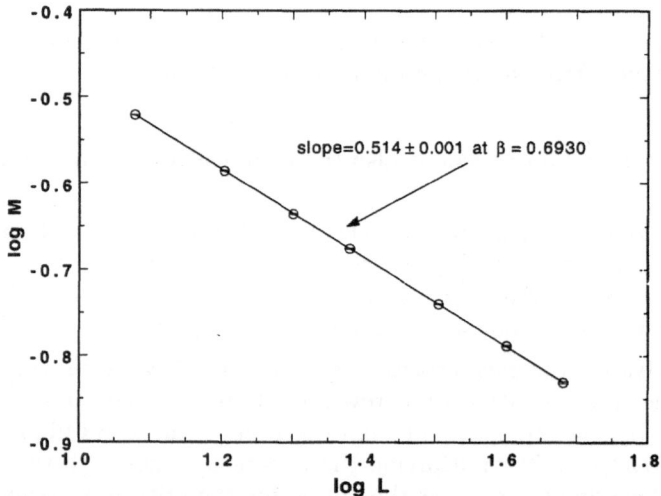

Fig. 14. FSS of the magnetization at $\beta = 0.6930 \approx \beta_c$. The linear fit yields an estimate of the exponent ratio, $\beta/\nu = 0.514(1)$

6   Concluding Remarks

The intention of this chapter was to give an elementary introduction to the basic concepts of modern Monte Carlo simulations and to illustrate their usefulness by applications to one typical model. Since the choice of the 3D Heisenberg model was obviously biased by my own work in this field, I want to conclude with at least a few remarks on the 3D Ising and 3D XY model, for which quite a few

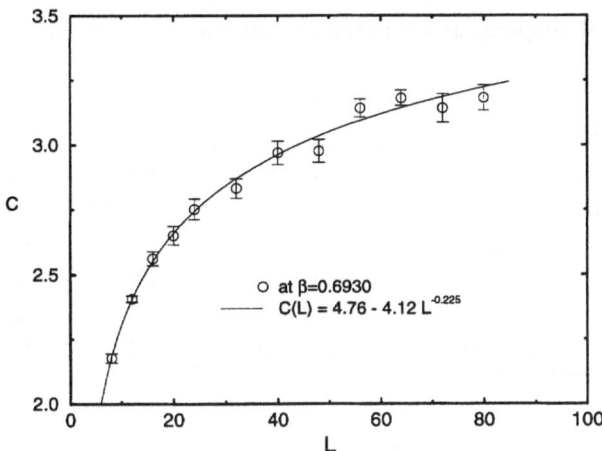

Fig. 15. FSS of the specific heat at $\beta = 0.6930 \approx \beta_c$. The non-linear fit taking into account a regular background term yields an estimate of the exponent ratio, $\alpha/\nu = -0.225(80)$

high-precision simulations have also been performed.

Due to its relative simplicity, the 3D Ising model is the best-studied model of all $O(n)$ spin systems. Apart from finite-size scaling analysis as described here, many other techniques have also been applied, including the Monte Carlo renormalization group (MCRG) and the finite-size scaling of partition function zeros. This has led to quite a few very accurate estimates of critical exponents. Some of them are compiled in Table 3, where for comparison field theory and series expansion estimates are given. As an amusing side remark it is worth mentioning the Rosengren [113] conjecture that the critical coupling of the 3D Ising model is given by $\beta_c = \tanh^{-1}[(\sqrt{5} - 2)\cos(\pi/8)] = 0.221\,658\,637\ldots$ – a value which is indeed in impressive agreement with the most precise Monte Carlo estimates!

A similar accuracy could also be reached for the 3D XY model, the simplest model of the $O(2)$ universality class which governs the critical behavior of the λ-transition in liquid helium. Some recent results of Monte Carlo simulations employing the Swendsen Wang and Wolff cluster update algorithm, as well as estimates using series expansions and field theory methods are compiled in Table 4.

By comparing the various Monte Carlo estimates collected in Tables 2–4 with results from field theory and series expansions it is fair to conclude that for $O(n)$ spin models modern Monte Carlo techniques are at the present time superior to series expansion analyses. The recently derived critical exponents are in fact competitive in accuracy with estimates obtained with the best and very elaborate methods of field theory.

Table 3. Critical coupling and selected critical exponents of the 3D Ising ($n = 1$) model

Method	β_c	ν	γ	Reference
g-expansion	–	0.6300(15)	1.241(2)	[102, 103]
ϵ-expansion	–	0.6310(15)	1.2390(25)	[104]
MCRG	0.221 652(3)	0.624(2)	–	[109]
MC	0.221 6595(26)	0.6289(8)	1.239(7)	[110]
MC	0.221 6546(10)	0.6301(8)	1.237(2)	[111]
series	0.221 655(5)	0.631(4)	1.239(3)	[112]

Table 4. Critical coupling and selected critical exponents of the 3D classical XY ($n = 2$) model

Method	β_c	ν	γ	Reference
^{4}He experiment	–	0.6705(6)	–	[114]
g-expansion	–	0.669(2)	1.3160(25)	[102, 103]
ϵ-expansion	–	0.671(5)	1.315(7)	[104]
MC	0.45421(8)	–	1.327(8)	[75]
MC	0.45408(8)	0.670(2)	1.316(5)	[76]
MC	0.45417(1)	0.662(7)	1.324(1)	[115]
series	0.45406(5)	0.67(1)	1.315(9)	[116]
series	0.45414(7)	≈0.673	≈1.325	[107]
series	0.45420(6)	0.679(3)	1.328(6)	[108]

Acknowledgements

The results for the 3D Heisenberg model are taken from joint work with C. Holm. It is a pleasure to thank him for a very fruitful and enjoyable collaboration. I am also grateful to S. Kappler and T. Sauer for numerous stimulating discussions and collaborations on various aspects of Monte Carlo simulations described here. Along the way, discussions with B. Berg, D. Johnston, and R. Villanova and their comments helped a lot. Last but not least I would like to thank K. Binder and D.P. Landau for sharing their experience and insight in many of the problems touched in these lecture notes. Finally, I gratefully acknowledge a Heisenberg fellowship from the Deutsche Forschungsgemeinschaft.

Appendix: Program Codes

The accompanying diskette contains five FORTRAN programs (is_clu.f, reweight.f, rew_his.f, 1dis_ex.f, 2disnm_ex.f) and two data files (3d_e004.plo, 3d_c004.plo) which can be used to reproduce Figs. 3, 7, and 8, and to generate an Ising model analog of Fig. 9. The output files denoted by ...plo are kept as simple as possible to allow easy plotting, e.g., with the standard utility gnuplot.

is_clu.f is a Monte Carlo simulation program for the nearest-neighbor Ising model with subroutines for the Wolff single-cluster (sc), Swendsen–Wang multiple-cluster (sw), Metropolis (me), and heat-bath (hb) update algorithm, which can be selected in the main parameter statement. All subroutines are set up to work for general D-dimensional (hyper-) cubic lattices of size L^D with periodic boundary conditions. The dimension, lattice size, simulation temperature, and the statistics parameters are also defined in the main parameter statement (the dimension and lattice size have to be changed globally in all subroutines as well). With the choice of parameters as given in is_clu.param_example, the 2D Ising (multiple-cluster) Monte Carlo data of Figs. 7 and 8 can be reproduced within about one minute run-time. Of course the detailed timing depends on the type of PC or workstation, but it should always be possible to run the simulation interactively. The average energy, magnetization, specific heat, susceptibility, Binder parameter, and cluster averages are written on standard output. Furthermore, the energy and magnetization histograms, and the "microcanonical" magnetization averages $\langle\langle|m|\rangle\rangle$ and $\langle\langle m^2\rangle\rangle$ (cp. (46)) at the simulation temperature are saved in the files ehis_b0.plo, mhis_b0.plo, malis_b0.plo, and m2lis_b0.plo for easy plotting, and in e_b0.his for further reweighting analyses (containing the necessary parameter informations). With these output files it is straightforward to produce for the Ising model a plot similar to Fig. 9. If desired the time evolution of the energy and magnetization measurements can be saved in the files e_series.plo and m_series.plo, respectively, by turning on the corresponding logical switches in the parameter statement.

For simplicity the standard UNIX random number generator RAND( ) is called in the MC program. For illustration purposes this generator is good enough, but for a serious simulation study it should at any rate be replaced by a more reliable random number generator. Again only for simplicity all statistical error analysis subroutines are omitted. For a sensible MC study this is clearly unacceptable.

reweight.f takes as input the energy histogram and the "microcanonical" magnetization lists stored in e_b0.his and computes by reweighting the energy and specific heat, the susceptibilities χ/β and χ'/β, and the Binder parameter as a function of inverse temperature β. The desired β-range can be set in the parameter statement, and the dimension and lattice size parameters must be the same as in is_clu.f. The results are written into the files e016_mc.plo, c016_mc.plo, sus016_mc.plo, chi016_mc.plo, and U016_mc.plo, respectively, where 016 indicates the linear lattice size.

The MC data for the one-dimensional Ising chain can be tested against the exact results provided by 1dis_ex.f for any chain length. The two-dimensional MC results can be used together with the output from 2disnm_ex.f to reproduce Fig. 7. Further comparison data for a 16^2 lattice from high-statistics single-cluster simulations at $\beta_c = \ln(1+\sqrt{2})/2$ are $2+e = 0.546\,85(10)$, $C = 1.4978(10)$, $\chi = 139.669(31)$, $\langle|C|\rangle = 139.656(29)$, and $U = 0.611\,537(50)$. In three dimensions the MC data can be compared with the exact energy and specific heat curves for a 4^3 lattice contained in the data files 3d_e004.plo and 3d_c004.plo.

rew_his.f reads again as input the energy histogram stored in e_b0.his, computes reweighted histograms, and stores them in, e.g., ehis_b4750.plo. The dimension and lattice size parameters must be the same as in is_clu.f. Here b4750 indicates that the histogram is reweighted to $\beta = 0.4750$. The new inverse temperature is inquired interactively by the program. In this way Fig. 8 can be reproduced. If gnuplot is used for plotting the histograms, then by using the escape character "!" the reweighting program can be called and the new histogram immediately displayed without leaving the plot session.

1dis_ex.f computes the exact temperature dependence of the energy, specific heat, and susceptibility of the one-dimensional Ising chain with periodic boundary conditions of arbitrary length L. For, e.g., $L = 16$, the results are stored in the files 1d_e016.plo, 1d_c016.plo, and 1d_chi016.plo.

2disnm_ex.f implements the exact solution of Ferdinand and Fisher (1969) for the 2D nearest-neighbor interaction Ising model on finite $L_x \times L_y$ lattices (L_x, L_y = even) with periodic boundary conditions. The desired lattice size and inverse temperature range can be chosen in the parameter statement of the main program. The output are two data files, e.g., 2d_e016.plo and 2d_c016.plo for $L_x = L_y = 16$, containing minus the internal energy per site, $-E/V$, and the specific heat per site, C, as a function of the inverse temperature β. By running this code for various lattice sizes, Fig. 3 can be reproduced.

References

[1] T.H. Berlin and M. Kac, Phys. Rev. **86**, 821 (1952)
[2] V.L. Berezinskii, Zh. Eksp. Teor. Fiz. **61**, 1144 (1971) [Sov. Phys. JETP **34**, 610 (1972)]
[3] J.M. Kosterlitz and D.J. Thouless, J. Phys. C **6**, 1181 (1973)
[4] J.M. Kosterlitz, J. Phys. C **7**, 1046 (1974)
[5] H. Kleinert, *Gauge Fields in Condensed Matter*, Vol. I (World Scientific, Singapore 1989)

[6] A. Patrascioiu and E. Seiler, Munich preprint MPI-PhT/95-73, hep-lat/9508014 (1995)

[7] R.B. Potts, Proc. Camb. Phil. Soc. **48**, 106 (1952)

[8] F.Y. Wu, Rev. Mod. Phys. **54**, 235 (1982)

[9] F.Y. Wu, Rev. Mod. Phys. **55**, 315(E) (1983)

[10] J.D. Gunton, M.S. Miguel, and P.S. Sahni, in *Phase Transitions and Critical Phenomena*, Vol. 8, eds. C. Domb and J.L. Lebowitz (Academic Press, New York 1983), p. 269

[11] K. Binder, Rep. Prog. Phys. **50**, 783 (1987)

[12] H.J. Herrmann, W. Janke, and F. Karsch, eds. *Dynamics of First Order Phase Transitions* (World Scientific, Singapore 1992)

[13] W. Janke, in *Computer Simulations in Condensed Matter Physics VII*, eds. D.P. Landau, K.K. Mon, and H.-B. Schüttler (Springer, Berlin 1994), p. 29

[14] H. Kleinert, *Gauge Fields in Condensed Matter*, Vol. II (World Scientific, Singapore 1989)

[15] W. Janke and H. Kleinert, Phys. Rev. B **33**, 6346 (1986)

[16] W. Janke and S. Kappler Nucl. Phys. B (Proc. Suppl.) **34**, 674 (1994)

[17] W. Janke and S. Kappler, Phys. Lett. A **197**, 227 (1995)

[18] W. Janke and S. Kappler, Nucl. Phys. B (Proc. Suppl.) **42**, 770 (1995)

[19] W. Janke and S. Kappler, Europhys. Lett. **31**, 345 (1995)

[20] W. Janke and S. Kappler, Mainz preprint KOMA-95-65 (September 1995), hep-lat/9509057, to appear in the proceedings of the conference *Lattice '95*, Melbourne, Australia, July 11–15, (1995)

[21] S. Kappler, Ph.D. Thesis, Johannes Gutenberg-Universität Mainz (1995)

[22] A.E. Ferdinand and M.E. Fisher, Phys. Rev. **185**, 832 (1969)

[23] K. Binder, in *Monte Carlo Methods in Statistical Physics*, ed. K. Binder (Springer, Berlin 1979), p. 1

[24] M.E. Barber, in *Phase Transitions and Critical Phenomena*, Vol. 8,eds. C. Domb and J.L. Lebowitz (Academic Press, New York 1983), p. 146

[25] V. Privman, ed., *Finite-Size Scaling and Numerical Simulations of Statistical Systems* (World Scientific, Singapore 1990)

[26] K. Binder, in *Computational Methods in Field Theory*, Schladming Lecture Notes, eds. H. Gausterer and C.B. Lang (Springer, Berlin 1992), p. 59

[27] M.E. Fisher and A.N. Berker, Phys. Rev. B **26**, 2507 (1982)

[28] V. Privman and M.E. Fisher, J. Stat. Phys. **33**, 385 (1983)

[29] K. Binder and D.P. Landau, Phys. Rev. B **30**, 1477 (1984)

[30] M.S.S. Challa, D.P. Landau, and K. Binder, Phys. Rev. B **34**, 1841 (1986)

[31] V. Privman and J. Rudnik, J. Stat. Phys. **60**, 551 (1990)

[32] C. Borgs and R. Kotecky, J. Stat. Phys. **61**, 79 (1990)

[33] C. Borgs and R. Kotecky, Phys. Rev. Lett. **68**, 1734 (1992)

[34] J. Lee and J.M. Kosterlitz, Phys. Rev. Lett. **65**, 137 (1990)

[35] J. Lee and J.M. Kosterlitz, Phys. Rev. B **43**, 3265 (1991)

[36] C. Borgs, R. Kotecky, and S. Miracle-Solé, J. Stat. Phys. **62**, 529 (1991)

[37] C. Borgs and W. Janke, Phys. Rev. Lett. **68**, 1738 (1992)

[38] W. Janke, Phys. Rev. B **47**, 14757 (1993)

[39] J.M. Hammersley and D.C. Handscomb, *Monte Carlo Methods* (London 1965)

[40] K. Binder and D.W. Heermann, *Monte Carlo Simulations in Statistical Physics: An Introduction* (Springer, Berlin 1988)

[41] D.W. Heermann, *Computer Simulation Methods in Theoretical Physics*, 2nd ed., (Springer, Berlin 1990)

[42] K. Binder, ed. *The Monte Carlo Method in Condensed Matter Physics* (Springer, Berlin 1992)

[43] N. Madras and A.D. Sokal, J. Stat. Phys. **50**, 109 (1988)

[44] A.M. Ferrenberg, D.P. Landau, and K. Binder, J. Stat. Phys. **63**, 867 (1991)

[45] W. Janke and T. Sauer, J. Stat. Phys. **78**, 759 (1995)

[46] N. Metropolis, A. Rosenbluth, M. Rosenbluth, A. Teller, and E. Teller, J. Chem. Phys. **21**, 1087 (1953)

[47] M. Creutz, Phys. Rev. D **36**, 515 (1987)

[48] S.L. Adler, Phys. Rev. D **37**, 458 (1988)

[49] H. Neuberger, Phys. Lett. B **207**, 461 (1988)

[50] R. Gupta, J. DeLapp, G.G. Battrouni, G.C. Fox, C.F. Baillie, and J. Apostolakis, Phys. Rev. Lett. **61**, 1996 (1988)

[51] J.D. Doll, R.D. Coalsen, and D.L. Freeman, Phys. Rev. Lett. **55**, 1 (1985)

[52] J. Goodman and A.D. Sokal, Phys. Rev. Lett. **56**, 1015 (1986)

[53] J. Goodman and A.D. Sokal, Phys. Rev. D **40**, 2035 (1989)

[54] G. Mack, in *Nonperturbative quantum field theory*, Cargèse lectures 1987, ed. G.'t Hooft et al. (Plenum, New York 1988), p. 309

[55] D. Kandel, E. Domany, D. Ron, A. Brandt, and E. Loh, Jr., Phys. Rev. Lett. **60**, 1591 (1988)

[56] D. Kandel, E. Domany, and A. Brandt, Phys. Rev. B **40**, 330 (1989)

[57] G. Mack and S. Meyer, Nucl. Phys. B (Proc. Suppl.) **17**, 293 (1990)

[58] A.D. Sokal, *Monte Carlo Methods in Statistical Mechanics: Foundations and New Algorithms*, Cours de Troisième Cycle de la Physique en Suisse Romande, Lausanne (1989)

[59] A.D. Sokal, *Bosonic Algorithms*, in *Quantum Fields on the Computer*, ed. M. Creutz (World Scientific, Singapore 1992), p. 211

[60] R.H. Swendsen and J.-S. Wang, Phys. Rev. Lett. **58**, 86 (1987)

[61] U. Wolff, Phys. Rev. Lett. **62**, 361 (1989)

[62] P.W. Kasteleyn and C.M. Fortuin, J. Phys. Soc. Japan **26** (Suppl.), 11 (1969)

[63] C.M. Fortuin and P.W. Kasteleyn, Physica **57**, 536 (1972)

[64] C.M. Fortuin, Physica **58**, 393 (1972)

[65] C.M. Fortuin, Physica **59**, 545 (1972)

[66] U. Wolff, Nucl. Phys. B **322**, 759 (1989)

[67] U. Wolff, Nucl. Phys. B **334**, 581 (1990)

[68] M. Hasenbusch, Nucl. Phys. B **333**, 581 (1990)

[69] C.F. Baillie, Int. J. Mod. Phys. C **1**, 91 (1990)

[70] R.H. Swendsen, J.-S. Wang, and A.M. Ferrenberg, in *The Monte Carlo Method in Condensed Matter Physics*, ed. K. Binder (Springer, Berlin 1992), p. 75

[71] U. Wolff, Phys. Lett. A **228**, 379 (1989)

[72] D.W. Heermann and A.N Burkitt, Physica A **162**, 210 (1990)

[73] P. Tamayo, Physica A **201**, 543 (1993)

[74] N. Ito and G.A. Kohring, Physica A **201**, 547 (1993)

[75] M. Hasenbusch and S. Meyer, Phys. Lett. B **241**, 238 (1990)

[76] W. Janke, Phys. Lett. A **148**, 306 (1990)

[77] C. Holm and W. Janke, Phys. Rev. B **48**, 936 (1993)

[78] K. Binder, Z. Phys. B **43**, 119 (1981)

[79] K. Binder, Phys. Rev. A **25**, 1699 (1982)

[80] A. Billoire, R. Lacaze, A. Morel, S. Gupta, A. Irbäck, and B. Petersson, Nucl. Phys. B **358**, 231 (1991)

[81] B.A. Berg and T. Neuhaus, Phys. Lett. B **267**, 249 (1991)

[82] B.A. Berg and T. Neuhaus, Phys. Rev. Lett. **68**, 9 (1992)

[83] W. Janke, B.A. Berg, and M. Katoot, Nucl. Phys. B **382**, 649 (1992)

[84] K. Rummukainen, Nucl. Phys. B **390**, 621 (1993)

[85] W. Janke and T. Sauer, Phys. Rev. E **49**, 3475 (1994)

[86] T. Sauer, Ph.D. Thesis, Freie Universität Berlin (1994)

[87] W. Janke and S. Kappler, Nucl. Phys. B (Proc. Suppl.) **42**, 876 (1995)

[88] W. Janke and S. Kappler, Phys. Rev. Lett. **74**, 212 (1995)

[89] A.M. Ferrenberg and A.M. Swendsen, Phys. Rev. Lett. **63**, 1658(E) (1989)

[90] A.M. Ferrenberg and A.M. Swendsen, Phys. Rev. Lett. **61**, 2635 (1988)

[91] N.A. Alves, B.A. Berg, and S. Sanielevici, Nucl. Phys. B **376**, 218 (1992)

[92] R.G. Miller, Biometrika **61**, 1 (1974)

[93] B. Efron, *The Jackknife, the Bootstrap and Other Resampling Plans* (SIAM, Philadelphia 1982)

[94] A.M. Ferrenberg and A.M. Swendsen, Phys. Rev. Lett. **63**, 1195 (1989)

[95] P. Peczak and D.P. Landau, Bull. Am. Phys. Soc. **35**, 255 (1990)

[96] P. Peczak and D.P. Landau, J. Appl. Phys. **67**, 5427 (1990)

[97] P. Peczak and D.P. Landau, Phys. Rev. B **47**, 14260 (1993)

[98] P. Peczak, A.M. Ferrenberg, and D.P. Landau, Phys. Rev. B **43**, 6087 (1991)

[99] C. Holm and W. Janke, Phys. Lett. A **173**, 8 (1993)

[100] C. Holm and W. Janke, J. Phys. A **27**, 2553 (1994)

[101] E. Brézin and J. Zinn-Justin, Nucl. Phys. B **257** [FS14], 867 (1985)

[102] J.C. Le Guillou and J. Zinn-Justin, Phys. Rev. Lett. **39**, 95 (1977)

[103] J.C. Le Guillou and J. Zinn-Justin, Phys. Rev. B **21**, 3976 (1980)

[104] J.C. Le Guillou and J. Zinn-Justin, J. Physique Lett. **46**, L137 (1985)

[105] I. Dimitrović, P. Hasenfratz, J. Nager, and F. Niedermayer, Nucl. Phys. B **350**, 893 (1991)

[106] K. Chen, A.M. Ferrenberg, and D.P. Landau, Phys. Rev. B **48**, 3249 (1993)

[107] J. Adler, C. Holm, and W. Janke, Physica A **201**, 581 (1993)

[108] P. Butera and M. Comi, Phys. Rev. B **52**, 6185 (1995)

[109] C.F. Baillie, R. Gupta, K.A. Hawick, and G.S. Pawley, Phys. Rev. B **45**, 10438 (1992)

[110] A.M. Ferrenberg and D.P. Landau, Phys. Rev. B **44**, 5081 (1991)

[111] H.W.J. Blöte, E. Luijten, and J.R. Heringa, J. Phys. A **28**, 6289 (1995)

[112] J. Adler, J. Phys. A **16**, 3585 (1983)

[113] A. Rosengren, J. Phys. A **19**, 1709 (1986)

[114] L.S. Goldner and G. Ahlers, Phys. Rev. B **45**, 13129 (1992)

[115] A.P. Gottlob and M. Hasenbusch, Physica A **201**, 593 (1993)

[116] P. Butera, M. Comi, and A.J. Guttmann, Phys. Rev. B **48**, 13987 (1993)

Metastable Systems and Stochastic Optimization*

Karl Heinz Hoffmann

Institut für Physik, Technische Universität, D-09107 Chemnitz, Germany
e-mail: hoffmann@physik.tu-chemnitz.de

Abstract. Metastable systems such as spin glasses show a wealth of interesting relaxation phenomena. Stochastic optimization procedures such as simulated annealing help to solve a number of industrially important minimization problems. Here we show that the two fields are intimately connected by the thermally activated relaxation dynamics of complex energy landscapes. The numerical as well as the analytical tools to analyse it are discussed. Finally two applications, aging phenomena in spin glasses and adaptive simulated annealing procedures, are presented.

1   An Introduction to Complex Systems

Metastable systems and stochastic optimization – at first sight one wonders what these two fields have in common. Why is it worth discussing them together?

The aim of this paper is to show that both fields are intimately connected, and that concepts from one field can be put to use in the other. Let us start by first describing what these fields deal with.

Metastable systems are characterized by states which decay very slowly over long periods of time. The usual idea is that some mechanism prevents the system from reaching its true equilibrium easily, instead the system might be caught in some state which can only be left very slowly – a metastable state. An example is a system with deep wells in its free energy function, from which an escape can take a very long time. Such a system should reveal itself when, for instance, its thermal relaxation is studied.

Stochastic optimization procedures on the other hand try to provide solutions to optimization problems which have many local minima in their objective function. For these optimization problems the usual steepest descent algorithms fail as they get easily caught in local minima. For such problems stochastic optimization procedures, and especially simulated annealing, have been used with growing success – not to determine the global minimum but to provide "good" solutions with values of the objective which are not too "far" apart from the desired global minimum. Very often the global minimum can only be determined by an enumeration of all feasible solutions, and their number will increase exponentially with the problem size. Thus these problems are sometimes called NP-hard.

---

* Software included on the accompanying diskette.

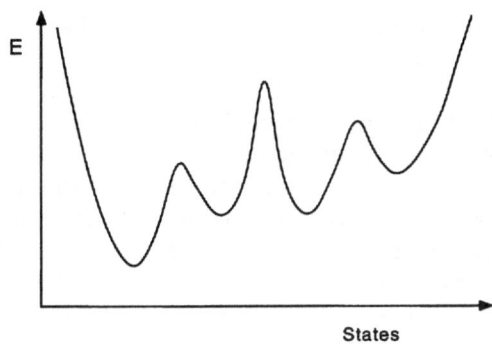

Fig. 1. A sketch of a complex state space. The energy is depicted as a function of a sequence of neighboring states. It resembles a cut through a mountainous landscape and thus the energy function is sometimes refered to as the energy landscape

From this short characterization of the two fields it becomes clear that in both cases the existence of a function on a high-dimensional state space which exhibits many local minima is of great importance. Such a state space is often called complex. Figure 1 shows a sketch of a such a complex state space. It is important to note that in order to draw a figure like that one needs a definition of what the neighboring states are. Otherwise one would not know what the abscissa means. From a mathematical point of view one needs this definition to define a local minimum (a state where all *neighbors* have higher energy).

A typical example of a system with a complex state space is an Ising spin glass [1]. Its energy is given by

$$E = \sum_{i,j} J_{ij} s_i s_j, \tag{1}$$

where the Ising spins s_i can only take the values $+1$ or -1, and the coupling constants J_{ij} are random quantities, which can take positive and negative values. Here the states are defined by the configuration of all spins $\{s_i\}$, and neighboring states are obtained from each other by flipping one of the spins.

An often-used picture for complex state spaces is that of a mountainuous landscape, where the heights of the mountains represent the energy and the two horizontal axis have to mimic two of the many dimensions of the physical system. Even though this picture is very suggestive, one should bear in mind that the high dimensionality of the physical system might not be properly taken care of. So generally speaking this picture can lead the intuition, but the consequences should be mathematically checked.

We shall now proceed by discussing the dynamics in such complex state spaces. We are particularly interested in the thermal relaxation dynamics, i.e., the dynamics which describe the equilibration of the system in contact with a heat bath. It will turn out that there are two major tools to analyse the thermal relaxation. One will come from computational physics, namely simulation methods, and the other will come from the realm of theoretical physics, namely the theory of Marcov processes [2].

After these two tools have been presented in Sect. 2, we shall use them to discuss two particular problems occuring in complex systems. The first problem (Sect. 3) comes from the field of metastable systems and deals with the way the above-described techniques can be used to model the relaxation of complex systems in contact with a constant temperature heat bath. This problem will also serve as an example for the application of Marcov processes.

The second problem (Sect. 4) comes from the field of stochastic optimization and deals with finding the global minimum of such complex systems, or – in physical terms – the ground state. This problem will make use of simulation techniques, and will thus provide some insight into the problems there.

2   Dynamics in Complex Systems

The link between the two topics of this paper is the complex state space. For both fields it is important to understand the dynamics in these state spaces. For the physical systems the dynamics of interest is the thermal relaxation in contact with a heat bath. For the stochastic optimization schemes it will turn out that the same dynamics can be very profitably used to obtain solutions to complex optimization problems. So let us first turn to the thermal relaxation in contact with a heat bath.

We will view it as a thermally induced hopping process between the configurations of the system. The energy fluctuations between system and bath will allow transitions between different states or configurations with varying probability depending on their connectivity and energy.

There are now two tools available to analyse the behavior of the system. One way to proceed is to use simulations of the thermal hopping process, the other is to use the theory of Marcov processes. Interestingly, it turns out that the latter can also be used as a theory to describe the stochastic simulations.

2.1   Thermal Relaxation Dynamics: The Metropolis Algorithm

Let us first deal with the algorithmic approach using simulations. The aim in simulating a system in contact with a heat bath is to create a Boltzmann distribution in the state space according to the temperature of the heat bath. This problem is solved by the Metropolis algorithm [3].

Let $\Omega = \{\alpha\}$ represent this state space, let $E(\alpha)$ be the energy defined on the state space, and let T be the temperature of the heat bath in which the physical system is immersed. In addition a so-called neighborhood relation or move class is needed, which typically takes the form of an undirected graph structure on the state space. We will denote by $N(\alpha)$ the set of neighbors of a state α in this graph. As an example remember the above-mentioned case of an Ising spin glass, where two states (= spin configurations) usually are neighbors if they differ by one spin flip. We remark that for certain problems one may have several alternative move classes available, for instance for the spin-glass case one could also use states as neighbors which differ by more than one spin flip.

To start the algorithm a state ω_0 is chosen at random. Then at each step of the algorithm a neighbor ω' of the current state ω_k is selected at random to become the candidate for the next state. It actually becomes the next state only with probability

$$P_{\text{acceptance}} = \begin{cases} 1 & \text{if } \Delta E \leq 0 \\ \exp(-\Delta E/T) & \text{if } \Delta E > 0, \end{cases} \tag{2}$$

where $\Delta E = E(\omega') - E(\omega_k)$, and $k_B = 1$ by choice of units. If this candidate is accepted, then $\omega_{k+1} = \omega'$, otherwise the next state is the same as the old state, $\omega_{k+1} = \omega_k$.

This probabilistic decision rule is implemented by choosing a random number r, uniformly distributed between 0 and 1 and compare it with the Boltzmann factor $\exp(-\Delta E/T)$. If $r < \exp(-\Delta E/T)$, then $\omega_{k+1} = \omega'$, else $\omega_{k+1} = \omega_k$.

2.2   Thermal Relaxation Dynamics: A Marcov Process

A different way of studying thermal relaxation is by modeling it as a discrete time Marcov process. The thermally induced hopping process induces a probability distribution in the state space. The time development of $P_\alpha(k)$, the probability to be in state α at step k, can then be described by a master equation [2]

$$P_\alpha(k+1) = \sum_\beta \Gamma_{\alpha\beta}(T)P_\beta(k). \tag{3}$$

The transition probabilities $\Gamma_{\alpha\beta}(T)$ depend on the temperature T and can be chosen in a number of ways, but always they have to insure that the stationary distribution is the Boltzmann distribution $P_\alpha^{\text{eq}}(T) = g_\alpha \exp(-E_\alpha/T)/Z$, where $Z = \sum_\alpha P_\alpha^{\text{eq}}$ is the partition function and g_α is the degeneracy of state α. The latter is needed if the states α already represent quantities which include more than one microstate.

One of the possible choices for the transition probabilities is the Glauber dynamics [4], which we here just mention. Another one is the Metropolis dynamics, which turns out to be also the one which describes the process induced by the above-described Metropolis algorithm. Its transition probabilities are defined as follows.

First one defines the infinite-temperature transition probabilities $\Pi_{\beta\alpha} = \Gamma_{\beta\alpha}(\infty)$ from state α to β by

$$\Pi_{\beta\alpha} = \begin{cases} 0 & \text{if } \beta \notin N(\alpha) \\ 1/\mid N(\alpha) \mid & \text{if } \beta \in N(\alpha), \end{cases} \tag{4}$$

where $\mid N(\alpha) \mid$ is the number of neighbors of α. These are the transition probabilities if the algorithm automatically accepts each attempted move, i.e., if $T = \infty$.

At finite temperature the acceptance decision is superimposed on Π in (4) to give $\Gamma(T)$ defined by

$$
\Gamma_{\beta\alpha} = \begin{cases} \Pi_{\beta\alpha} \exp(-\Delta E/T) & \text{if } \Delta E > 0, \ \alpha \neq \beta \\ \Pi_{\beta\alpha} & \text{if } \Delta E \leq 0, \ \alpha \neq \beta \\ 1 - \sum_{\xi \neq \alpha} \Gamma_{\xi\alpha} & \text{if } \alpha = \beta, \end{cases} \tag{5}
$$

where now $\Delta E = E(\beta) - E(\alpha)$.

2.3   Thermal Relaxation Dynamics: A Simple Example

A simple example will show, why the Metropolis algorithm creates a Boltzmann distribution as the stationary solution of (3) in state space. Let us consider a (rather artificial) state space with states numbered by natural numbers i and a neighborhood relation such that states with successive numbers are neighbors $N(i) = \{i - 1, i + 1\}$ and $N(0) = \{1\}$. The energy is $E(i) = iE_0$. Thus the transition probabilities are 0, apart from

$$
\Gamma_{i+1,i} = \frac{1}{2} \exp(-E_0/T),
$$

$$
\Gamma_{i-1,i} = \frac{1}{2}, \tag{6}
$$

$$
\Gamma_{i,i} = 1 - \frac{1}{2}(\exp(-E_0/T) + 1).
$$

State 0 has to be treated differently, as it has no lower neighbor we find $\Gamma_{1,0} = \exp(-E_0/T)$ and $\Gamma_{0,0} = 1 - \exp(-E_0/T)$.

In the stationary solution P_0^{eq} is constant, thus the probability flow $\Gamma_{1,0}P_0^{eq}$ from state 0 to state 1 has to equal the reverse flow $\Gamma_{0,1}P_1^{eq}$ from state 1 to state 0. Then the same must be true for the flow between state 1 and state 2, and so forth. So we find

$$
\Gamma_{i+1,i}P_i^{eq} = \Gamma_{i,i+1}P_{i+1}^{eq} \tag{7}
$$

and thus for $i > 0$

$$
\frac{\Gamma_{i+1,i}}{\Gamma_{i,i+1}} = \exp(-E_0/T) = \frac{\exp(-(i+1)E_0/T)}{\exp(-(i)E_0/T)} = \frac{P_{i+1}^{eq}}{P_i^{eq}}, \tag{8}
$$

as expected for the Boltzmann distribution.

For $i = 0$ we find

$$
\frac{\Gamma_{1,0}}{\Gamma_{0,1}} = \frac{|N(1)|}{|N(0)|} \exp(-E_0/T)
$$

$$
= \frac{2}{1} \exp(-E_0/T) = \frac{P_1^{eq}}{P_0^{eq}}, \tag{9}
$$

which shows that the infinite temperature equilibrium distribution gives twice the probability to state 1 as it gives to state 0: $g_1/g_0 = 2$.

Usually statistical mechanics requires for a microscopic state space that the stationary distribution $P_\alpha^{eq}(\infty) = g_\alpha/Z$ of $\Pi = \Gamma(\infty)$ has to be uniform [5]. An easy way to achieve this is to make $| N(\alpha) |$ constant over α, i.e., to make each vertex have the same number of neighbors. If however nonuniform degeneracies g_α are needed, then the transition rates in the Metropolis algorithm need to be adjusted accordingly by choosing an appropriate move class. Note that one has only one free choice: either one defines a move class, which then fixes the degeneracies, or one requires certain degeneracies and that puts constraints on the move class.

If in a given system (7) holds for all neighbors,

$$\Gamma_{\beta,\alpha}P_\alpha^{eq} = \Gamma_{\alpha,\beta}P_\beta^{eq}, \tag{10}$$

then the system is said to obey detailed balance, as the probability flows along each bond of the neighborhood graph balance out separately. While in the above example detailed balance holds due to its structure, in more complicated systems one has to prove or assume that it holds.

3   Modeling Constant-Temperature Thermal Relaxation

In this section we want to study the thermal relaxation of a complex system at constant temperature by means of a Marcov process model. We are then immediately confronted with a major problem common to macroscopic physical systems described on a microscopic level: the enormous number of states. Some simplification is asked for and a starting point for such an operation is given by the rough qualitative picture of the thermal relaxation process painted as a random hopping process in a mountain range (or diffusion if considered in a continuous time limit).

This picture suggests that – as in real mountains – the movement within one valley is easy compared to the movement from one valley into another, which involves the crossing of a pass. In terms of the thermal relaxation this means that within such a valley in state space the relaxation proceeds quite fast and local equilibrium is obtained after a short while. Then the barrier crossing between two valleys takes much longer and – as there are valleys inside larger valleys inside larger valleys etc. – we have to expect a whole spectrum of longer and longer relaxation times.

Note that, again as in real mountains, the transition of a pass might not only be influenced by the height (the energy) of the pass alone, there might be also a dynamical restriction (for instance the width of a pass) which needs its own modeling. Nonetheless it is important to realize that already the complex energy landscape leads to a slow down of the relaxation.

3.1 Coarse-Graining a Complex State Space

In this subsection we show how the above concept can be used in a more technical fashion.

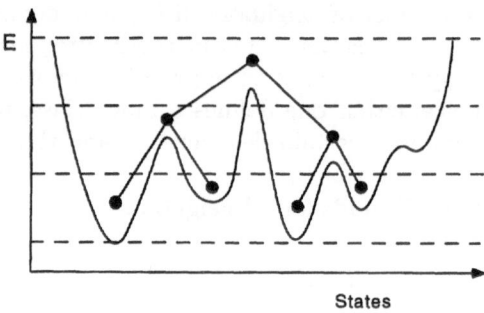

Fig. 2. The coarse-graining of a complex state space is shown. All connected states within one energy band (here indicated by the *dashed lines*) are lumped into the nodes of a tree

Figure 2 depicts the underlying idea [6]. Suppose one introduces hyperplanes with energy

$$E(l) = \Delta E_t \cdot l \tag{11}$$

into the state space and calls all states between two neighboring hyperplanes an energy band. The connected (in the sense of the neighborhood graph introduced above) parts of the state space within an energy band are then lumped together to represent the nodes of a new coarse-grained structure.

If ΔE_t is large compared with the energy change $\Delta E_{\mathrm{micro}}$ of a transition between microstates, then the connectivity between the microstates induces a unique connectivity between the nodes as indicated in Fig. 2. The resulting structure has a tree topology. Each node has only one connection to a higher energy node, this is called the mother node. The connected nodes at lower energy are called daughter nodes. The number of daughter nodes will vary from node to node. Also the number of states lumped into a node will vary. In general, we expect that the higher the nodes are in energy the more states they will include, i.e., the degeneracy of a node will increase with its energy.

There are also other coarse-graining procedures [6] which lead to trees with nodes not separated by equal energy intervals. The important point is that a hierarchical tree-like structure is the result, and thus trees can be regarded as a generic coarse-grained structure for complex state spaces.

An even coarser lumping would collect all states belonging to one energy band into one node. This leads directly to the simple example presented in Sect. 2.3. However, this would lump [7] states together which could reach each other in the underlying complex state space only via higher barriers, i.e., states belonging to a higher energy band. Thus the essential feature of a complex state space, the crossing of barriers, would be neglected and so all the complexity of the state space would be lost in the model! It is thus important that only the connected parts of the state space are lumped together.

Consider now the thermal hopping in the complex energy landscape. Rather than calculating the full-time dependence of the probability distribution in the state space, we can choose to monitor the presence or absence from a node of the corresponding tree. We have hereby defined a stochastic process, which in general will not be a Marcov process because the induced transition probability from node to node might depend on the internal (microscopic) distribution within one node.

However, it turns out [8] that inside a coarse-grained area a kind of local equilibrium distribution is very quickly established, which then makes the coarse-grained relaxation process (at least approximately) Marcovian. The result is that Marcov processes on tree structures are good modeling tools for the thermal relaxation of complex systems [6, 9, 10].

3.2 Tree Dynamics

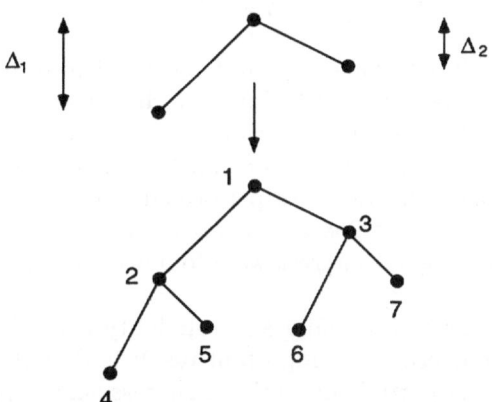

Fig. 3. The construction of an LS tree

Thermal relaxation dynamics have been studied for a number of tree structures. We present here one example: the LS tree. Following [11, 12] the tree is constructed as shown in Fig. 3. The building block is a 'mother' node connected to two 'daughters' of lower energy, the energy differences I being Δ_1 and Δ_2 respectively. The tree is constructed iteratively out of one initial block by (a) duplicating the already existing structure, and (b) identifying the top nodes of the two resulting twins with the daughters of the building block. For $\Delta_1 = \Delta_2$ the structure is regular and coincides with that studied, e.g., in [9].

A level in the tree is the set of nodes connected by the same number of bonds to the root (the top node of the tree). Each node has thus a unique level index l, with the local energy minima being at level 0, their mothers at level 1 and so forth up to the root node being at level l_{max}, which is the height of the tree. Each node is contained in a unique subtree of height m, dubbed the mth subtree containing that node. The whole tree contains $N = 2^{l_{max}+1} - 1$ nodes. For each node α we introduce a degeneracy $g(\alpha)$, which can account for the number of microstates lumped into the node in a coarse-graining procedure. We assume a level-dependent degeneracy $g(\alpha) = g(l) = \kappa^l$ for all nodes at level l. κ is a free model parameter.

For the dynamics it is more convenient to use a continuous time description

$$\partial_t P_\alpha(t) = \sum_\beta \Gamma^c_{\alpha,\beta} P_\beta(t), \tag{12}$$

with the transition matrix $\Gamma^c_{\alpha,\beta} = \Gamma_{\alpha,\beta} - \delta_{\alpha,\beta}$. Specifically the transition probability Γ^c is chosen to be:

$$\Gamma^c_{mother,daughter} = s^{(I)}\kappa e^{-I/T}, \tag{13}$$

$$\Gamma^c_{daughter,mother} = s^{(I)} \quad , \tag{14}$$

where $I = \Delta_1, \Delta_2$ indicates the energy difference between mother and daughter nodes. Note that the parameter $s^{(I)}$ does not change the (Boltzmann) equilibrium distribution on the tree, it only defines the time scale.

The time development of the above-defined model has been thoroughly studied by numerical diagonalization. We performed most calculations on a tree of height $l_{max} = 7$, leading to a 255×255 transition matrix Γ. Finite size effects for the temperature range of interest were found to be quite negligible for this system size.

Consider the case where initially all probability is concentrated within one local minimum. The overall and important result is that the equilibration procedes by a sequence of partial equilibria in successively larger subtrees. Figure 4 shows for instance the time development of the probability within subtrees of different heights. One very characteristic feature in this figure is the power-law decay of the probability, which can be traced back to the regular hierarchical structure of the model.

In order to see the effect of the parameter $s^{(I)}$, let us consider the typical situation in which a complex system is quenched into a low energy-state. If the low-energy part of the phase space can meaningfully be modeled by a tree, highly asymmetric downward rates imply that the initial condition for the thermal relaxation process following the quench can be concentrated into a small region. In our simple model, an arbitrarily sharp initial condition can be obtained for instance by choosing s^{Δ_2} sufficiently large. More generally we expect the values of $s^{(I)}$ to play a role when temperature variations are important [13].

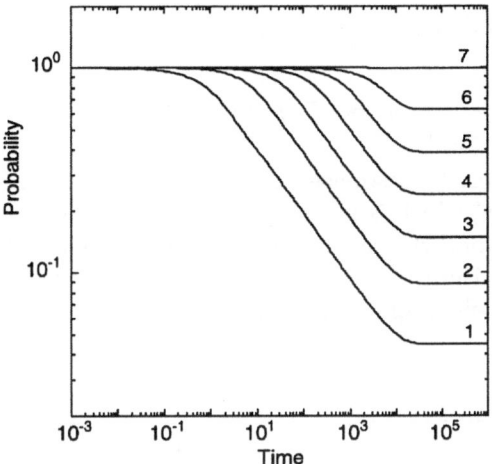

Fig. 4. The probability contents in subtrees of different height (1 to 7) are shown as a function of time. Note the power-law dependence

3.3 A Serious Application: Aging Effects in Spin Glasses

We now present an application of the above ideas to demonstrate the power of the modeling approach. We show that a number of highly interesting experimental results for spin glasses can be explained by a very simple relaxation model based on a tree structure.

First we describe briefly the experiments. Aging effects were first observed in spin glass systems [14, 15, 16, 17, 18, 19] and have also been measured in high-T_c superconductors [20] and CDW systems [21] as well. In the so-called Z(ero)F(ield)C(ooled) experiments a sample is cooled to a low temperature and 'aged' for a 'waiting time' t_w, without field. Thereafter a small field is applied, and the response of the magnetization to the perturbation is measured. Contrary to one's naive expectations, the response depends both on the time during which the system has been acted upon by the field and on the waiting time. This situation persists through many decades, and indicates that the system never reaches thermodynamic equilibrium during the observation time.

In the spin glass ZFC magnetization experiments the applied field H is kept constant, and the salient feature of the data is a kink in the magnetization $M(t, t_w)$ plotted as a function of logarithmic time at $t = t_w$ or equivalently, a maximum in the derivative $S(t, t_w)$ of the magnetization with respect to the logarithm of the time at $t = t_w$. This feature is only present for temperatures below the critical spin glass temperature T_g.

Rather than trying to coarse-grain the state space of a microscopic spin glass model, the starting point for the model is a tree structure as presented in the previous subsection. We now show that the model is able to reproduce the main aging features of the experiments. In order to calculate response properties, we use linear response theory

$$M(t, t_{\rm w}) = \int_0^t {\rm d}t' R(t', t_{\rm w}) H(t'), \qquad (15)$$

with the nonequilibrium response function $R(t', t_{\rm w})$ [22] rather than the usual correlation function, which is not appropriate for this nonequilibrium situation.

It would be beyond the scope of this paper to discuss the response function $R(t', t_{\rm w})$ in detail. Here it is important that, apart from magnetic properties which have to be defined in addition, $R(t', t_{\rm w})$ depends crucially on the solution $P_\alpha(t)$ of (12).

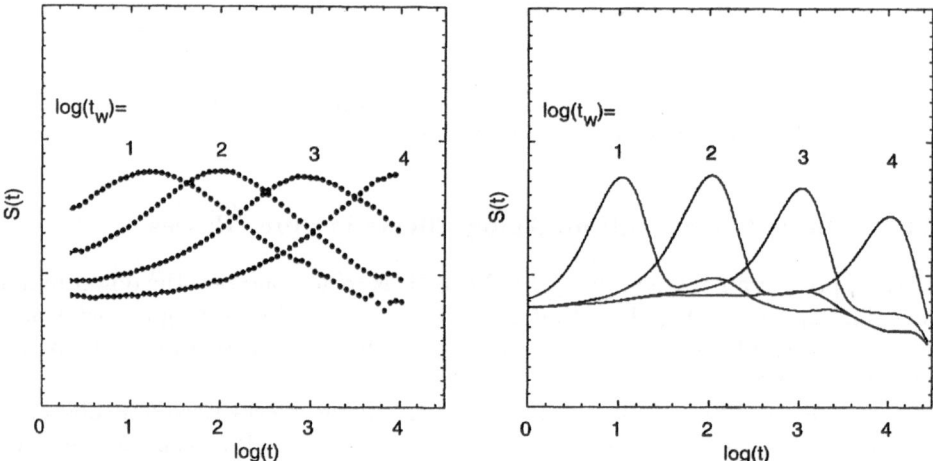

Fig. 5. A comparison between the experimental data (*left*) for ZFC experiments and the tree-model data (*right*)

Figure 5 shows a comparison between the experimental data and the model result. Note that the latter reproduces very well the maxima in the relaxation rate at times which correspond to the waiting times [23, 24].

There are a number of further experiments which measure the response as temperatures are changed during the waiting time. This leads to partial reinitialization effects which can also be reproduced very well by this simple model. For an in-depth discussion see [13].

Finally we note that a theoretical understanding of the physical mechanism behind the aging phenomenon of spin glasses has important implications for the more general issue of describing relaxation phenomena in complex systems such as glasses and polymers, as well as some nonphysical systems which are currently treated by statistical mechanical methods, for instance Boltzmann machines and simulated annealing schemata [25].

4   Stochastic Optimization:
How to Find the Ground State of Complex Systems

Whereas the previous section was devoted to the thermal relaxation of complex systems in contact with a fixed-temperature bath, we shall now turn towards the relaxation process under the influence of temperature changes. The reason for doing so is the problem of finding the ground state in complex systems. By now it is certainly clear that the multivalley state space structure with its many local minima makes it quite complicated to find the ground state. Simple descent methods which rely on following the energy function down to states with lower and lower energy will invariably lead into one of the many local minima, but rarely into the global one. Recent investigations [26] show that the number of local minima can grow exponentially with the system size, thus just hoping to find the ground state by chance or a few repeated trials does not work. So, other methods are needed.

4.1   Simulated Annealing

One idea for solving this problem was born in the early eighties: simulated annealing [27, 28]. It is based on the observation that a careful annealing of physical systems brings them closer to equilibrium than a quenching process. This is well known and widely used in the preparation of single crystals or in the manufacturing of telescope mirrors. Careful annealing of a real physical system brings it into its equilibrium with the ambient temperature T and thus for $T \to 0$ the system moves into its groundstate(s).

But how can this fact be used for finding ground states of spin glass *models*?

The answer is simple. Simulate the thermal relaxation of the system by the Metropolis algorithm and slowly turn down the temperature parameter which enters in the transition rate: simulated annealing!

One quickly realizes that this procedure has a much wider application range than physical models. Given a state space with an (energy) function with many local minima for which the global minimum is sought, all one needs to do is to artificially introduce a neighborhood relation on the states and then perform a random hopping process according to the Metropolis rules, where the temperature is now just an external parameter which has to be lowered properly. Simulated annealing can thus be used as a general stochastic optimization tool.

Simulated annealing has been applied to a wide range of problems with a complex state space structure. Apart from finding the ground state of a spin glass [29] simulated annealing has also proved a useful tool in the design of integrated circuits [27, 30, 31] for partitioning, routing, and placement [32]. It has been applied to many other problems including the traveling salesman [33, 34], graph partitioning [35], restoration of images [36], and parameter estimations [37]. While this list is far from exhaustive, it shows that the problems attacked by simulated annealing are of great scientific and industrial importance.

A thorough analysis showed [36] that the simulated annealing procedure indeed finds the ground state with a probability of 1. However, the annealing schedule, i.e., the way in which the temperature parameter T is lowered as a function of time, needs to be very slow and is

$$T(t) \sim \frac{1}{\ln t}. \tag{16}$$

Note that to reach the ground state with a probability of 1 one needs infinite time.

Thus for the finite time usually available to humans one has to live with the fact that the ground state is not found with a probability of 1. That brings us to the problem of finding that schedule which provides the "best" solution we can get under the restriction of finite (computer) time, which in our case translates into a finite number of Metropolis steps. In other words the question is, what is the "optimal" schedule?

Before this question can be answered the yardstick with respect to which the optimality is determined must be defined. Indeed several criteria are possible, the two commonly used are: (a) *the final energy*, (b) *the BSF energy* $E_{\text{BSF}}(k) = \min_{0 < k' < k} E(k')$, i.e., the lowest energy seen up to a certain time k.

The final and the BSF energies are stochastic quantities [38]. Their probability distribution evolves with time and is induced by the underlying random walk in the state space governed by the Metropolis algorithm. The distribution as such cannot be optimized, only certain aspects (for instance its mean, its median, or its mode). The choice between the different criteria has to be made externally. After a choice has been made, the determination of the optimal schedule becomes a new optimization problem, which can be attacked analytically or numerically.

4.2   Optimal Simulated Annealing Schedules: A Simple Example

Below we show a very simple example for which the optimal schedule has been determined. The example system consists of only three states $\Omega = \{1, 2, 3\}$ and shows how the crossing of a single barrier is optimized. The states have energies $E(1) = 0, E(2) = 1$, and $E(3) = B > 1$, and the move class is $N(1) = \{3\}$, $N(2) = \{3\}, N(3) = \{1, 2\}$. Thus 1 is the global minimum, 2 is the local one, and 3 represents the barrier in between.

For this simple model optimal schedules for the mean final energy were determined [39]. The knowledge of the optimal schedule allows a comparison between different schedules. For instance it was shown that the optimal schedule performs much better than any exponential schedule $T(t) = T_0 \nu^t$, or linear schedule $T(t) = T_0 - \epsilon t$, thus providing some indication for the potential gains. In Fig. 6 this is demonstrated. The mean final energy for the optimal, the linear, and the exponential schedules are compared as a function of the total available annealing time τ. The linear and the exponential schedules use the best possible values for

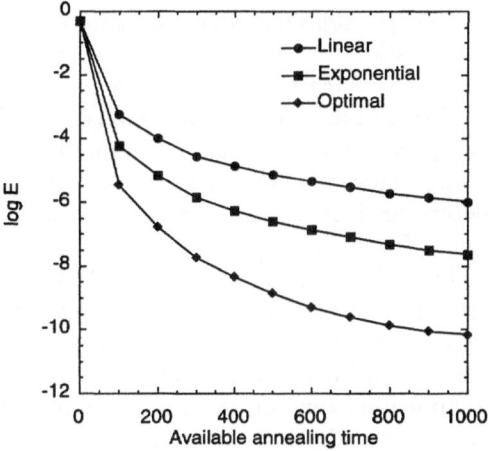

Fig. 6. A comparison between the mean final energy as a function of the available annealing time for the best linear, the best exponential and the optimal schedules

ϵ and ν. Note how much better the optimal schedule performs compared to the other two.

Analysing this simple model shows that the barrier height enters the optimal schedule as an essential parameter. Later optimal schedules [40, 41] for larger systems were determined numerically. It turned out that the optimal schedules are dominated by a single barrier during certain time intervals.

Summarizing these studies of optimal schedules one sees it is essential to hold the system close enough to equilibrium in order to prevent getting trapped in local minima. On the other hand one has yet to maintain a certain disequilibrium in order to anneal as quickly as possible.

4.3 Adaptive Annealing Schedules and the Ensemble Approach to Simulated Annealing

Investigations of truly optimal schedules for simple systems [39, 41, 42] have shown that the schedule depends critically on the barrier height which has to be overcome to leave a local minimum. In the usual optimization problem these barrier heights are unknown, moreover they differ from problem to problem. Thus, the schedule has to be adapted to the problem.

Adaptive schedules using information gathered during the annealing have already been suggested before [43, 44]. Here we present as a simple example an adaptive schedule which has been proven to work well on spin glass and standard travelling salesman problems. The schedule is easy to implement and has only negligible computational overheads.

It is based on the ensemble approach to simulated annealing, in which a collection of copies of the system, rather than just one, is annealed according to the same schedule [37, 38, 45, 46]. One of the most important advantages of the ensemble approach is that statistical information about the ensemble can be used to adjust the schedule with which the temperature is lowered. In principle the temperature T can be lowered after every Metropolis step. We decided however to use the more usual approach in which a certain number of steps is performed at the same temperature before T is lowered by a certain amount. This number of steps can be predetermined or it can be set adaptively during the annealing using the statistical data from the ensemble.

The philosophy behind the schedule [47] is to hold the ensemble fairly close to the equilibrium corresponding to the annealing temperature. As an indicator for this, the ensemble average of the energy $\langle E \rangle$ is monitored. For an ensemble close to equilibrium, $\langle E \rangle$ will fluctuate around the thermodynamic equilibrium corresponding to the annealing temperature, whereas for an out of equilibrium situation $\langle E \rangle$ will move towards that value. The adaptive schedule is implemented as follows:

1. Set an initial temperature T_0.
2. Perform m Monte Carlo steps per ensemble member with the ensemble. We will call this a Monte Carlo sweep.
3. Let $\langle E \rangle_{(j)}$ be the ensemble average after the jth Monte Carlo sweep. IF $\langle E \rangle_{(j)} < \langle E \rangle_{(j-1)}$
   THEN   GOTO 2,
   ELSE   GOTO 4.
4. Reset the temperature $T_{j+1} = cT_j$ with $c < 1$.
5. IF the maximum number of Monte Carlo sweeps has not been reached
   THEN   GOTO 2,
   ELSE   end of annealing run.

Note that even though the temperature is always lowered by a factor c, the schedule is not exponential as the number of Monte Carlo sweeps spent at each temperature varies. Due to the finite ensemble size, fluctuations will mask the "true" behavior of the energy average and our criterion for being close to equilibrium will lower the temperature even though some of the ensemble members are stuck in local minima. Thus we expect that sooner or later the ensemble will fall out of equilibrium.

From a technical point of view the ensemble version of the simulated annealing algorithm proceeds by selecting randomly n (due to the size of the state space usually) different initial states, where n is the ensemble size. Then all these states are subjected to the same annealing schedule and evolve according to the Monte Carlo dynamics.

We implemented the simulated annealing ensemble algorithms on a Parsytec parallel computer using the PARIX environment and on high-performance RISC workstations. The program consists of n equivalent work processes and a master process. The latter is responsible for the input/output, the evaluation of the data received from the workers and for their control. Each work process handels the Monte Carlo algorithm for one member of the ensemble. So each process has to communicate only with the master process and only once after each Monte Carlo sweep.

The configuration of the processor array was chosen to be a simple tree. The master process was placed on the root of the tree. Each of the other processors had to accomodate a certain number of work processes depending on the ensemble size used and the number of processors available.

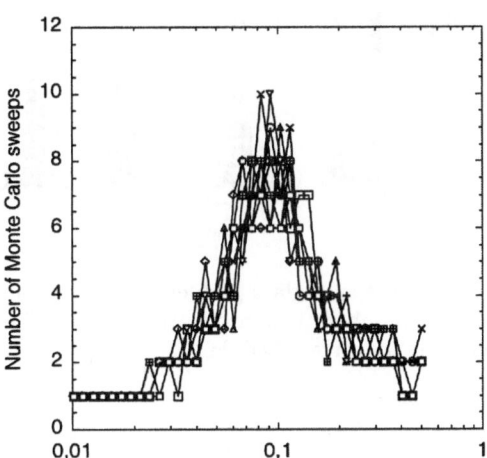

Fig. 7. The adaptive annealing using the ensemble approach at work. The time (i.e., the number of Monte Carlo sweeps) spent at that temperature varies automatically during the adaptive annealing process. At lower temperatures the ensemble has fallen out of equilibrium. Here the results of several runs are shown

Figure 7 demonstrates how the adaptive schedule works. It shows for a particular optimization problem the number of Monte Carlo sweeps as a function of the temperature T. One clearly sees that with decreasing T the time (i.e., the number of Monte Carlo sweeps) spent at that temperature increases first, and then starts to decay once the trapping of the ensemble members has started.

Figure 8 [48] shows how during the annealing the ensemble is transported towards lower energies. While initially the distribution moves down easily it starts to sharpen when it gets close to the ground state. In this example a spin glass system was annealed which has an approximate ground state energy around $E_{\min} = -780$ in the units used for Fig. 8. The simulated annealing procedure was carried out adaptively with $c = 0.9$ and $T_0 = 14.0$.

Finally we remark that in a number of cases the time dependence of the moments $< (E_{\text{BSF}} - E_{min})^n >$ of the BSF energy can be reasonably well fitted by power laws for a certain time span [48, 49]. This shows again the connection to the thermal relaxation in tree models, where these power laws also occur.

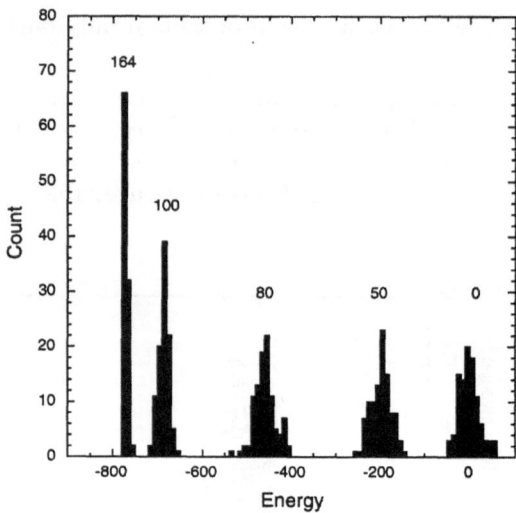

Fig. 8. Energy distribution for a spin glass problem during the annealing run at different times, indicated by the number of Monte Carlo sweeps. The simulated annealing procedure was used with an adaptive schedule

5   Summary

In this paper we have shown that metastable systems and stochastic optimization are fields that are intimitely connected through the concept of a complex state space, i.e., a space with a function with many local minima. The dynamics in this state space, which is induced by a contact with a heat bath – real or artificial – can be analysed by simulations based on the Metropolis algorithm or by a Marcov-process description. This was demonstrated with a simple example.

We have then presented two serious applications: the thermal relaxation behavior of spin glasses and optimized simulated annealing procedures. In the first application the coarse-graining of the complex state space leading to a tree structure is an important intermediate step. In the second application the need for adaptive annealing procedures was pointed out and a first solution was presented.

Appendix: Examples and Exercises (with S. Schubert)

As an example for stochastic optimizaton we study the traveling salesman problem (TSP). The aim of the TSP is to find the shortest tour connecting a given number of points in a x–y plane, which possibly represent towns that a salesman has to visit.

In order to show how this problem falls within the formal framework set up in Sect. 2.1. we note that a single tour, i.e., a sequence of towns, constitutes a state α in the state space Ω. The next step is to define a neighborhood relation. One possible method to create a neighbor of a given tour is to choose two connections along the tour randomly, to cut them, and to insert one part of the tour in reversed order. Another move class chooses two towns randomly and exchanges their position in the tour. Finally the tour length is the equivalent of the energy and is then the objective function, which is to be minimized. The TSP is known to be NP-hard, and thus stochastic optimization procedures such as simulated annealing or threshold accepting [50] have been used to search for the minimal tour.

Simulated Annealing. The program `tsp_sa.c` shows a simple simulated annealing (SA) procedure for the TSP. The number of towns is N (take $N = 6$ initialy) and they are supposed to be on the circumference of a circle. Thus the shortest tour is well known and we are able to compare the result of the program with the real minimum state.

The tour is represented by a one-dimensional integer field `tour[i]` which contains the numbers of towns in the order in which they are visited.

In the outer loop the temperature is decreased exponentially after each Monte Carlo sweep. Here the number of Monte Carlo steps per sweep is set equal to the number of towns.

One Monte Carlo step consists of the following steps:

1. propose a move;
2. calculate the cost $\Delta r = r_{new} - r_{old}$ where r denotes the length of the tour;
3. draw a random number z, uniformly distributed between 0 and 1;
4. accept the new tour if $\exp(-\Delta r/T) > z$, else reject the new move.

Threshold Accepting. The program `tsp_ta.c` shows a simple example of a traveling salesman problem using another technique, namely threshold accepting (TA). TA differs from SA in the decision rule for accepting or rejecting a neighbor. In TA a neighbor is accepted if the cost function decreases or if it increases by less than a given external parameter, which plays the role of the temperature. The computation advantage of TA lies in that no calculation of an exponential function and no further random number is needed.

Exercise. Write a program with many more and randomly spread towns (graphical output). Try out different annealing schedules (linear, exponential, adaptive) and try to find the optimal range for the parameter T. See the self-explaining code for the two example programs on the enclosed diskette.

References

[1] K.H. Fisher and J.A. Hertz, *Spin Glasses* (Cambridge University Press, Cambridge 1991)

[2] G. van Kampen, *Stochastic Processes in Physics and Chemistry* (North Holland, Amsterdam 1991)

[3] N. Metropolis, A. Rosenbluth, M. Rosenbluth, A. Teller, and E. Teller, J. Chem. Phys. **21**, 1087 (1953)

[4] K. Binder and D.W. Heermann, *Monte Carlo Simulation in Statistical Physics* (Springer, Heidelberg 1992)

[5] K. Binder, ed., *Monte Carlo Methods in Statistical Physics* (Springer, New York 1979)

[6] K.H. Hoffmann and P. Sibani, Phys. Rev. A **38**, 4261 (1988)

[7] B. Andresen, K.H. Hoffmann, K. Mosegaard, J. Nulton, J.M. Pedersen, and P. Salamon, J. Phys. (France) **49**, 1485 (1988)

[8] H.A. Kramers, Physica (The Hague) **7**, 284 (1940)

[9] K.H. Hoffmann, S. Grossmann, and F. Wegner, Z. Phys. B **60**, 401 (1985)

[10] P. Sibani, Phys. Rev. B **35**, 8572 (1987)

[11] P. Sibani and K.H. Hoffmann, Europhys. Lett. **16**, 423 (1991)

[12] C. Uhlig, K.H. Hoffmann, and P. Sibani, Z. Phys. B **96**, 409 (1995)

[13] K.H. Hoffmann, S. Schubert, and P. Sibani, preprint: *Age reinitialization in spin glass dynamics and in hierarchical relaxation models*, Institute of Physics, Technical University of Chemnitz-Zwickau (1995)

[14] L. Lundgren, P. Svedlindh, P. Nordblad, and O. Beckmann, Phys. Rev. Lett. **51**, 911 (1983)

[15] M. Ocio, H. Bouchiat, and P. Monod, J. Physique Lett. **46**, L-647 (1985)

[16] P. Nordblad, P. Svedlindh, J. Ferre, and M. Ayadi, J. Magn. Magn. Mater. **59**, 250 (1986)

[17] P. Svedlindh, P. Granberg, P. Nordblad, L. Lundgren, and H.S. Chen, Phys. Rev. B **35**, 268 (1987)

[18] N. Bontemps and R. Orbach, Phys. Rev. B **37**, 4708 (1988)

[19] J. Hammann, M. Ocio, and E. Vincent, in *Relaxation in Complex Systems and Related Topics*, ed. I.A. Campbell and C. Giovannella (Plenum Press, New York 1990), p. 11

[20] C. Rossel, Y. Maeno, and I. Morgenstern, Phys. Rev. Lett. **62**, 681 (1989)

[21] K. Biljakovic, J.C. Lasjaunias, P. Monceau, and F. Levy, Phys. Rev. Lett. **62**, 1512 (1989)

[22] K.H. Hoffmann, T. Meintrup, P. Sibani, and C. Uhlig, Europhys. Lett. **22**, 565 (1993)

[23] P. Sibani and K.H. Hoffmann, Phys. Rev. Lett. **63**, 2853 (1989)

[24] C. Schulze, K.H. Hoffmann, and P. Sibani, Europhys. Lett. **15**, 361 (1991)

[25] E. Aarts and J. Korst, *Simulated Annealing and Boltzmann Machines* (Wiley, Chichester 1989)

[26] P. Sibani, C. Schoen, P. Salamon, and J.-O. Andersson, Europhys. Lett. **22**, 479 (1993)

[27] S. Kirkpatrick, C.D. Gelatt Jr., and M.P. Vecchi, Science **220**, 671 (1983)

[28] V. Cerny, JOTA **45**, 41 (1983)

[29] R. Ettelaie and M.A. Moore, J. Physique Lett. **46**, L-893 (1985)

[30] P. Slarry and G. Dreyfus, J. Physique Lett. **45**, L-39 (1983)

[31] S.R. White, Proceedings of the ICCD '84 (IEEE), 646 (1984)

[32] M.P. Vecchi and S. Kirkpatrick, Proceedings of the IEEE Trans. Computer-Aided Design CAD **2**, 215 (1983)

[33] E. Bonomi and S.L. Lutton, SIAM Rev. **26**, 551 (1984)

[34] R. Durbin and D. Willshaw, Nature **326**, 689 (1987)

[35] P. Salamon, J. Nulton, J.R. Harland, and J. Pedersen, Comput. Phys. Commun. **49**, 423 (1988)

[36] S. Geman and D. Geman, IEEE, PAMI **6**, 721 (1984)

[37] M.O. Jakobsen, K. Mosegaard, and J.M. Pedersen, in *Model Optimization in Exploration Geophysics 2*, ed. A. Vogel (Friedr. Vieweg and Son, Braunschweig/Wiesbaden 1988), p. 361

[38] K.H. Hoffmann, P. Sibani, J.M. Pedersen, and P. Salamon, Appl. Math. Lett. **3**, 53 (1990)

[39] K.H. Hoffmann and P. Salamon, J. Phys. A: Math. Gen. **23**, 3511 (1990)

[40] M. Christoph and K.H. Hoffmann, J. Phys. A: Math. Gen. **26**, 3267 (1993)

[41] K. Ergenzinger, K.H. Hoffmann, and P. Salamon, J. Appl. Phys. **77**, 5501 (1995)

[42] K.H. Hoffmann, M. Christoph, and M. Hanf, in *Parallel Problem Solving from Nature*, eds. H.P. Schwefel and R. Maenner (Springer, Berlin 1991)

[43] I. Morgenstern and D. Wuertz, Z. Phys. B **67**, 397 (1987)

[44] S. Rees and R.C. Ball, J. Phys. A **20**, 1239 (1987)

[45] G. Ruppeiner, J.M. Pedersen, and P. Salamon, J. Phys. I **1**, 455 (1991)

[46] SDSC, *EBSA C Library Documentation, Version 2.1* (San Diego Supercomputer Center, 1994)

[47] K.H. Hoffmann, D. Wuertz, C. de Groot, and M. Hanf, in *Parallel and Distributed Optimization*, eds. M. Grauer and D.B. Pressmar (Springer, Heidelberg 1991), p. 154

[48] R. Tafelmayer and K.H. Hoffmann, Comput. Phys. Commun. **86**, 81 (1995)

[49] P. Sibani, J.M. Pedersen, K.H. Hoffmann, and P. Salamon, Phys. Rev. A **42**, 7080 (1990)

[50] G. Dueck and T. Scheuer, J. Comput. Phys. **90**, 161 (1990)

Modelling and Computer Simulation of Granular Media

Dietrich E. Wolf

Höchstleistungsrechenzentrum, Forschungszentrum Jülich (KFA), D-52425 Jülich, Germany, e-mail: d.wolf@kfa-juelich.de

Abstract. The purpose of this chapter is twofold; to give an introduction into the physics of granular media, emphasizing modern concepts and research topics, and to review the models most pertinent to the computer simulation of these phenomena. Soft-particle molecular dynamics, event-driven molecular dynamics, the contact dynamics of Moreau and Jean, and the bottom-to-top-restructuring model are discussed in detail. Their range of applicability is carefully assessed, artefacts are pointed out, and key results obtained with the respective methods are presented.

1 The Physics of Granular Media

1.1 What are Granular Media?

Sand, pellets, coal, grains all have an important property: they can flow, for example, through hoppers. However, in contrast to a fluid they also form piles, in fact compact granular material can be hard like a solid. One doesn't have to worry that one's child may drown in the sandbox. (It may drown in a silo filled with corn however, such accidents have happened.) This duality between fluid- and solid-like behavior has been the reason why granular media have been of central technological importance since the beginnings of civilization: solids can only be processed in the granular form. The term "granular media" is also used, for example, for suspensions (grains in fluid) or pastes (cohesive). However, in this chapter I want to restrict myself to dry, cohesionless granular media.

Besides the technological importance and the exotic phenomena, some of which will be described below, there are basic physical properties which make granular materials an interesting research subject [1, 2, 3, 4]. Important key words are disorder, threshold dynamics and dissipative interactions, and their consequences, stick-and-slip motion, pattern formation, and fluctuations on various scales. In particular the passage from microscopic laws to macroscopic behavior, i.e. the prediction of characteristic length and time scales pertinent to the dynamic behaviour still poses many fundamental questions. Many of these questions became tractable only because of the new research tool: high performance computers.

It is instructive to define granular media in terms of characteristic energies (see Table 1). For clusters up to ca. 10^6 molecules (such as soot particles), the thermal energy at room temperature is more important than gravitational energy

Table 1. Characteristic properties of granular media

	radius [m]	number of molecules
molecule	10^{-10}	1
soot particle	10^{-8}	10^6
dust, powder	10^{-6}	10^{12}
sand	10^{-4}	10^{18}
gravel	10^{-2}	10^{24}

differences on the length scale of the cluster. For granular media this is not the case. Typical Van der Waals energies are of the same order as room temperature so that cohesion also becomes weak for particles larger than $1\,\mu$m. However, this estimate is rather rough as other attractive forces between the particles may become important. Cohesion is not negligible if powders are a little wet, or even among pebbles, if they are covered with glue!

Particles in a granular medium interact when they collide. (Long-range interactions due to electrostatic charging will not be considered here.) For the dynamics of granular media it is essential that the interaction is *dissipative* and *nonlinear*. The dissipation is due to viscoelasticity (plastic deformation) and to Coulomb friction, caused by the surface roughness of the grains;

$$F_t = -\mu_d F_n \mathrm{sign}(v_t), \quad \text{for} \quad v_t \neq 0, \tag{1}$$

where μ_d is the coefficient of dynamic friction, F_n the normal force at the contact, and v_t the relative tangential velocity of the two particles. The nonlinearity is also of twofold origin: the Coulomb law is discontinuous at zero velocity. In fact, the static friction force $\mu_s F_n$ with $\mu_s > \mu_d$ has to be overcome to trigger relative motion of the particles in contact. Moreover the elastic restoring force leading to a reversal of normal (and to some extent also tangential) relative velocities depends on the shape of the particles and is in general different from the linear Hooke's law.

1.2   Stress Distribution in Granular Packing: Arching

In 1852 Hagen [5] discovered that, as an immediate consequence of Coulomb friction , the pressure at the bottom of a container filled with sand is essentially independent of the filling height. This phenomenon is due to so-called arching, which means that forces are transmitted to the side walls such that lower parts of the filling do not have to carry the weight of the parts above. This is the reason why the flow velocity in an hour glass does not depend on the filling height. The constant pressure is in marked contrast to an incompressible liquid for which the pressure increases linearly.

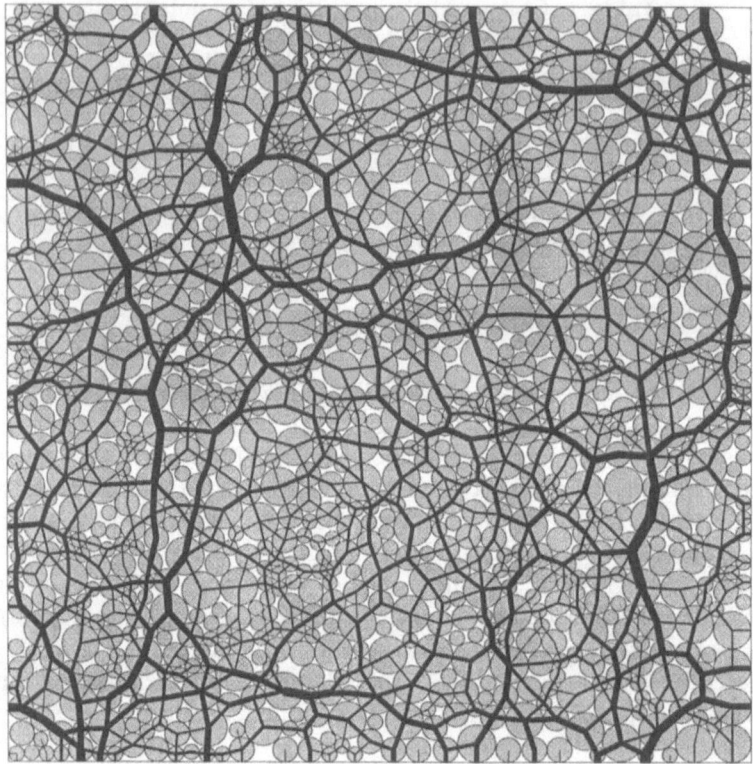

Fig. 1. Two-dimensional static assembly of discs in a horizontal container with three fixed walls and one piston, on which a fixed force is applied. Network of contacts with non-vanishing normal forces, whose strength is indicated by the width of the connecting lines. Contact dynamics simulation as described in Sect. 4. From [7]

The simple derivation of this result [6] for a cylindrical container of radius w starts from calculating the pressure increase from the top to the bottom of a horizontal slice of thickness Δx. It is due to the weight $G = \pi w^2 m \rho g \Delta x$ of this slice, where m is the average mass, ρ the number density of the grains, and g the gravitational acceleration. The weight is partly compensated by the friction at the container walls. This friction is proportional to the normal force $F_n = 2\pi w \Delta x p(x)$, where $p(x)$ denotes the pressure a distance x away from the surface of the filling. As the static friction force is indetermined within the bounds $\pm \mu_s F_n$, the value of the proportionality constant μ depends on the way in which the container was filled. (Actually, in pathological cases, the friction force may even have the same direction as G, in which case Hagen's observation is no longer correct. This happens when the packing is under compression stabilized by "upside-down" arches.) The pressure change, $\Delta p = (G - \mu F_n)/\pi w^2$ can easily be integrated to give

$$p(x) = p_\infty(1 - \exp(-2\mu x/w)), \quad \text{where} \quad p_\infty = \rho g w/2\mu. \tag{2}$$

Near the surface, $x = 0$ and the pressure increases linearly as in an incompressible fluid, however at a depth comparable to the radius of the container the pressure approaches a constant value p_∞.

This model calculation explains only the height dependence of the average pressure. As a granular medium is not an unstructured continuum but a disordered discrete packing, pressure fluctuations are significant. Experiments and computer simulations show that the forces are concentrated on an irregular network of contacts (Fig. 1) and actually have a power-law distribution of normal (and tangential) components smaller than their average values [7],

$$w(F_\mathrm{n}) \sim F_\mathrm{n}^{-0.3} \quad \text{(in two dimensions)}. \tag{3}$$

For larger contact forces the distribution decays exponentially.

1.3 Dilatancy, Fluidization and Collisional Cooling

In flowing granular matter of high density the arches constituting the force network break and reassemble differently all the time. These processes are governed by two basic phenomena, dilatancy and collisional cooling.

Dilatancy was discovered by Reynolds 1885 who observed that deformation of a dense granular packing without an increase of its volume is impossible. The particles sit in "cages" formed by their neighbours, and relative displacements require that these cages break up, necessarily an increase in the pore volume between the particles. The dilatancy phenomenon can be spectacularly demonstrated by putting a non-varnished wooden stick into a bottle which is then filled with sand. By tapping at the bottle the sand is compactified. If one now tries to pull the stick out, the whole bottle is lifted and can be carried around on the stick. The reason is that the walls suppress the volume increase necessary to allow the shearing exerted by the rough surface of the stick when pulled out.

In a more general sense the dilatancy phenomenon also occurs in flowing granular materials. Consider for example Fig. 2, which shows results of a computer simulation by Thompson and Grest, 1991 [8]. The horizontal motion of the upper wall with a constant velocity U leads to the coexistence of a compact lower part of the granular medium between the plates and a fluidized upper part that acts as a "lubricant" for the motion of the wall. An increase of U leads to the fluidization of more material. For high enough velocities the compact lower part will vanish entirely, and the fluidization will be complete. In addition to the shear-induced fluidization we observe that the width of the fluidized region is proportional to U, such that the average shear rate $\partial_y v_x$ is approximately constant. As a result, an increased shear rate implies a larger volume. This is due to the constant confining pressure on the walls, chosen as the boundary condition here.

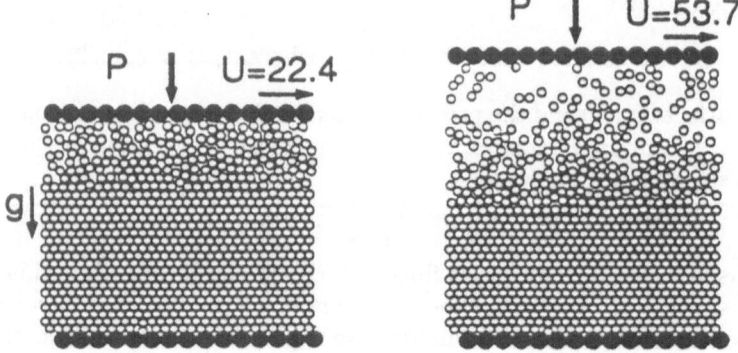

Fig. 2. Two-dimensional vertical system of discs subject to a horizontal shearing. P is the pressure at the upper wall, and U its velocity relative to the fixed lower wall. Molecular dynamics simulation as described in Sect. 2 with linear spring-dashpot model. From [8]

In order to keep a fully fluidized granular medium at constant volume, Bagnold (1954) showed that the confining pressure must increase like with shear rate squared. This is also true for the shear stress:

$$\sigma_{xy} \propto (m/d)(\partial_y v_x)^2 \operatorname{sign}(\partial_y v_x) \tag{4}$$

with a dimensionless proportionality constant; m is the grain mass and d is the mean distance between the grains (in the two-dimensional system considered by Bagnold the factor $1/d$ is absent in (4)). This shows that a fluidized granular medium is a non-Newtonian fluid. In Newtonian fluids the internal friction is proportional to the shear rate,

$$\sigma_{xy} = \eta \partial_y v_x, \tag{5}$$

where η is the viscosity.

Bagnold gives a simple derivation for (4), which I rephrase here as a dimensional argument: σ_{xy} has the dimension of a force per area, i.e. [mass/time^2 length] and describes the momentum transfer per unit time and unit area between adjacent volume elements. Now Bagnold argues that the momentum transfer is exclusively due to particle collisions induced by the shear rate $\partial_y v_x$ with the dimension of an inverse time. Let us assume that the distance d between the particles and the particle radii R are of the same order of magnitude. The only dimensional parameters which can possibly enter the problem are therefore m, d and the shear rate. Equation (4) is the only combination of these quantities with the proper dimensions. The proportionality constant will depend on the dimensionless ratio R/d. Why does this argument not hold for molecular fluids, where one finds (5) instead? The reason is that collisions of the molecules are mostly due to their thermal motion. The thermal energy therefore is another dimensional quantity which must enter the argument. Thus, a simple dimensional argument is no longer possible.[1]

As soon as one stops agitating the fluidized granular medium by vibration or shearing, it "freezes" into a metastable state. Two facts are responsible for this: first the negligibility of the thermal energy of room temperature compared to the energy barrier, if one wants to move a grain on the surface of a dense packing from one local minimum to the next, and second collisional cooling. The term "cooling" here refers to granular temperature rather than to the thermodynamic one. By analogy with the equipartition theorem in thermodynamics, the notion of granular temperature was introduced [10] to characterize the mean square deviation of particle velocities from their average:

$$T_\mathrm{g} = \langle \mathbf{v}^2 \rangle - \langle \mathbf{v} \rangle^2 \quad . \tag{6}$$

Hence, in a static pile $T_\mathrm{g} = 0$, whereas in the fluidized state $T_\mathrm{g} \neq 0$. After each collision the absolute value of the relative velocity between the two particles is reduced, because the dissipative nature of the interaction converts granular into thermodynamic temperature which henceforth is irrelevant for the dynamics of the granular medium. This is what is meant by collisional cooling.

In granular flow through a vertical pipe the collisional cooling leads to clustering of particles [11] in the pipe. For a sufficiently high density this can support the formation of arches, even in the absence of, for example, air flowing around the particles. The resulting kinematic waves are supposed to be analogous to spontaneous traffic jams on highways [12].

In this section I have discussed so far two flow regimes of granular media, one for high density (Hagen) and one for medium density (Bagnold). As a summary of this section I want to illustrate how they differ in the case of a vertical pipe. It is instructive to remind oneself first of the behaviour of a Newtonian fluid. In the steady state the divergence of the stress tensor (5), which is the internal friction force on a volume element, has to balance the gravitational acceleration,

$$\rho m g = \partial_y \sigma_{xy}, \tag{7}$$

where x is the vertical direction. Inserting (5) one finds that $\partial_y^2 v_x$ is constant. If $\bar{v}$ is the average velocity in the pipe and w its diameter, then the second derivative of v_x is of the order of magnitude $\bar{v}/w^2$. Hence, we find the well-known result of Hagen–Poiseuille flow, that

$$\bar{v} \propto w^2. \tag{8}$$

Granular flow has a much weaker dependence on the pipe diameter. First, for high density, gravitational accelerations are balanced by arching, which replaces the viscous internal friction (5). The physics of this balance are entirely

---

[1] A simple dimensional argument is, however, possible if one *postulates* that the viscosity of molecular gases does not depend on the shear rate. Then one obtains $\eta \propto \sqrt{mk_BT}/R^2$ [9].

determined by the volume fraction of the grains (dimensionless), the Coulomb friction coefficient (dimensionless), the diameter of the pipe and the gravitational acceleration. Again applying dimensional arguments, the average flow velocity must be

$$\bar{v} \propto (gw)^{1/2}. \tag{9}$$

Second, in the fluidized regime, when Bagnold's law (4) describes the internal friction which balances the gravitational acceleration, application of (7) shows that

$$\bar{v} \propto w^{3/2}. \tag{10}$$

Recently we [13] discovered a third regime for low density in which the balance of gravitational acceleration involves the production of granular temperature due to particle collisions with the wall. The steady state is reached when the production of granular temperature is balanced by the collisional cooling. In this regime,

$$\bar{v} \propto w. \tag{11}$$

This analytical prediction has been confirmed by computer simulations for which we used the event-driven algorithm described in Sect. 3.

1.4   Stick-and-Slip Motion and Self-Organized Criticality (with S. Dippel)

Coming back to Fig.2, the question arises, what are the dynamics close to the interface between the compact and the fluidized regions. One way to get insight into this is to relax the constraint of fixed speed U of the upper plate by pulling it slowly with a spring. Then one gets stick-and-slip motion (Fig.3). As long as the upper plate does not move, the spring tension increases linearly with time until a threshold is reached at which the plate starts sliding. The friction force exerted by the granular medium on the upper plate is smaller than the threshold force once it is in motion. Hence the plate is accelerated so that the spring tension drops sharply. However, then the friction wins and reduces the velocity of the plate to zero, and the plate sticks again until the spring tension exceeds the threshold the next time.

Stick-and-slip motion is usually viewed as the consequence of a static friction coefficient that is larger than the dynamic one. However, the results in Fig. 3 were obtained with a Molecular Dynamics simulation without any explicit implementation of static friction. In granular media the threshold dynamics leading to stick-and-slip phenomena can have a variety of origins. In the present example it is the dilatancy threshold which has to be overcome in order to move the upper plate: Fig. 3 shows that the distance between the plates is larger when there is motion and that gravitation and collisional cooling let the system freeze into a compact state when the upper plate is at rest. By contrast, in the case of static solid friction, microscopic asperities of the surfaces are interlocked. The simulation of Thompson and Grest in a way provides a macroscopic metaphor for this.

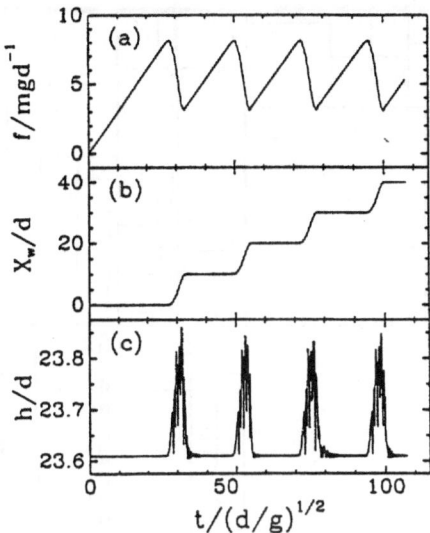

Fig. 3. Stick-and-slip motion: Time dependence of (a) the force per unit length, f, with which the top wall is pulled, (b) displacement of the wall, X_w, and (c) spacing between the walls, h, all in natural units (d is the grain diameter). From [8]

The regularity of the stick-slip sequence is due to the fact that the threshold dynamics manifests itself only in a single degree of freedom, the position of the upper plate. Dropping this restricition, an important concept was introduced by Bak, Tang and Wiesenfeld 1987 [14]: self-organized criticality. Without going into the details this concept can be illustrated in the present context by imagining the upper plate in Fig.2 consisting of many small elements coupled by springs. Then the system develops a non-trivial distribution of tensions among all the springs. Therefore, if one element starts sliding, it causes neighbouring elements to exceed the threshold, too, starting an "avalanche" which has been proposed as a model for earthquakes [15]. Self-organized criticality means that the avalanches have no characteristic size, i.e. their size distribution is a power law. Introductory reviews on the concept of self-organized criticality can be found in [16]. Originally, sand piles were proposed as a paradigm for self-organized criticality [14]. Adding grains to a pile, the surface becomes steeper and steeper until an avalanche starts. It turned out, however, that these avalanches have a power-law distribution only if inertia effects are avoided [17].

1.5   Segregation, Convection, Heaping (with S. Dippel)

Geologically most solid particles on earth were originally rocks. Sand is the result of a long fragmentation process. However, as Kolmogorov showed 1941 [18] in a study of the statistics of crushed-ore sizes, a fragmentation process usually leads to a much broader size distribution (ideally a log-normal distribution),

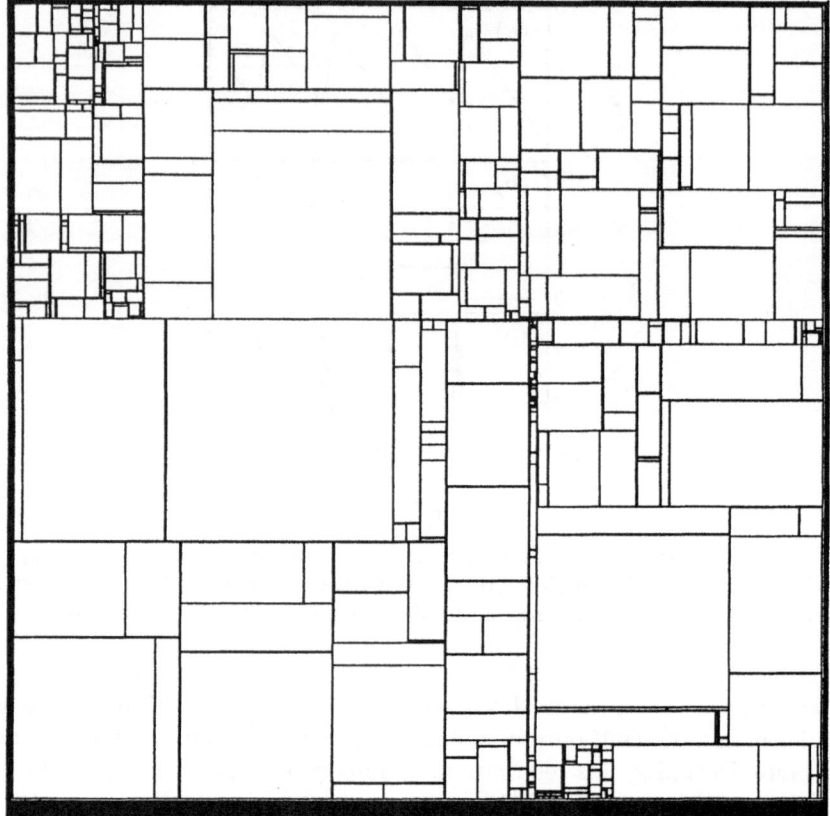

Fig. 4. Kolmogorov model of fragmentation

than that which we know from, for example, beaches. Figure 4 shows such a set of fragments with a log-normal size distribution. It was obtained by dividing a square by a straight line at an arbitrary position into two parts, then every part again into two parts and so forth.

Therefore the question arises: what is nature's sieve? Most importantly, nature separates particle sizes by selective transport by water and wind, and by subsequent sedimentation. However, even without interaction with a hydrodynamic medium, size segregation occurs in granular media. Three mechanisms are familiar from everyday experience. First, small grains can percolate under the influence of gravity downward through the gaps of a stable packing of larger particles [19]. Second, in granular flow along rough surfaces small particles are more easily trapped than big ones [20, 21, 22, 23]. This is why, for example, in rockslides and on mountain slopes the bigger rocks accumulate at the bottom of the slope. It also explains the radial segregation in a drum that rotates about a horizontal axis [21, 24, 25]. If it is less than half filled with a mixture of small

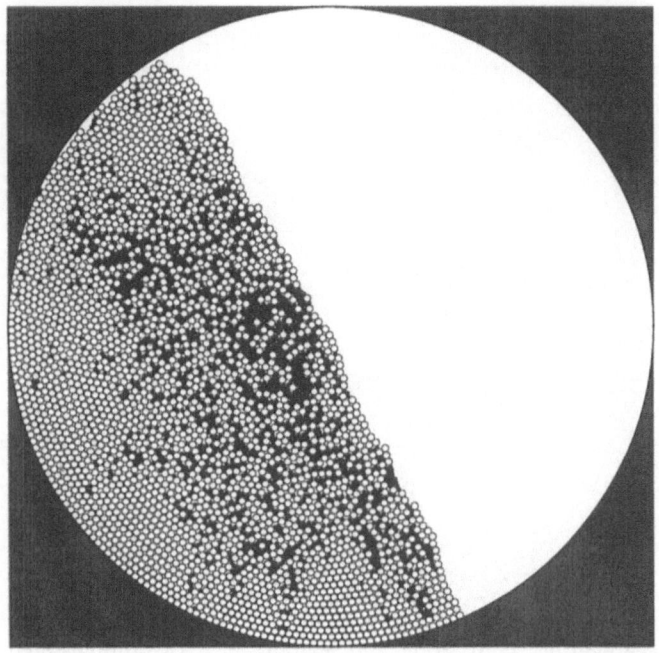

Fig. 5. Rotating two-dimensional drum with 2512 large (*white*) and 2483 (*black*) discs with half the diameter, after three total revolutions. Simulation with the BTR model described in Sect.5.2. From [59]

and big particles of equal mass density, the large ones accumulate in the outer regions within less than a full rotation (Fig. 5) .

A more intriguing (and also more controversial) mechanism of segregation can be observed when a box filled with particles of different sizes is shaken vertically. It has been termed "Brazil nut effect" [26, 27, 28], as in Brasilian fruit trucks the big nuts were found on top of the smaller fruit after transportation. Recently, it has been shown that in many cases this segregation is intimately connected to another phenomenon occuring in vibrated granular materials: convection. When rough particles in a container with rough walls are vibrated, a flux of particles in the middle of the container moves upwards, forming a heap on the surface (see Fig. 6), while at the side walls particles are dragged down into the bulk. However, the zone in which downflow occurs is quite narrow so that large particles having been carried up to the top of the pile by the wide upward stream cannot enter it again, thus staying on top [29, 30, 31].

However, if the box is not vibrated vigorously enough for convection to set in or to extend throughout the whole depth of the container (as it usually starts up in the higher, more easily fluidized regions of the piling), other mechanisms seem to be at work. Due to the vibrations, small local rearrangements of the particles can take place. A small particle can slip underneath a big one leading to the

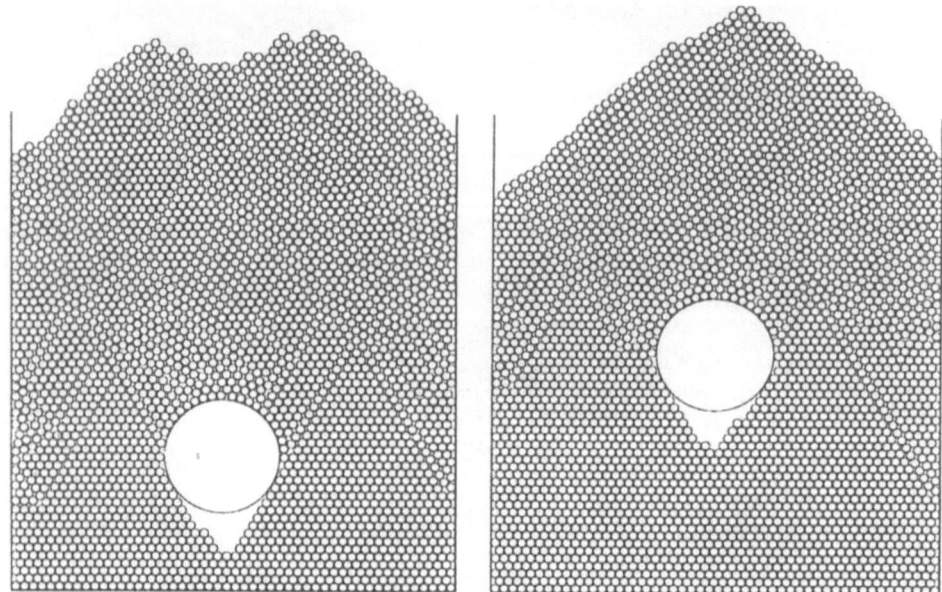

Fig. 6. Size segregation, convection and heaping in a two-dimensional packing [61]. Simulation with the BTR model described in Sect.5.1. Implementation of wall friction as in [58]. Ratio of particle radii is 13. *Left*: after 60 shakes. *Right*: after 180 shakes

upward motion of the big particle. This is more likely than a well-concerted giving way of all the small particles on which the big particle rests, which would be needed for downward motion. The character of the ascent of the big particles was found to depend on the ratio of the big and the small radii [27]. In two dimensions, if this ratio is smaller than about 12.9 [32, 28] a single large disc rises intermittently, depending on the amplitude of the shaking, whereas for larger ratios it rises continuously in every shaking cycle.

Among the less well understood phenomena related to heaping, segregation and pattern formation in general, we would like to mention two which are particularly intriguing. First, the dynamics of band segregation one observes in three dimensional drums (see for example [33]) and second, the cellular patterns one obtains, if a thin layer of grains is vibrated vertically, reminiscent of Rayleigh–Benard convection [34].

2   Molecular Dynamics Simulations I: Soft Particles

2.1   General Remarks

In molecular-dynamic simulations Newton's equations of motion

$$m_i \ddot{\mathbf{r}}_i = \sum_{j=1}^{N} \mathbf{F}_{ij} \tag{12}$$

are discretized and solved numerically to give the time evolution of a system of N particles. It should be borne in mind that this method is based on a *model* of the forces $\mathbf{F}$, so that the results have a range of validity which needs to be assessed carefully in order to avoid wrong interpretations. The computational challenge of this method lies in the fact that the hydrodynamic and agitation time scales which one wants to investigate are vastly larger than the duration τ of a collision or the time between two collisions.

In this section I use a fixed discretization time step Δt, which should be about $10^{-2}\tau$, to keep relative errors of physical quantities integrated over the whole collision time of order 10^{-4}. In contrast to Lennard–Jones fluids, where the characteristic time $\tau \sim 10^{-13}$s, the collision time here is typically of the order $\tau \sim 10^{-4}$s.

As a result of the integration over one time step two grains may turn out to overlap. Within the soft-particle model [35, 36] of granular media this overlap

$$\xi = R_i + R_j - |\mathbf{r}_{ij}| \tag{13}$$

(for spherical particles) is physically interpreted as the elastic deformation in the collision between the two grains i and j. Here R_i denotes the radius of the i-th particle, and $\mathbf{r}_{ij}$ is the difference between the centre of mass positions of the collision partners. The forces depend on ξ and the tangential component of the relative velocity, as explained in the next section.

For integrating (12) we have used the Gear predictor-corrector scheme of fifth order [37]: in every time step one calculates the position, velocity, acceleration and the 3rd and 4th time derivative of the position of each particle. The Taylor expansion of these five quantities gives a prediction $\mathbf{r}_p^{(i)}$, $i = 0, ...4$, for their values after the next time step. The predicted acceleration $\mathbf{r}_p^{(2)}$ is now compared with the force (divided by the mass) calculated for the predicted positions and velocities. The difference $\Delta \mathbf{r}_p^{(2)}$ is used to correct the predicted values:

$$\mathbf{r}^{(i)} = \mathbf{r}_p^{(i)} + c_i \Delta \mathbf{r}_p^{(2)}. \tag{14}$$

The coefficient c_0 depends on whether or not the forces depend on the velocities. Here they do, so that the coefficients are $(c_0, ..., c_4) = (19/90, 3/4, 1, 1/2, 1/12)$.

The reviews [38] and [39] contain many examples of where this type of simulation has been applied. Also the simulation results of Thompson and Grest [8] mentioned in Sect. 1 have been obtained with this model. The next few sections will address questions of implementation and validity of soft-particle molecular dynamics in the context of granular media.

2.2 Normal Force

Among the various different implementations used in the recent literature I want
to discuss only those of the form

$$F_{\mathrm{n}} = -k\xi^\alpha - \gamma\frac{d}{dt}(\xi^\beta). \tag{15}$$

The first term is the elastic restoring force and the second term describes vis-
coelastic dissipation. One has to imagine that (15) is the result of an integration
over the contact area, which increases with ξ. Therefore, even if the material of
the colliding particles is described by linear elasticity, in general the first term
deviates from Hooke's law $\alpha = 1$. Hertz (1882) showed that for elastic spheres
$\alpha = 3/2$ [40].

It is clear that $\alpha = \beta$ if the material of the colliding particles obeys linear
viscoelasticity [41] . Then the stress tensor is equal to a sum of two terms, one
proportional to the strain tensor and one proportional to its time derivative. The
integral over the contact area is then a corresponding sum of two terms, whereby
the second is proportional to the time derivative of the first one. Nonetheless,
frequently $\alpha = 3/2$ is used together with $\beta = 1$; however, we shall see that some
results are only applicable to rather exotic materials with nonlinear viscoelasticity.

As explained in the previous section we need to know the collision time in
order to choose the time step of the simulation appropriately. The collision time
for the normal force (15) can be estimated by a simple dimensional argument. I
neglect the damping term which can only increase the collision time. Then the
equation

$$\ddot{\xi} = -(k/m)\xi^\alpha \tag{16}$$

has to be solved with initial conditions $\xi(0) = 0$ and $\dot{\xi}(0) = v_n$, the normal
component of the relative velocity before the collision. The two parameters k/m
and v_n can be combined in unique ways to give a characteristic time and a
characteristic length, which have to be of the same order of magnitude as the
collision time and the maximal overlap ξ_{max}, respectively:

$$\tau \sim \xi_{\max}/v_{\mathrm{n}}, \quad \xi_{\max} \sim (mv_{\mathrm{n}}^2/k)^{1/(1+\alpha)}. \tag{17}$$

For $\alpha \neq 1$ the collision time depends on the impact velocity. Therefore the
simulation time step has to be chosen in accordance with the relative velocities
to be expected during the simulation.

The dissipation in a head-on collision is characterized by the coefficient of
normal restitution, an important material parameter, which is defined as the
ratio between the final and the initial normal component of the relative velocity,

$$e_{\mathrm{n}} = |v_{\mathrm{n}}^{(f)}/v_{\mathrm{n}}^{(i)}|. \tag{18}$$

It varies between 0 for completely inelastic and 1 for perfectly elastic collisions.
Experimentally one finds for a large class of materials [42, 43, 41, 44] that the

restitution coefficient decreases with increasing v_n, i.e. the more violent the impact, the more dissipative it is.

With (15) and (17) one can easily estimate the energy dissipation in one collision, and hence the restitution coefficient. The typical value of the damping force is $\gamma\xi_{\max}^\beta/\tau$, and the dissipated energy gets an extra factor $\xi_{\max}$,

$$E^{(f)} - E^{(i)} \sim -\gamma\xi_{\max}^{\beta+1}/\tau. \tag{19}$$

With $E^{(i)} = mv_n^{(i)2}/2$ the restitution coefficient can be expanded for weak dissipation as

$$e_n = (E^{(f)}/E^{(i)})^{1/2} \approx 1 + (E^{(f)} - E^{(i)})/2E^{(i)}. \tag{20}$$

Inserting (19) and (17) results in

$$1 - e_n \sim (\gamma/m)(m/k)^{\beta/(1+\alpha)}v_n^{(2\beta-\alpha-1)/(1+\alpha)}. \tag{21}$$

Figure 7 shows $1-e_n$ as a function of v_n in a double logarithmic plot [45]. The data were obtained by implementing (15) for $\alpha = \beta = 1$ (linear spring-dashpot model), for $\alpha = \beta = 3/2$ (Hertz–Kuwabara–Kono model), and for $\alpha = 3/2$, $\beta = 1$ (Hertz linear dashpot model). The results are in very good agreement with (21).

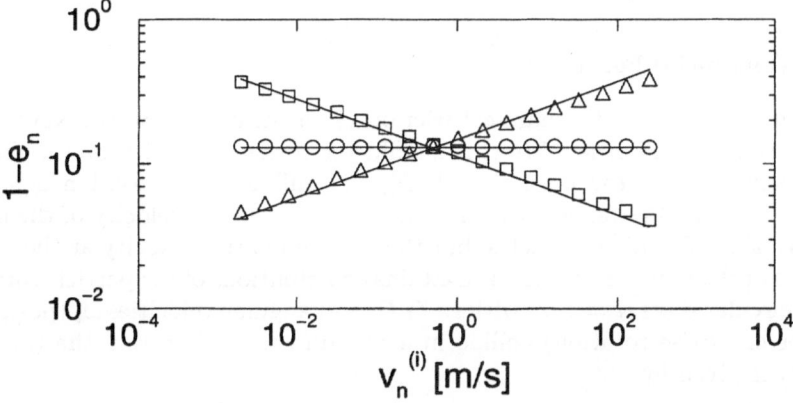

Fig. 7. Log-log plot $1 - e_n$ versus the normal component of the relative velocity. $\alpha = \beta = 1$ (*circles*), $\alpha = \beta = 3/2$ (*triangles*), and $\alpha = 3/2$, $\beta = 1$ (*squares*). The power laws predicted by (21) are indicated by the *lines*

The Hertz linear dashpot model gives the result that e_n decreases with v_n, i.e. more violent impacts are more elastic. The same is true whenever $\beta < (\alpha + 1)/2$, according to (21). Materials are known which become more elastic for high frequencies and are easily deformed plastically at low frequencies. Equation (reftau) implies that the collision time decreases with increasing impact velocity, as $\alpha > 1$ normally, and hence the collision may well become more elastic for those "exotic" materials. This does not apply, however, to most "normal" materials like metals, glasses, hard plastic etc. Nevertheless the Hertz linear dashpot model has been used in many recent simulations. The results should reflect the correct physical behaviour of normal materials, provided the relative velocities do not vary over a wide range. On the other hand, all these simulations show a tendency towards fluidization . One of the reasons for this could be that large relative velocities are damped out very slowly and contribute strongly to the granular temperature. Other reasons will be discussed in sections 2.4 and 2.5.

In any model with $\alpha = \beta > 1$ large relative velocities are damped out faster than slow relative velocities, and this trend is the stronger the larger α is. For the linear spring-dashpot model the restitution coefficient does not depend on the relative velocity. Together with the fact that the collision time does not depend on the velocity either, and that the linearity makes this law very well behaved numerically, the simple linear spring-dashpot model is a good compromise between physical accuracy and numerical efficiency. However, there are physical phenomena for which the nonlinearity of the Hertz law is important. The global elastic properties of a granular packing are one example [46]. In a dense system the overlap depends on the pressure. Linearizing the Hertz law for small fluctuations of the overlap around a non-zero average shows that the effective spring constant increases with the external pressure. This explains why shock waves spread faster through a granular medium the more compressed it is.

2.3   Tangential Force

The implementation of dynamical friction (1) is straight forward. Nevertheless two remarks are in order. First, it should be borne in mind that the different implementations of the normal force F_n also influence the friction behaviour. Second, v_t is *not* the tangential component of the relative velocity of the centres of mass of the colliding particles, but that of the relative velocity at the particle surfaces at the point of contact. It contains contributions of the particle rotations. If two circular discs ($i = 1, 2$) with radii R_i and angular velocities ω_i (positive for counter clockwise rotation) collide in a two-dimensional system the tangential velocity is given by

$$v_t = (\mathbf{v}_2 - \mathbf{v}_1) \cdot \mathbf{t} + \omega_1 R_1 + \omega_2 R_2, \qquad (22)$$

where $\mathbf{t}$ denotes the unit tangential vector. The equation of motion for the centres of mass (12) has to be supplemented by that for the spin of the particles

$$\dot{\omega}_i = R_i F_t / I_i, \qquad (23)$$

where I_i denotes the moment of inertia.

Figure 8 shows the final tangential velocity as a function of the initial tangential velocity [45], both normalized by the initial normal velocity for $\mu_d = 0.25$, and compares the curves with experimental data obtained for cellulose acetate

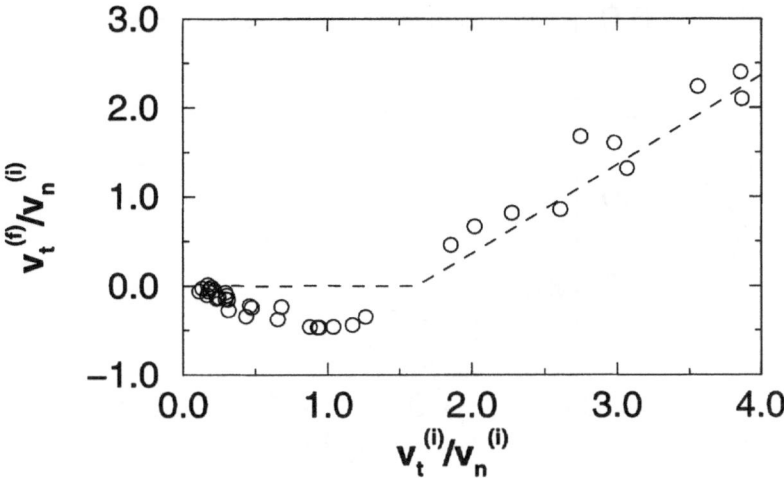

Fig. 8. Final versus initial tangential velocity, both normalized with the initial normal velocity. *Dashed line*: simulation with (1) [45]. Experimental data from [47]

spheres [47, 48]. The behaviour for sufficiently large initial tangential velocities can be well fitted by adjusting μ_d. However for small values the dynamic friction simply reduces the tangential velocity to zero (strictly speaking the finite simulation time step always leads to an overshooting so that the final velocity has tiny oscillations around zero, not observable in Fig.8). In the experiment, by contrast, the tangential velocity is reversed. The physical reason is a tangential elastic restoring force which builds up in a sticking contact and hence is connected with static friction.

The simplest implementation of tangential elasticiy is due to Cundall and Strack (1979) [35]. As soon as two particles touch ($t = t_0$), an imaginary spring is attached to the points of contact. During the collision the tangential displacement ζ of the points of first contact is recorded and identified with an elastic stretching of the spring,

$$\zeta(t) = \int_{t_0}^{t} v_t(t')dt'. \tag{24}$$

The friction law (1) is then usually replaced by

$$F_t = - \min(|k_t\zeta|, |\mu_d F_n|)\text{sign}(\zeta). \tag{25}$$

Figure 9 shows that with this model one can fit the experimental data.

In order to explain what (25) means, let us consider a collision with initial tangential velocity v_t. First the imaginary spring gets stretched, and if the tangential velocity reaches zero before $|F_t|$ reaches $|\mu_d F_n|$, the contact should be viewed as nonsliding. Then one gets an (undamped) fictitious tangential oscillation as long as the contact exists. As soon as the particles loose their overlap due

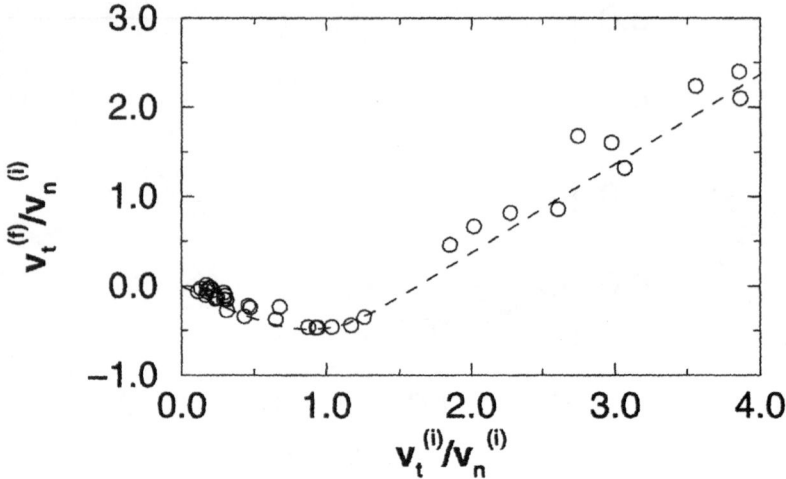

Fig. 9. Same as Fig. 8. Simulation with (25), $k_t/k_n = 1/5$ [45]

to normal motion, the spring vanishes. Depending on the phase of the tangential oscillation when this happens, the final tangential velocity can be reversed compared to the initial one. If the contact persists, the tangential spring can resist a finite external tangential force and thus models static friction. If $|k_t\zeta|$ exceeds $|\mu_d F_n|$ the force no longer increases. In this case one has to distinguish two scenarios. If the tangential velocity never reaches zero before the two particles separate again, the energy spent to stretch the spring is lost and the contact had been sliding. Note that there is no difference between the static and dynamic friction coefficient in this model force. The second scenario is that the tangential velocity reaches zero while the contact still exists. Then the particle starts an imaginary oscillation as above and the contact is interpreted as sticking. However, now the tangential oscillation is not harmonic: as long as $|k_t\zeta| > |\mu_d F_n|$ the force does not depend on ζ. Note that in this case $F_t = -|\mu_d F_n|\text{sign}(\zeta)$ cannot be interpreted as sliding friction, which never would lead to a reversal of tangential motion, and the work done by this force is not dissipated. Again the tangential velocity at the end of the collision may be opposite to its initial direction.

2.4   Detachment Effect

Equation (17) shows that the collision time τ and hence the simulation time step is proportional to $k^{-1/(1+\alpha)}$. As the phenomena one is interested in, such as segregation or convection, take place on much larger time scales, one common simulation strategy has been to choose k much smaller than in real materials, thereby allowing larger time steps. This seemed to be a good trade off for the artificially enlarged collision times. The restitution coefficient can be kept at a realistic value by choosing γ accordingly.

Luding [49] showed that this strategy only works for low densities, i.e. in the fully fluidized state. If the collision time is enlarged so much that it becomes comparable to the time between successive collisions (still much shorter than the times one is interested in) then the dynamical behaviour of the system changes drastically. Instead of having many successive binary collisions, one gets many-particle collisions in which clusters of several particles overlap at the same time. Luding argues that this reduces dissipation enormously.

He considers a simple example of a one-dimensional equidistant arrangement of N grains that move all with the same velocity v towards a plate from which they are going to be reflected. If the distance is much smaller than $v\tau$ then there is essentially a single multi-particle collision of all grains. The whole cluster can be viewed as a single elastic particle with a total deformation $N\xi_{max}$. The collision time of the cluster is then N times the collision time of a binary collision. For the linear spring-dashpot model this implies that the dissipated energy per particle is the same as in a single binary collision. By contrast, if the distance between the particles initially was much larger than $v\tau$ then one gets a sequence of about $N^2/2$ binary collisions until all particles are reflected. The dissipated energy per particle is significantly increased.

This example shows that dissipation is suppressed in dense systems if one artificially increases τ in order to be able to simulate larger time intervals. This leads to unrealistically strong fluidization. Event driven simulations (see Sect. 3) avoid the detachment effect.

2.5 Brake Failure Effect (with J. Schäfer)

The molecular dynamics scheme described in this section gives rise to an artefact, the so-called brake failure effect. Consider two equal spheres of radius R that collide with initial relative velocity $\mathbf{v}^{(i)}$ under an angle ϑ. We choose cartesian coordinates such that the x-axis is in direction of $\mathbf{v}^{(i)}$.

A braking function can be defined as the change of v_x caused by the interaction between the colliding spheres,

$$\Delta v_x \propto \int_{t_{cont}} F_x(t)\mathrm{d}t, \tag{26}$$

where t_{cont} is the time for which the spheres overlap. If they could freely penetrate each other without feeling any interaction, they would overlap for a time

$$\tau_0 = 4R\cos\vartheta/v^{(i)}. \tag{27}$$

As long as $\tau \ll \tau_0$, the simulation gives correct results: the contact time is determined by the collision time (17), $t_{cont} \approx \tau$. Then the spheres are reflected from each other, i.e.

$$\Delta v_x \propto v^{(i)} \tag{28}$$

with geometrical factors depending on ϑ. The braking function increases linearly with the initial velocity.

However, for $\tau_0 \ll \tau$, the braking function behaves unphysically. The duration of contact becomes $t_{\text{cont}} = \tau_0$. The integral (26) can be approximated by $\int F_x(t)\mathrm{d}t \approx \bar{F}_x \tau_0$, where $\bar{F}_x$ is a mean force in x-direction. Using (27) we obtain: $\Delta v_x \propto 1/v^{(i)}$. This decrease of the braking function with increasing initial velocity is meant by the term *brake failure*.

The transition between both regimes takes place at $\tau_0/\tau \approx 1$, such that we obtain a critical velocity for brake failure v_c:

$$v_c \approx \frac{4R\cos\vartheta}{\tau},\tag{29}$$

being smaller the more oblique the impact and the higher the collision time τ is. The brake failure effect therefore can be stated in the following terms: Time-step-driven simulations of collisions exhibit an unrealistically small dissipation when the impact velocity exceeds the critical velocity v_c.

In principle this may even happen for a frontal collision $\vartheta = 0$ if the impact velocity is so large that the maximal overlap $\xi_{\text{max}} > 4R$. For a reasonable simulation setup this can be avoided. But as one approaches grazing incidence, $\vartheta = \pi/2$, brake failure is bound to set in. To illustrate this effect, Fig.10 shows the braking function for two spheres.

A possible artifact due to brake failure is bistability in simulated granular flow through a vertical pipe [51]: instead of a unique steady-state velocity one obtains

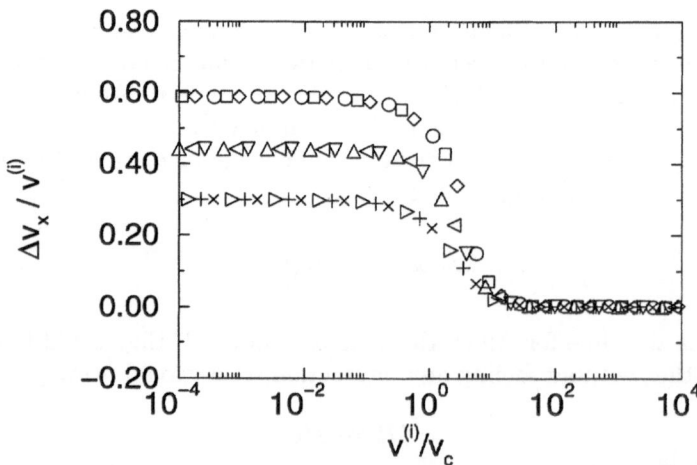

Fig. 10. Brake failure: for velocities larger than v_c collisions have vanishing effect. The relative change of smaller velocities decreases the more grazing the collision is. From *top to bottom*: $\vartheta = 64.2°, 71.8°$, and $77.2°$. Different symbols indicate data collapse for various different collision times at fixed ϑ. From [50]

a physical or an unphysical steady state depending on the initial conditions. Like the detachment effect, brake failure is avoided in event-driven simulations (next section).

3   Molecular Dynamic Simulations II: Hard Particles (with J. Schäfer)

3.1   Event-Driven Simulation

In systems in which the particle contacts can be considered as instantaneous binary collisions and only the contact forces act between the grains, the evolution of the system need not be integrated numerically: between collisions, the analytical solution of the equations of motion can be used. Then two tasks remain to be done: the calculation of the time of the next collision in the system, and the definition of a collision operator that describes post-collisional velocities and angular velocities as a function of the pre-collisional ones and the impact geometry. The former can be done using simple algebra and some intelligent bookkeeping [37]; the latter involves some definite (and simplified) model of the physics of colliding bodies.

We use the assumption that a collision of two grains can be entirely described by a coefficient of normal restitution e_n plus two coefficients related to tangential motion: the coefficient of friction μ, which applies for high impact angles where throughout the collision dynamic friction occured, and the coefficient of tangential restitution which applies for low impact angles . On the basis of this simplified picture, the following collision operator can be derived [52, 47].

3.2   Collision Operator

The collision operator is completely determined if we know the momentum change **J** that occurs during the contact. Decomposing **J** into normal and tangential components, J_n and J_t, we have for the difference of final and initial velocities of grains i and j ($\Delta \mathbf{v}_i = \mathbf{v}_i^f - \mathbf{v}_i^i$ etc.):

$$m_i \Delta \mathbf{v}_i = -m_j \Delta \mathbf{v}_j = \mathbf{J} \tag{30}$$

$$(I_i/R_i)\,\Delta \omega_i = (I_j/R_j)\,\Delta \omega_j = J_t. \tag{31}$$

Here, R is the radius and $I = 2/5\,mR^2$ the rotational inertia of the grains. The relative velocity at the contact points has the form

$$\mathbf{v} = \mathbf{v}_i - \mathbf{v}_j + (R_i \omega_i + R_j \omega_j)\mathbf{t}. \tag{32}$$

Using (30) and (31), we obtain:

$$\Delta \mathbf{v} = \mathbf{v}^f - \mathbf{v}^i = m_{\text{eff}}^{-1}\mathbf{J} + \tfrac{5}{2} m_{\text{eff}}^{-1} J_t \mathbf{t}. \tag{33}$$

The three coefficients that describe the impact are the coefficient of normal restitution $e_n = -v_n^f/v_n^i$, the friction coefficient $\mu = |J_t|/|J_n|$, and the coefficient of tangential restitution $e_t = v_t^f/v_t^i$. Using these definitions together with the normal and tangential component of Eqn. (33) yields the following components of $\mathbf{J}$:

$$J_n = -m_{\text{eff}}(1 + e_n)v_n^i, \tag{34}$$

$$J_t^{(\mu)} = -\mu\, m_{\text{eff}}\,(1 + e_n)v_t^i/\tilde{\psi}^i, \tag{35}$$

and

$$J_t^{(e_t)} = \tfrac{2}{7}m_{\text{eff}}(e_t - 1)v_t^i. \tag{36}$$

Here, $\tilde{\psi}^i = -|v_t^i|/v_n^i$ is a parameter measuring the obliqueness of the impact. The transition between the cases (35) and (36) occurs when they take equal values, i.e. at an obliqueness $\tilde{\psi}_0^i$ of

$$\tilde{\psi}_0^i = \tfrac{7}{2}\mu\frac{1 + e_n}{1 - e_t}. \tag{37}$$

3.3 Limitations

When particle densities are low, e.g. in a dilute gas, the event-driven simulation scheme works very fast compared to the time-step-driven method. However, the explicit determination of the time of the next collision is the most expensive part of the calculation, and for dense systems in which the frequency of collisions is high, the method becomes less advantageous. A real limitation arises from the assumption of instantaneous impacts. The dynamics of dissipative systems such as granular materials tend to produce clustered states [11, 53], and in this situation long-lasting contacts, not contained in the concepts of the method, are frequent. Much as Achilles in Zeno's famous paradox can never overtake the tortoise, the simulation time then converges towards the time when a long-lasting contact closes, whereas the collision frequency between the respective particle pair diverges. This is called "inelastic collapse" [53]. In principle, an algorithm handling clusters as a distinct class of particles could circumvent this problem; such a routine exists for 1D systems [54], but its extension to two dimensions is non-trivial and has not yet been undertaken.

4 Contact Dynamics Simulations (with L. Brendel and F. Radjai)

4.1 General Remarks

One of the most interesting theoretical challenges is to understand the emergence of characteristic lengths and times relevant for the collective behaviour of a packing with lasting contacts from microscopic contact interactions. One simple example is an array of parallel cylinders of equal radius R on a horizontal plane (Fig.11). It is pushed with a constant force by a block and terminated by a

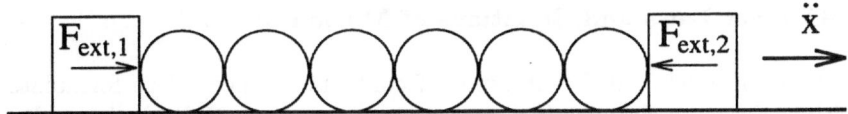

Fig. 11. System with many lasting contacts, showing self-organization of rotational degrees of freedom

similar block, such that the cylinders stay in contact with each other and with the plane. The important feature of this arrangement is that the rotational degrees of freedom are frustrated. This is illustrated in Fig.12: because of Coulomb friction, a contact wants to be rolling ($v_t = 0$) instead of sliding ($v_t \neq 0$), i.e. neighbouring cylinders want to have opposite spin. (Spin here is meant in the mechanical sense, but the analogy with an antiferromagnet is perfectly correct.) If there are three contacts forming a closed loop, at least one of them has to be sliding. The question then arises, how do the rotational degrees of freedom organize themselves into a collective structure in a simple situation like Fig.11.

On a macroscopic level the whole array can be viewed as one "block" sliding over the plane. Although on this level the rotations are internal degrees of freedom of the block, they do have important implications for the macroscopic friction which is no longer a Coulomb law. Another example is the calculation of the force network as in Fig.1. Clearly an accurate implementation of static friction is needed to study these questions.

Reading Sect. 2.3 about the implementation of Coulomb friction in the soft-particle model one can hardly avoid some feeling of dissatisfaction about the treatment of static friction: it is really rather ad hoc, the representation of microscopic dynamics by the imaginary tangential spring seems a bit unnatural and is only justified by the results at the macroscopic level. Lasting contacts involve undamped fictitious oscillations on the microscopic level. In the event-driven algorithm (Sect. 3) lasting contacts cannot be treated at all. In this section an exact implementation of Coulomb friction will be described, based on the work of J. J. Moreau and M. Jean [55, 56].

Fig. 12. Frustration of rotational degees of freedom in the presence of friction

4.2   Contact Laws and Equations of Motion

Some basic notions of different types of contacts are needed to formulate the algorithm. First the classification according to the tangential velocity and acceleration:

- A contact is called *sticking* if $v_t = \dot{v}_t = 0$. The tangential force at such a contact can be anything between $-\mu_s F_n$ and $\mu_s F_n$. Its precise value depends on the other forces acting on the particles in contact: the static friction force has to compensate them such that $\dot{v}_t$ is indeed zero.
- A contact is *activated* if $v_t = 0$, but $\dot{v}_t \neq 0$. Then $F_t = -\mu_s F_n \mathrm{sign}(\dot{v}_t)$.
- A contact is *sliding* if $v_t \neq 0$. Then $F_t = -\mu_d F_n \mathrm{sign}(v_t)$.

The allowed combinations of $(\dot{v}_t, F_t)$ for the case $v_t = 0$ are plotted in Fig.13, the *Coulomb graph*. Because of the multi-valuedness at $\dot{v}_t = 0$ this is not a function.

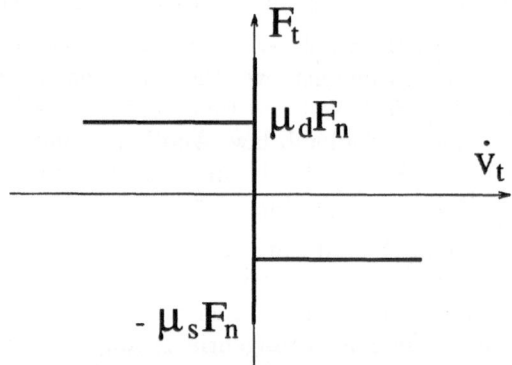

Fig. 13. Coulomb graph of allowed pairs $(F_t, \dot{v}_t)$

Similarly for the allowed combinations of $(\dot{v}_n, F_n)$ for $v_n = 0$ one obtains the so called *Signorini graph* (Fig.14) if one considers perfectly rigid particles. One distinguishes the following:

- *Closed* contacts; $v_n = \dot{v}_n = 0$. For them the normal force can have any value ≥ 0. The value adjusts itself such that it compensates all forces which would lead to an interpenetration of the particles.
- *Opening* contacts; $v_n = 0, \dot{v}_n > 0$. For such a contact F_n has to be zero.

Finally, if the normal component of the relative velocity $v_n \neq 0$, there cannot be a contact between the perfectly rigid particles.

The equations of motion give in addition a dynamic relationship between the accelerations and forces. For every contact the relative normal and tangential

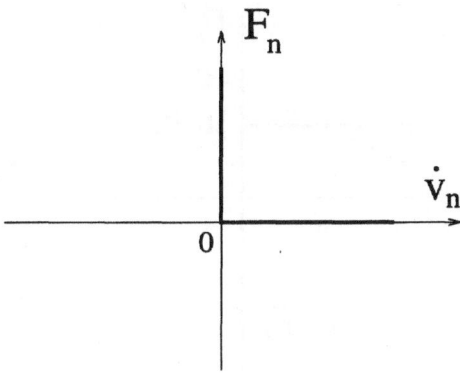

Fig. 14. Signorini graph of allowed pairs $(F_n, \dot{v}_n)$

 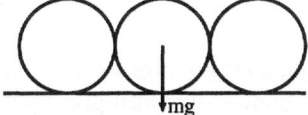

Fig. 15. Two examples, where $\dot{v}_n = F_n = 0$ is violated for the contact between the middle particle and the support; see text

accelerations are given by

$$\dot{v}_n = \frac{1}{m^*} F_n + A_n, \tag{38}$$

$$\dot{v}_t = \left(\frac{1}{m^*} + \left(\frac{R^2}{I} \right)^* \right) F_t + A_t. \tag{39}$$

The first terms on the right-hand side are the usual two body interactions with the reduced mass $1/m^* = 1/m_1 + 1/m_2$, and $(R^2/I)^* = R_1^2/I_1 + R_2^2/I_2$, where m_i, R_i and I_i are the masses, radii and moments of inertia of the two particles in contact. A_n and A_t are accelerations due to contacts with further particles or due to external forces such as gravity. They are linear functions of the forces acting at the other contacts.

Without A_n the only solution $(\dot{v}_n, F_n)$ of (38) allowed by the Signorini graph is $\dot{v}_n = F_n = 0$. However it is easy to imagine situations in which $F_n = 0$ and $\dot{v}_n \neq 0$. Consider, for example, three cylinders as in Fig.15. In the left example the middle one is accelerated upward due to the frictional force exerted by the rotating cylinders to the left and to the right. The normal force at the opening contact with the plane is zero. Similarly, the right example shows a situation in which the normal force at the contact between the middle cylinder and the plane is directed upward but does not lead to an acceleration because it is compensated by the weight.

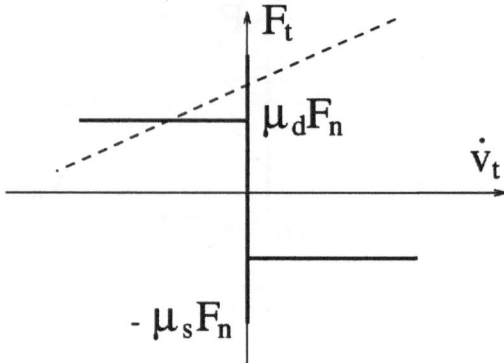

Fig. 16. The line $(1/m^* + (R^2/I)^*)^{-1}(\dot{v}_t - A_t)$ has two intersections with the Coulomb graph: the tangential acceleration at the contact in indetermined. It can be zero or negative in this example

In general the intersection between the linear function $F_n(\dot{v}_n) = m^*(\dot{v}_n - A_n)$ and the Signorini graph gives a solution for F_n and $\dot{v}_n$ simultaneously. Similarly the intersections of the corresponding linear function $F_t(\dot{v}_t)$ obtained from (39) with the Coulomb graph are solutions for F_t and $\dot{v}_t$ simultaneously. In fact, as A_n and A_t depend on all the other contact forces, one has to determine $(F_n, \dot{v}_n, F_t, \dot{v}_t)$ for all contacts simultaneously. One important fact is illustrated by Fig.16: as $\mu_s > \mu_d$ the solution obviously is not always unique, the straight line can have two intersections with the Coulomb graph. Both pairs $(F_t, \dot{v}_t)$ are solutions of the dynamic equations and the contact laws simultaneously. Which of the solutions is realized, depends on the history. This, of course, reminds us of the hysteretic behaviour in stick-and-slip motion .

Radjai, Brendel and Roux [57] showed that indeterminacy of forces is a generic feature of the contact laws: even if one assumes $\mu_s = \mu_d$ the number of possible static solutions grows linearly with the number of particles in a dense two-dimensional system. In the one-dimensional example (Fig.11) with $\mu_s = \mu_d$ this does not happen. Then the solution can be found by an iterative procedure described in the next section. In the case of indeterminacy this does not suffice and additional specifications, described in [57] have to be made.

4.3   Iterative Determination of Forces and Accelerations

In order to determine the accelerations and the forces for the example system (Fig.11) one starts with an arbitrary guess about the status of all contacts. For example one can assume that all contacts are closed and sliding (with randomly chosen initial rotation velocities). Then $\dot{v}_n = 0$ and $F_t = -\mu_d F_n \text{sign}(v_t)$ are given for all contacts. F_n and $\dot{v}_t$ are then calculated from the equations of motion. If all pairs $(F_n, \dot{v}_n)$ and $(F_t, \dot{v}_t)$ are allowed by the Signorini- and Coulomb-graphs, respectively, one would have found the solution. If not, one has to correct the initial guess about the status of the contacts. For example, if one of the F_n turns out to be negative, one drops the assumption that this contact was closed and lets it be opening, i.e. one sets $F_n = 0$ instead of $\dot{v}_n = 0$. After a few iterations a valid solution will be found. Then one knows all tangential accelerations and hence can calculate the time at which one of the tangential relative velocities drops to zero. Then this contact becomes sticking, but in the course of further iterations it may be activated again. In this way the time evolution of the system can be calculated.

4.4 Results

The simple example system (Fig.11) shows a surprisingly rich behaviour. No matter what the initial rotation of the cylinders was, sliding and sticking contacts organize themselves such that a unique state with constant acceleration is reached. This state can consist of up to three spatial domains. The first, next to the pushing block, has all contacts between cylinders sticking and all contacts between cylinders and the plane sliding. This means that the cylinders are either not rotating or counterrotating. In the third domain, at the end of the array, all cylinders roll, i.e. their contacts with the plane are sticking, whereas the contacts among them are sliding. In both the first and the third domain the absolute values of the angular accelerations of the cylinders are constant. In between there is the second domain in which all contacts are sliding, the ones among the cylinders as well as the ones with the plane. The angular accelerations increase from cylinder to cylinder as one goes from domain 1 to 3.

The lengths of the three domains depend on the pushing force and the friction coefficients for the contacts between cylinders or with the plane, respectively. The first domain grows with increasing pushing force (and hence increasing normal force between the cylinders).

The linear acceleration $\ddot{x}$ of the whole array of N cylinders of mass m is given by

$$F_{\text{ext},1} - F_{\text{ext},2} - F_{\text{friction}} = Nm\ddot{x} \tag{40}$$

where $F_{\text{ext},1}$ is the pushing force and $F_{\text{ext},2}$ is the friction force of the terminating block with the plane. Equation (40) *defines* the global friction force F_{friction}. Since the first domain, where the cylinders slide over the plane, grows with increasing pushing force, F_{friction} increases, too, see Fig.17. This means that on the macroscopic scale the Coulomb law, stating that the friction force only depends on the normal force Nmg, is no longer valid in granular materials.

5 The Bottom-to-Top Restructuring Model

5.1 The Algorithm and its Justification (with E. Jobs)

The bottom-to-top restructuring model (short BTR Model) is an algorithm designed for high-speed simulation of granular systems which periodically return

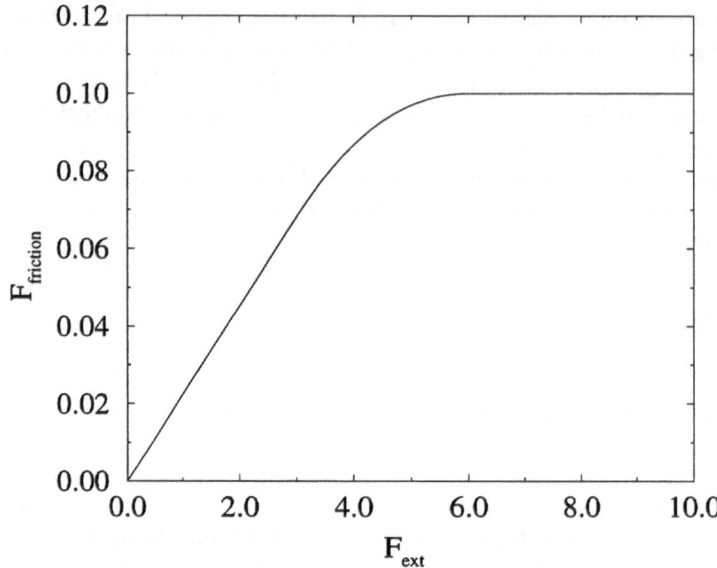

Fig. 17. The global friction force as defined in (40) grows with increasing pushing force $F_{ext,1}$ ($F_{ext,2} = 0.01$), as fewer and fewer cylinders are rolling

to a static packing in a container of any shape and exposed to an external, directed force field (such as gravity). It was first applied by Jullien, Meakin and Pavlovitch [27] to study size segregation under vertical shaking. Figures 5 and 6 were obtained with this model.

Although it is possible to simulate a whole class of granular systems with this algorithm [58], the most intuitive example is a vertically vibrating box filled with particles. Imagine that the time Δt between two shakes is long enough that all particles settle in a static packing. The idea of the BTR model is to formulate transition rules from one static configuration to the next.

For the vertically shaken box the following transition rule has been suggested by Jullien, Meakin and Pavlovitch [27]: all particle positions are updated one by one in the order of ascending height. Starting with the lowest particle each particle seeks its nearest local minimum, ignoring all higher particles.

The physical picture behind this is that each shake dilates the packing. The particles then fall, one after the other, in the direction of the gravitational field until they hit the surface of already settled particles or the walls of the container. Then, each particle follows the path of steepest descent until it reaches a local minimum, while higher ones are still in free flight. Once the particle has reached this position it will stay there until the end of the restructuring cycle. After all particles have found their minima they are again collectively displaced or dilated and a new restructuring cycle starts.

Extreme material properties are implicitly assumed in this rule. First, the coefficient of restitution has to be close to zero so that one can neglect that particles bounce back. Second, the dynamic and static friction coefficients have to be large so that particles cannot escape from shallow minima and settled particles are not kicked out of their positions. In contrast to molecular dynamics calculations, the algorithm needs no force calculations but only geometrical considerations (to find the path of steepest descent). Thus the BTR model is particularly adapted to study geometrical effects. Whereas molecular dynamics often show a tendency towards fluidization, the BTR model by construction lets a configuration freeze in after each cycle.

5.2 Simulation of a Rotating Drum
(with T. Scheffler and G. Baumann)

In order to understand the basic properties of the behavior in a slowly rotating drum, we adapted the BTR model to this situation. As above, we consider only the case, where after each avalanche the packing adopts a static configuration. This configuration is then rigidly rotated by a small angle. Then, starting with the lowest particle, the same restructuring rule as in Sect. 5.1 is applied to find the new configuration.

However, now the justification of the algorithm has to be reconsidered. Is it really allowed to ignore the higher particles when one particle relaxes into the nearest local minimum, even if there is no significant dilation of the packing, in contrast to the shaking? Figure 18 shows the traces of all particles during one update: although in principle allowed, unphysical restructurings occur only extremely seldom. The reason is that one has to divide the filling into a bulk and a surface region. In order to find the new position of a bulk particle all higher particles may be ignored, because they move essentially on parallel trajectories and therefore cannot change the result. In the avalache zone, however, higher particles easily give way, and again do not significantly influence the rolling down of particles below.

Using this algorithm the first convincing simulation of radial size segregation was made [58], which compares well with experiments [25], see Fig. 5. It was shown that segregation occurs for all ratios of the disc radii without changing its character [21], in contrast to segregation due to vertical shaking. Ergodicity of the trajectories of a tagged particle in the drum has been shown [59]. An interesting prediction of the BTR model is that a monodisperse filling generally has a lower (dynamic) angle of repose than a bidisperse one, where the material of all particles is the same. This was found in simulations of a two-dimensional drum [60]. For smaller drums the difference between the angles of repose for the two fillings decreases and vanishes eventually. An experimental check of this prediction should be a test of the applicability of the BTR model. Also the fluctuations of the surface angle have been investigated in this model. Their power spectrum decreases with the inverse frequency squared [21].

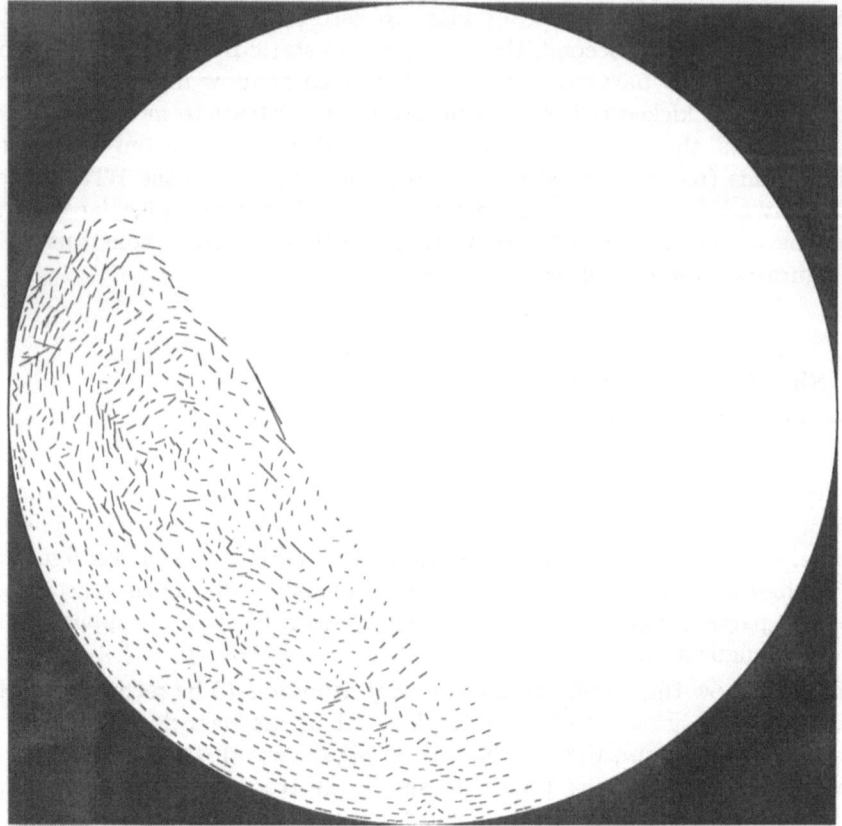

Fig. 18. Traces of the particles in one elementary rotation step. From [21]

6   Conclusion

The last four sections give an overview over some models used to simulate granular media. Each one has its particular strengths and limitations. All of them have been successfully applied to study phenomena described in the first section. It has been one purpose of this chapter to point out that the simulation model has to be chosen such that the phenomenon one wants to study is not outside the range of applicability of the model.

The most versatile model uses soft-particle molecular dynamics with forces that depend on the overlap. It is simple and can be used over a wide range of densities. However, it is not the most efficient model, and it is hampered by an inherent trend towards fluidization. The artefacts due to the detachment effect and brake failure have to be considered if one wants to simulate dense systems or rapid flows with this model.

In the fluidized state event driven simulations are more efficient. They avoid the artefacts of the soft-particle model, however they fail due to the inelastic collapse, if the system develops long lasting contacts.

By contrast, contact dynamics has its most spectacular applications just for systems with long-lasting contacts. It should be kept in mind that this simulation method can be generalized to allow for the opening and closing of contacts [7, 56, 55], and it seems to me that this method can be developed into a more versatile model than ordinary soft-particle molecular dynamics. However, the model uses rigid particles, i.e. elasticity is only represented by a coefficient of restitution, which is a concept of binary collisions. Therefore one can easily construct special multiple collisions, in which rigid particles behave differently from, for example, steel spheres [62]. The efficient implementation of this method is not easy.

Finally, the bottom-to-top-restructuring model turned out to give unexpectedly good results in situations, in which the system evolves from one static configuration to the next under the influence of gravity. The transition rule can be justified in the limiting case of vanishing restitution coefficient and large friction. Among the discussed models this one is certainly computatinally most efficient. On the other hand, it cannot be used to study fluidized granular media.

Some simulation techniques have not been addressed in this review such as cellular automata [63, 64], which are also extremely efficient, and finite element methods [65]. So far, cellular automata have only been applied in two dimensions in the context of granular media. The three-dimensional case is expected to require an embedding into a higher-dimensional space, as is known from hydrodynamic cellular automata. Finite element methods do not deal with particle dynamics, as the methods discussed here, but treat the medium on a coarser scale.

Acknowledgements

These lecture notes would never have been finished without the help of my collaborators on this subject George Batrouni, Gerald Baumann, Lothar Brendel, Sabine Dippel, Imre Jánosi, Elmar Jobs, Franck Radjai, Tim Scheffler and Jochen Schäfer, some of whom are co-authors on several sections as indicated. With all of them I had many discussions and shared the excitement of research. This work was supported by the European Community within the network CHRX-CT93-0354 "Cooperative Structures in Complex Media" and by the DFG grant no. Wo 577/1-1.

References

[1] H.M. Jaeger and S.R. Nagel, Science 255, 1523 (1992); S.R. Nagel, Rev. Mod. Phys. 64, 321 (1992)

[2] A. Hansen and D. Bideau (eds.), *Disorder and Granular Media* (North-Holland, Amsterdam 1992)

[3] J. Rajchenbach, E. Clément, J. Duran, and T. Mazozi, in: *Scale Invariance, Interfaces, and Non-equilibrium Dynamics*, eds. A. McKane, M. Droz, J. Vannimenus, D.E. Wolf (Plenum, New York 1995), p. 313

[4] H.J. Herrmann, Phys. Bl. **51**, 1083 (1995)

[5] G. Hagen, Monatsberichte (Preuß. Akad. d. Wiss., Berlin 1852), p. 35

[6] H.A. Janssen, Zeitschrift VDI **39**, 1045 (1895)

[7] F. Radjai: *Dynamique des rotations et frottement collectif dans les systèmes granulaires*, PhD-thesis, (Université de Paris-Sud XI, Orsay 1995)

[8] P.A. Thompson and G.S. Grest, Phys. Rev. Lett. **67**, 1751 (1991)

[9] L.D. Landau and E.M. Lifshitz, *Lehrbuch der Theoretischen Physik* Vol. X: Physikalische Kinetik (Akademie-Verlag, Berlin 1983) p. 27

[10] S. Ogawa, in: *Proc. US–Japan Seminar on Continuum-Mechanical and Statistical Approaches in the Mechanics of Granular Materials*, eds. S.C. Cowin and M. Satake, (Gakujutsu Bunken Fukyukai, Tokio 1978)

[11] I. Goldhirsch, in [12], p. 251; I. Goldhirsch, G. Zanetti, Phys. Rev. Lett. **70**, 1619 (1993)

[12] D.E. Wolf, M. Schreckenberg, and A. Bachem (eds.), *Traffic and Granular Flow* (World Scientific, Singapore 1996)

[13] J. Schäfer and D.E. Wolf, in [12], p. 311

[14] P. Bak, C. Tang, and K. Wiesenfeld, Phys. Rev. Lett. **59**, 381 (1987)

[15] J.M. Carlson and J.S. Langer, Phys. Rev. Lett. **62**, 2632 (1989)

[16] P. Bak and K. Chen, Sci. Am. **264**, 46 (1991); P. Bak and M. Paczuski, Phys. World **6**, 39 (1993); G. Grinstein, in: *Scale Invariance, Interfaces, and Non-equilibrium Dynamics*, eds. A. McKane, M. Droz, J. Vannimenus, D.E. Wolf (Plenum, New York 1995), p. 261

[17] V. Frette, K. Christensen, A. Malthe-Sørenssen, J. Feder, T. Jøssang, and P. Meakin, Nature **379**, 49 (1996)

[18] A.N. Kolmogorov, Doklady Akad. Nauk. SSSR **31**, 99 (1941); T. Ishii and M. Matsushita, J. Phys. Soc. Japan **61**, 3474 (1992)

[19] D.R. Wilkinson and S.F. Edwards, Proc. R. Soc. Lond. A **381**, 33 (1982)

[20] P. Meakin, Physica A **163**, 733 (1990)

[21] G. Baumann, I.M. Jánosi, and D.E. Wolf, Phys. Rev. E **51**, 1879 (1995)

[22] D. Bideau, I. Ippolito, L. Samson, G.G. Batrouni, S. Dippel, A. Aguirre, A. Calvo, and C. Henrique, in [12], p. 279

[23] S. Dippel, L. Samson, and G.G. Batrouni, in [12], p. 353

[24] E. Clément, J. Rajchenbach, and J. Duran, Europhys. Lett. **30**, 7 (1995)

[25] F. Cantelaube and D. Bideau, Europhys. Lett. **30**, 133 (1995)

[26] A. Rosato, F. Prinz, K.J. Strandburg, and R. Swendsen, Phys. Rev. Lett. **58**, 1038 (1987)

[27] R. Jullien, P. Meakin, and A. Pavlovitch, Phys. Rev. Lett. **70**, 2195 (1993)

[28] S. Dippel and S. Luding, J.Phys. I France **5**, 1527 (1995)

[29] J.B. Knight, H.M. Jaeger, and S.R. Nagel, Phys. Rev. Lett. **70**, 3728 (1993)

[30] J. Duran, T. Mazoni, E. Clément, and J. Rajchenbach, Phys. Rev. E **50**, 5138 (1994)

[31] T. Pöschel and H.J. Herrmann, Europhys. Letters **29**, 123 (1995)

[32] J. Duran, J. Rajchenbach, and E. Clément, Phys. Rev. Lett. **70**, 2431 (1993)

[33] O. Zik, D. Levine, S.G. Lipson, S. Shtrikman, and J. Stavans, Phys. Rev. Lett. **73**, 644 (1994)

[34] F. Melo, P. Umbanhowar, and H.L. Swinney, Phys. Rev. Lett. **72**, 172 (1994)

[35] P.A. Cundall and O.D.L. Strack, Géotechnique **29**, 47 (1979)
[36] P.K. Haff and B.T. Werner, Powder Technol. **48**, 239 (1986)
[37] M.P. Allen and D.J. Tildesley, *Computer Simulation of Liquids* (Clarendon Press, Oxford 1987)
[38] H.J. Herrmann, in: *Lectures in Computational Physics*, eds. P.L. Garrido and J. Marro (Springer, Heidelberg 1995)
[39] G. Ristow, in: *Annual Reviews of Computational Physics I*, ed. D. Stauffer (World Scientific, Singapore 1994)
[40] K.L. Johnson, *Contact Mechanics* (Cambridge University Press, Cambridge 1989); L.D. Landau and E.M. Lifshitz, *Lehrbuch der Theoretischen Physik* Vol.VII: Elastizitätstheorie (Akademie-Verlag, Berlin 1989)
[41] G. Kuwabara and K. Kono, Jap. J. Appl. Phys. **26**, 1230 (1987)
[42] W. Goldsmith, *Impact* (Edward Arnold, London 1960)
[43] F.G. Bridges, A. Hatzes, and D.N.C. Lin, Nature **309**, 333 (1984)
[44] R. Sondergaard, K. Chaney, and C.E. Brennen, J. Appl Mech. **57**, 694 (1990)
[45] J. Schäfer, S. Dippel, and D.E. Wolf, J. Physique I (1996)
[46] S. Melin, Phys. Rev. E **49**, 2353 (1994); C.-H. Liu and S.R. Nagel, Phys. Rev. Lett. **68**, 2301 (1992)
[47] S.F. Foerster, M.Y. Louge, H. Chang, and K. Allia, Phys. Fluids **6**, 1108 (1994)
[48] T.G. Drake, J. Geophysical Research **95**, 8681 (1990)
[49] S. Luding, E. Clément, A. Blumen, J. Rajchenbach, and J. Duran, Phys. Rev. E **50**, 4113 (1994)
[50] J. Schäfer, PhD-thesis, (Gerhard-Mercator-Universität, Duisburg 1996)
[51] J. Schäfer and D.E. Wolf, Phys. Rev. E **51**, 6154 (1995)
[52] J.T. Jenkins, J. Appl. Mech. **59**, 120 (1992)
[53] S. McNamara, W.R. Young, Phys. Rev. E **50**, R28 (1994)
[54] S. Luding, E. Clément, A. Blumen, J. Rajchenbach, and J. Duran, Phys. Rev. E **49**, 1634 (1994)
[55] J.J. Moreau, Eur. J. Mech. A/Solids **13**, 93 (1994)
[56] M. Jean in: *Mechanics of Geometrical Interfaces*, eds. A.P.S. Selvadurai and M.J. Boulon (Elsevier, Amsterdam 1995)
[57] F. Radjai, L. Brendel, S. Roux, *Non-smoothness, Indeterminacy and Friction in Granular Systems*, HLRZ-preprint 80/95
[58] G .Baumann, E. Jobs, and D.E. Wolf, Fractals **1**, 767 (1993)
[59] G. Baumann, I.M. Jánosi, and D.E. Wolf, Europhys. Lett. **27**, 203 (1994)
[60] G. Baumann, T. Scheffler, I.M. Jánosi, and D.E. Wolf, in [12], p. 347
[61] E. Jobs, unpublished
[62] D.E. Wolf and F. Radjai, unpublished
[63] G. Peng and H.J. Herrmann, Phys. Rev. E **48**, R1796 (1994)
[64] A. Károlyi and J. Kertész, in [12], p. 359
[65] S. Roux, in: *Statistical Models for the Fracture of Disordered Media*, eds. H.J. Herrmann and S. Roux, (Elsevier, Amsterdam 1990)

Algorithms for Biological Aging*

Dietrich Stauffer

Institut für Theoretische Physik, Universität zu Köln, D-50923 Köln, Germany
e-mail: `stauffer@thp.uni-koeln.de`

Abstract. There are many theories about why we age; this chapter will present some theories that are suited for simulation without claiming that they are the truth. Perhaps Darwinian selection leads to an accumulation of bad mutations in old age since those which happened in young age and prevented procreation were weeded out much more strongly. This balance of mutations and selection pressure has been dealt with in many Monte Carlo simulations since 1993. Particularly efficient is the simulation of Thadeu Penna which stores the mutations in the 32 bits of one computer word, with each bit position corresponding to one year in life, and on-bits signaling the presence of the mutation, off-bits the absence.

1 Introduction

Except for this writer, everybody gets old. Even Brigitte Bardot (or Ronald Reagan, or Marxism...). Why? How can we avoid that? Is there eternal youth outside of the movie "Death becomes her"? I don't know. But I know how to program computers. Thus I present here one of the many aging theories which is particularly suited for computers: evolutionary population dynamics [1].

This is part of a widespread field, for which Ausloos coined the acronym BEER: biological evolution engineering research. The aim of modeling biology is not to be as realistic as possible but to be as simple as is compatible with the aspects one wants to study. A single cell is too complicated to be put completely onto today's computers, just as a glass of beer contains too many molecules to be simulated. Nevertheless we understand today quite well the transition between liquid and vapor, through simple approximations such as the van der Waals equation, or simple models like the Ising model, proposed by Ern(e)st Ising's advisor Lenz 75 years ago at a conference close to the region of Chemnitz. A water molecule is much more complicated than can be described by a single-bit variable via occupied and empty. Nevertheless, for 20 years we have believed that water and all other simple fluids have the same critical exponents at the liquid vapour critical point as the three-dimensional Ising model. So the Ising model does not describe hydrogen bonding and the anomalous properties of water near $4^o C$, but it does describe the behavior near the critical point.

So what is the analog of the Ising model for aging? I still don't know but perhaps we will know ten years from now. There are major hurdles to a simple understanding: The boiling of water is as common place as is biological aging,

---

* Software included on the accompanying diskette.

but experiments are clear, reproducible, and give a few numbers of interest such as the densities on the coexistence curve or the vapor pressure as a function of temperature. What is aging? For lack of better definitions we defined it as the reduction of survival probabilities with advancing age of the individual. Clearly, this is not what we see in Brigitte Bardot. Also, experiments are difficult to reproduce, and biologists have not yet agreed on the major cause of aging. Similarly, in other fields of biology the experiments are difficult and seldom clear-cut, and thus do not easily differentiate between good and bad theories. For an analogy with physics history, imagine you should have simulated fluids before it was known that they consist of molecules and that heat is not a separate substance called phlogiston. You then might have made some nice phlogiston simulations which gave good results but had nothing to do with the truth. In this sense, biologically motivated simulations may also turn out to be based on completely wrong ideas, even if the results are correct. This also holds for aging theories.

This chapter is not about evolutionary or genetic algorithms. These are names for computational techniques which are inspired by mother nature, like sex or neural networks. Such simulations can then be applied to solve the spin glass ground state, or the traveling salesman problem. We, instead, use methods known from computational physics to model biology.

2   Concepts and Models

The survival probability S is the number of living beings of age $t+1$, in relation to the number for age t one time unit earlier. Whether our time unit is one year, or some smaller interval, depends on the biological application. For simplicity we always talk about years and have in mind fish, or human beings, and not so much fruit flies or nematode worms. Aging then is the decay of fitness with age, as is observed after the particular dangers for newborn children are overcome. The death rate $1-S$ increases with age roughly exponentially, as found by Gompertz in the last century. Of course, there are important exceptions: two world wars produced huge gaps in the European population whereas our model neglects such phenomena. Mathematically the death rate cannot become bigger than one, even though this was asserted in the January 1995 issue of Scientific American. The Gompertz law is thus valid for not too young and not too old people, in stable populations like that of Taiwan at the beginning of this century. All other aspects of aging, except this reduction of survival rate with increasing age, are neglected now.

Why does this survival rate go down? Actions of oxygen radicals may be responsible but are hardly suited for simulations similar to Ising models. Sex has been found to be dangerous to your health by some: Van Voorhies [2] found a significantly faster decrease of survivability in mated males compared with unmated males (also known as the Duran-Duran effect from the movie "Barbarella".) Sure it is fun if I can blame women for aging, but I doubt that a proposal to abolish sex has much chances for being funded or followed; even if implemented, who would then cite our papers 100 years from now? Much more similar to traditional physics simulations is the idea that random mutations move our genes away from the ideal configurations and thus cause aging, just like random thermal motion moves atoms away from their ideal lattice sites and produces liquids and gases.

If mutations are random they happen for all ages, and thus at first they might be thought to give a constant death rate independent of age. But this is not true due to Darwinian selection of the fittest: a mutation endangering the life of a person below the reproductive age reduces the number of offspring much more than a mutation affecting us only late in life when we barely get any children. Thus after some generations, the mutation for early life is restricted to a very small fraction of the population whereas the mutation for late age can be very widespread. In this sense, selection pressure tries to keep us close to the ideal genetic makeup whereas random mutations move us away from it, just like energy minimization tries to keep an equilibrated physics system close to the ground state while entropy (random thermal fluctuations) move it away. The balance of selection and mutation gives the equilibrium aging curve $S(t)$ just as the balance of energy and entropy gives the free energy minimum.

Thus an important ingredient of such an aging theory is an age structure with no reproduction in young age. In the simplest case, proposed by Partridge and Barton [1], we take just two age intervals: juveniles from age 0 to age 1, and adults from age 1 to age 2. Reproduction is possible at age 1 and age 2 only, and is followed by death after age 2. We then have the juvenile survival rate J and the adult survival rate A, and aging simply means

$$J > A \quad .$$

Mutations in such models were simulated by Monte Carlo methods in [3].

More realistic models use many aging steps, such as 32 or 64 years, and then it is tempting to store the whole genetic setup in a single computer word consisting of 32 or 64 bits. This is the case in the Penna model [4], and mutations in such models have also been studied by Monte Carlo [5]. Bit position t in such models then indicates whether this individual at age t and thereafter will suffer from the bad effects of a particular mutation.

In all cases one has to be careful to distinguish between hereditary and somatic mutations. The first ones are passed on to the offspring, like hemophilia, the second ones are not, like skin cancer from sun bathing. One of the crucial questions in any model is whether or not due to the accumulation of bad hereditary mutations from one generation to the next the whole population finally dies out: "mutational meltdown" [6].

3   Techniques

Some two-age models are reviewed in [7], and the program on the diskette thus gives only a one-age model from which the reader can construct a two-age program. This one-age model does not explain aging, but explains how the accumulation of hereditary bad mutations is programmed and may destroy the population. Initially, our (juvenile) survival rates juven are all set to unity and of course there are no adult survival rates in this one-age simplification. Then we make max iterations during which we not only calculate the average survival rate av but also check for each individual if it dies. This is done in loop 2: if a random number rand(0) between 0 and unity is greater than the survival rate, the individual dies; otherwise it survives and suffers a random mutation which reduces the survival rate by a random amount between 0 and ϵ. Now the survivors reach reproductive age, each one gets m children in loop 5, and each of these children inherits the (mutated) survival probability of its parent. Sex is avoided here to prevent any violation of pornography laws by this article; thus bacteria might be better suited than *homo sapiens* for this simple program. In the same way, a second age interval, adulthood, can be programmed with its own mutations added to those of youth.

This simulation shows in a few seconds of computer time how for $m = 2$ the population first grows since each individual gets two children. However, after some time the hereditary bad mutations have accumulated to such an amount that the survival rate sinks below 1/2, and the population shrinks instead of growing until it vanishes completely: mutational meltdown. The same effect also occurs in two-age models and can be avoided if we also allow positive mutations, or assume that these mutations are somatic and not inherited, as discussed extensively in [3]. The aging condition $J > A$ is often but not always fulfilled.

More realistic, of course, is a model with many ages. A Fortran program for the Penna bit string model was listed and explained by Penna and Stauffer [5]. Basically, each bit is set if at the age corresponding to this bit position a bad mutation becomes active. The individual feels all bad mutations of the present and all previous years. If this total number of bad mutations is larger than a fixed threshold or larger than the average number for the whole population (Thoms et al [5]), the individual dies; otherwise it lives at least a year longer. After some minimum age of reproduction, each individual (again asexually) can get children at each age, and each child differs by one bit from the genes of the parent. Thus the children in the same family have slightly different genes, and mutational meltdown can be avoided. The advantage of this model is that all genes can be stored in a single computer word, saving lots of storage and some computer time. (Bernardes [5] showed that replacing bits by small integers having more than two possible values does not drastically change the results.)

4 Results

With this bit-string model [5] one may find populations in reasonable agreement with natural age distributions. The survival rate is close to unity in youth, decays somewhat at middle age, and decays much more rapidly at old age. However, exceptions exist with an additional minimum in the survival rate at young age (Thoms et al). The mutations accumulate mostly in old age where their frequency can be ten times higher than in youth, even if initially all mutations were distrubuted randomly over all ages, or if we started with no mutations at all. In other words, families with lots of bad mutations in young age have mostly died out whereas those with many mutations affecting old age have survived and proliferate.

This balance of selection pressure (energy) and mutation accumulation (entropy) and its relation to aging becomes particularly clear if we simulate pacific salmon, a fish which dies soon after reproduction ("catastrophic senescence"). If we assume that reproduction only happens at some specific age, then all animals beyond that age no longer contribute to the population growth. (Parental care, and grandmotherly help, is ignored in this model.) Thus mutations acting on ages beyond this reproductive age can accumulate without being weeded out by selection pressure. After some time practically all fish above reproductive age are killed by their inherited mutations, whereas the survival rates below that reproductive age are close to unity [8]. The simulation results agree well with analytical solutions [9]. This first-order transition from a survival rate close to unity to one close to zero indicates how important reproduction is in biology. The publish or perish philosophy is a sociological example in a similar sense for universities. (The first-order transition claimed by Partridge and Barton to explain pacific salmon turned out to be an artifact of their choice of variables.)

The Penna–Moss theory [9] gives an exact solution for large populations for the special case where reproduction occurs only at one precise age R, and when one bad mutation acting against the animal is enough to kill it. For simplicity we assume an infinite supply of food and space. Then, in the stationary state N_0 babies give $N_1 = 31/32N_0$ animals of age 1, since their randomly mutated bit can be anywhere except in the first position, N_1 animals of age 1 will give $N_2 = 30/31N_1 = 30/32N_0$ animals of age 2, since there mutation is allowed to be anywhere in bits 3 to 32, etc. For age $k \leq R$ we have thus $N_k/N_0 = (32 - k)/32$ and older animals die. The birth rate b thus must obey $N_0 = bN_R$ or $b = 32/(32 - R)$ in a 32-bit model. Mutational meltdown is avoided for larger b since then enough children get their new mutation in the already mutated bits relevant for older ages $k > R$ where these mutations do not change anything.

If we can learn survival strategies from pacific salmon, can we also help fish to survive against human overfishing? Early simulations [10] used a birth rate fluctuating by a factor of nearly 30 to explain strong fluctuations in the fish populations, but such fluctuations are hardly realistic. The Penna model allows the simulation of age-dependent fishing, and then one can show that a slight increase of fishing may destroy the whole population (as happened with northern cod off Newfoundland in 1993, and is happening there now with Greenland halibut). If, on the other hand, young fish are allowed to live, then the fish population can survive [11]. Perhaps in the future, human effort in computers can balance better the human stupidity in overfishing.

It is often asserted that mutational meltdown is relevant only for asexual or very small sexual populations. This opinion is too optimistic according to

Bernardes. Sex often helps but does not always ensure the survival of the population [12]. One author [13] even claimed that there are no advantages in sex and that nature would be better off without men. This should be a challenge for male scientists to investigate in greater detail the effects of sexual reproduction on aging.

Acknowledgements

I thank Naeem Jan, Subinay Dasgupta, Stefan Vollmar, Suzana Moss de Oliveira, Thadeu Penna, and Americo Bernardes for many discussions and for collaboration.

References

[1] M.R. Rose, *Evolution Biology of Aging*, (Oxford University Press, Oxford 1991); B. Charlesworth *Evolution in Age-Structured Populations*, 2nd ed, (Cambridge University Press, Cambridge 1994); L. Bartridge and N.H. Barton, Nature **362**, 305 (1993)

[2] W.A. van Voorhies, Nature **360**, 456 (1992)

[3] T.S. Ray, J. Stat. Phys. **74**, 929 (1994); N. Jan, J. Stat. Phys. **77**, 915 (1994); S. Dasgupta, J. Physique I **4**, 1563 (1994); M. Heumann and M. Hotzel, J. Stat. Phys. **79**, 483 (1995)

[4] T.J.P. Penna, J. Stat. Phys. **78**, 1629 (1995)

[5] T.J.P. Penna and D. Stauffer, Int. J. Mod. Phys. C, **6**, 233 (1995); A.T. Bernardes and D. Stauffer, Int. J. Mod. Phys. C **7** (1996); J. Thoms, P. Donahue, and N. Jan, J. Physique I **5**, 935 (1995); J. Thoms, P. Donahue, D.L. Hunter, and N. Jan, J. Physique. I **5**, 1689; P.M.C. de Oliveira, and D. Stauffer, Physica A **221**, 453 (1995), H. Puhl, D. Stauffer, and S. Roux, Physica A **221**, 445 (1995); S.Goldberg, preprint for Physica A

[6] M. Lynch and W. Gabriel, Evolution **44**, 1725 (1990); M. Lynch, R. Burger, D. Butcher, and W. Gabriel, Journal of Heredity, **84**, 339 (1993)

[7] D. Stauffer, Braz. J. Phys. **24**, 900 (1994)

[8] T.J.P. Penna and S. Moss de Oliveira, and D. Stauffer, Phys. Rev. E **52**, 3309

[9] T.J.P. Penna and S. Moss de Oliveira, J. Physique I **5**, 1697 (1995)

[10] P.M. Allen and J.M.M. McGlade, Can. J. Fish Aquat. Sci. **43**, 1187 (1986); O. Fisken, J. Giske, and D. Slagstad, preprint (1995), C. Walsh, T.S. Ray, and N. Jan, preprint (1994), for models with spatial variation.

[11] S. Moss de Oliveira, T.J.P. Penna, and D. Stauffer, Physica A **215**, 298 (1995).

[12] A.T. Bernardes, preprint for Physica A

[13] R.J. Redfield, Nature **369**, 145 (1994)

Simulations of Chemical Reactions

Alexander Blumen[1], Igor Sokolov[1], Gerd Zumofen[2], and Joseph Klafter[3]

[1] Theoretische Polymerphysik, Universität Freiburg, Rheinstr 12,
 D-79104 Freiburg, Germany, email: blumen@tpoly1.physik.uni-freiburg.de
[2] Physical Chemistry Laboratory, ETH-Zentrum, CH-8092 Zurich, Switzerland
 email: gezu@msp.chem.ethz.ch
[3] School of Chemistry, Tel-Aviv University, Tel Aviv, 69978 Israel,
 email: klafter@chemsg1.tau.ac.il

Abstract. Chemical reactions seldom obey the "well-stirred-reactor" scheme, so that the actual spatial distribution of reactants is of importance at all times. Recent advances in the field are due to improved numerical techniques, which often paved the way for a deeper analytical treatment. We survey the knowledge on the A+B →0 reaction, paying special attention to mixing techniques (stirring transformations as well as Lévy-processes).

1 Introduction

Model chemical reactions are a very specific object of study, since despite their simply structured "rules of the game" they display very rich temporal patterns of behavior. From the point of view of simulations, model chemical reactions present the important advantage that in several particular cases exact solutions (or at least, asymptotic behaviors) are known; this provides then an excellent test for the numerically obtained results. On the other hand, as will be discussed in the following, analytical approximations may be quite misleading, when pushed beyond their (often poorly known) limits of validity. In several, very important instances, computer simulations have played a decisive role in showing that well-established patterns of thought were incorrect. In many cases the plotted results of simulations are intuitively easy to grasp; this has helped very much in establishing new ways of understanding the underlying phenomena.

In this chapter we will present several cases in which simulations were exceedingly helpful in advancing the analytical approach. We will start from simple, bimolecular reactions on simple lattices; this will help in keeping the basic picture simple, without being immediately submerged by realistic details.

2 The Basic Kinetic Approach

Let us start from the basic kinetic scheme [1]. Interestingly, this procedure in widespread use in physical chemistry is not adequate in describing many intriguing relaxation forms, as we proceed to show.

General irreversible reactions indexIrreversible reactions are of the type:

$$A_1 + A_2 + \ldots + A_n = \sum_{i=1}^{n} A_i \rightarrow 0 \ . \tag{1}$$

Assuming the reaction rate to be k, the classical description in the kinetic scheme leads to the following system of (in general) nonlinear differential equations:

$$\frac{\mathrm{d}A_i(t)}{\mathrm{d}t} = -k \prod_{i=1}^{n} A_i(t) \ . \tag{2}$$

In (2) $A_i(t)$ denotes the concentration of the ith molecular species, whose initial value is A_{i0}. The simplest case of (1) is the unimolecular reaction ($n = 1$) of the form $A \rightarrow 0$, whose solution is exponential:

$$A(t) = A_0 e^{-kt} \ . \tag{3}$$

Bimolecular reactions for which $n = 2$ are of the type

$$\frac{\mathrm{d}A(t)}{\mathrm{d}t} = -kA(t)B(t) = \frac{\mathrm{d}B(t)}{\mathrm{d}t} \ . \tag{4}$$

We set $C = B_0 - A_0$ and have as a general solution of (4):

$$\frac{1 + C/A(t)}{1 + C/A_0} = e^{Ckt} \ . \tag{5}$$

From (5) we infer for $B_0 \gg A_0$ that $C \simeq B_0$ and thus $C/A(t) \gg 1$. Hence for $B_0 \gg A_0$ the decay of the minority species is quasiexponential:

$$A(t) \simeq A_0 e^{-B_0 kt} \ . \tag{6}$$

On the other hand, if $A_0 = B_0$ then $C = 0$ in (5). An expansion in small C leads to the decay

$$A(t) = \frac{A_0}{1 + A_0 kt} \ , \tag{7}$$

from which at longer times, $t \gg (A_0 k)^{-1}$, an algebraic time dependence emerges:

$$A(t) \sim \frac{1}{kt} \ . \tag{8}$$

We pause to note that a very similar behavior is also obtained for the $A+A \rightarrow 0$ reaction, whose kinetic equation is:

$$\frac{1}{2}\frac{\mathrm{d}A(t)}{\mathrm{d}t} = -k[A(t)]^2 \ . \tag{9}$$

Separation of the variables in (9) and integration lead to:

$$A(t) = \frac{A_0}{1 + 2A_0 kt} \ , \tag{10}$$

a form very akin to (7). The long-time behavior obeys here

$$A(t) \sim \frac{1}{2kt} \ . \tag{11}$$

Thus, from unimolecular and from bimolecular reactions one has as long-time decays that are either exponential or $1/t$ algebraic dependent.

3 Numerical and Analytical Approaches for Reactions Under Diffusion

The problem with the $1/t$ dependencies of bimolecular reactions is that the finding is incorrect. Starting from simulations and from the better analytical understanding of the problem, as initiated in [2–6] we nowadays know that the A+B $\rightarrow$ 0 reaction obeys asymptotically

$$A(t) \sim \frac{1}{t^{d/4}} \ , \tag{12}$$

as long as $d \leq 4$. This is due to the fact that even under homogeneous initial

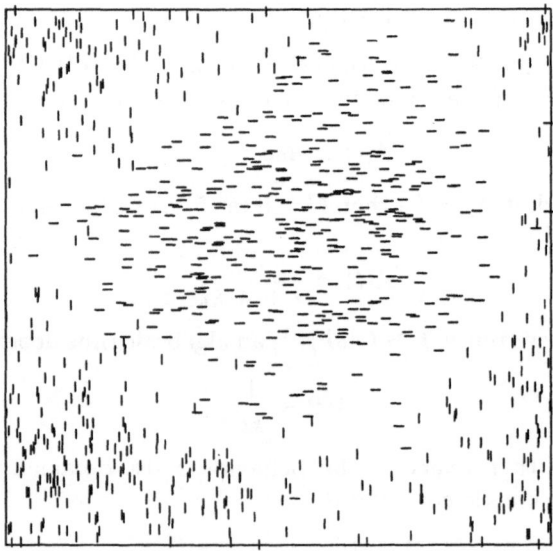

Fig. 1. Segregation in the A+B $\rightarrow$ 0 reaction. The reaction takes place on a square lattice of 1000 × 1000 sites and with periodic boundary conditions. A full lattice is considered as initial condition. The picture shows a snapshot after 10^5 time steps; roughly 1000 particles are displayed. The positions of the A and B particles are indicated by *vertical* and *horizontal bars*, respectively

conditions the microscopic reaction steps by themselves create nonhomogeneities and enhance already-existing density fluctuations. During the reaction large regions (clusters) containing only A or only B particles appear, see Fig. 1. This many-particle aspect of the problem was not recognized in the classical treatments of the problem, e.g., [7–10], which focused on the derivation of the reaction rate k from binary collisions. What then is the reason behind the simple finding of (11)? The basic assumption underlying the general kinetic scheme of (2) is the "well-stirred reactor" model, in which all spatial dependencies due to the positions of discrete particles are neglected. Thus the use of (2) implies a homogeneous spatial distribution of particles during the whole course of the reaction. As discussed above, such an assumption is untenable in general, since nonhomogeneous conditions are widespread. The diffusion can only partly wipe out density fluctuation effects [2–6,11–17] but only when the diffusion length is large compared to the mean cluster size. At low particle densities, diffusion (or stirring) cannot create a homogeneous background.

What about introducing the local particle densities $A(\mathbf{x}, t)$ and $B(\mathbf{x}, t)$ into the analytical picture? As we proceed to show, this indeed helps reproduce (12), but the results are still only qualitative. The idea is to start [2,3] from the following coupled diffusion-reaction equations:

$$\dot{A}(\mathbf{x}, t) = D\nabla^2 A(\mathbf{x}, t) - \kappa A(\mathbf{x}, t)B(\mathbf{x}, t) \tag{13}$$

and

$$\dot{B}(\mathbf{x}, t) = D\nabla^2 B(\mathbf{x}, t) - \kappa A(\mathbf{x}, t)B(\mathbf{x}, t) \ , \tag{14}$$

where D is the diffusion coefficient, and κ denotes the local bimolecular reaction rate. The connection to $A(t)$ and $B(t)$ is given by $A(t) = \langle A(\mathbf{x}, t)\rangle_{\mathbf{x}}$ and $B(t) = \langle B(\mathbf{x}, t)\rangle_{\mathbf{x}}$, i.e., by the spatial average. Equations (13) and (14) are far from being exact, since they are restricted to first-order density functions. Consequently, the reaction term is only approximate, since –at least– the joint probability density of A-B pairs is needed for a correct description [18–20]. We focus here on the case $A_0 = B_0$, which implies $A(t) = B(t)$ at all times. The analysis of (13) and (14) is simplified by setting $q(\mathbf{x}, t) = A(\mathbf{x}, t) - B(\mathbf{x}, t)$ and $s(\mathbf{x}, t) = A(\mathbf{x}, t) + B(\mathbf{x}, t)$, which leads to

$$\dot{q}(\mathbf{x}, t) = D\nabla^2 q(\mathbf{x}, t) \tag{15}$$

and to

$$\dot{s}(\mathbf{x}, t) = D\nabla^2 s(\mathbf{x}, t) - \frac{\kappa}{2}\left[s^2(\mathbf{x}, t) - q^2(\mathbf{x}, t)\right] \ . \tag{16}$$

We point out that (15) holds exactly, irrespective of the approximation introduced in the description of the reaction term. One has now $A(t) = \langle A(\mathbf{x}, t)\rangle_{\mathbf{x}} = \frac{1}{2}\langle s(\mathbf{x}, t)\rangle_{\mathbf{x}}$.

It is furthermore of interest to have the expressions corresponding to (15) and (16) also for discrete lattices, since this allows the generalization of the procedure to any connected underlying system and especially to fractal lattices. The discrete version of (15) and (16) is the following system of coupled differential-difference equations [19,20]:

$$\dot{q}(\mathbf{x}_j, t) = \Gamma \sum_{i \in \sigma_j} [q(\mathbf{x}_i, t) - q(\mathbf{x}_j, t)] \tag{17}$$

and

$$\dot{s}(\mathbf{x}_j, t) = \Gamma \sum_{i \in \sigma_j} [s(\mathbf{x}_i, t) - s(\mathbf{x}_j, t)] - \frac{\kappa}{2} [s^2(\mathbf{x}_j, t) - q^2(\mathbf{x}_j, t)] \quad . \tag{18}$$

In (17) and (18) $\mathbf{x}_j$ are the sites, σ_j denotes the set of nearest neighbors of site j, and Γ is the hopping rate between nearest-neighbor sites. Note that (17) is linear, and its solution can be expressed through the Green's function $G(\mathbf{x}_j, t; \mathbf{x}_0, 0)$, the conditional probability to be at $\mathbf{x}_j$ at time t having started at $\mathbf{x}_0$ at time zero. One has

$$q(\mathbf{x}_j, t) = \sum_i G(\mathbf{x}_j, t; \mathbf{x}_i, 0) q(\mathbf{x}_i, 0) \quad , \tag{19}$$

where $q(\mathbf{x}, 0)$ denotes the initial random configuration. From (19) all moments follow readily. Thus $\langle q(\mathbf{x}, t) \rangle_{\mathbf{x}}$ is zero at all times; the second moment, $\langle q^2(t) \rangle$, obtained by averaging $q^2(\mathbf{x}, t)$ over the lattice sites and over all initial configurations (ic) is

$$\langle q^2(t) \rangle = \langle q^2(\mathbf{x}, t) \rangle_{\mathbf{x}} = N^{-1} \sum_j \left\langle \left[\sum_i G(\mathbf{x}_j, t; \mathbf{x}_i, 0) q(\mathbf{x}_i, 0) \right]^2 \right\rangle_{\text{ic}} . \tag{20}$$

Here N denotes the number of sites of the lattice considered. Letting $q_0/2$ be the initial occupation probability for A or B particles, one has as initial distribution

$$q(\mathbf{x}_j, 0) = \begin{cases} +1 & \text{with probability } q_0/2 \\ -1 & \text{with probability } q_0/2 \\ 0 & \text{with probability } 1 - q_0 \end{cases} \quad . \tag{21}$$

It is now straightforward to calculate $\langle q^2(t) \rangle$ using (20) and (21), since the $G(\mathbf{x}_j, t; \mathbf{x}_i, 0)$ are independent of the initial configuration:

$$\langle q^2(t) \rangle = \frac{1}{N} \sum_{j,i,k} G(\mathbf{x}_j, t; \mathbf{x}_i, 0) G(\mathbf{x}_j, t; \mathbf{x}_k, 0) \langle q(\mathbf{x}_i, 0) q(\mathbf{x}_k, 0) \rangle_{\text{ic}}$$

$$= \frac{q_0}{N} \sum_{j,i,k} G(\mathbf{x}_j, t; \mathbf{x}_i, 0) G(\mathbf{x}_j, t; \mathbf{x}_k, 0) \delta_{ik}$$

$$= \frac{q_0}{N} \sum_{j,i} G^2(\mathbf{x}_j, t; \mathbf{x}_i, 0) \quad . \tag{22}$$

To proceed, we consider the Chapman–Kolmogorov equation for $0 \le t' \le t$:

$$G(\mathbf{x}_j, t; \mathbf{x}_i, 0) = \sum_m G(\mathbf{x}_j, t'; \mathbf{x}_m, 0) G(\mathbf{x}_m, t'; \mathbf{x}_i, 0) \quad , \tag{23}$$

which is obeyed by all Markov processes (irrespective of lattice structure and dimension, i.e., it holds also for fractals) [19]. Noticing furthermore, that the Green's function is symmetrical, see [19],

$$G(\mathbf{x}_j, t; \mathbf{x}_i, 0) = G(\mathbf{x}_i, t; \mathbf{x}_j, 0) \ , \tag{24}$$

one finds from (22) to (24) that

$$\langle q^2(t) \rangle = \frac{q_0}{N} \sum_j G(\mathbf{x}_j, 2t; \mathbf{x}_j, 0) = q_0 G(0, 2t) \ , \tag{25}$$

where $G(0, t)$ is the probability of being at the origin after time t, averaged over all starting sites. Equation (25) relates $\langle q^2(t) \rangle$ to the well-understood autocorrelation function $G(0, t)$, whose leading behavior follows asymptotically [21,22] the power law: $G(0, t) \sim a_{\tilde{d}} t^{\tilde{d}/2}$, where $\tilde{d}$ is the spectral dimension (d for Euclidean lattices), and the prefactor $a_{\tilde{d}}$ is lattice dependent. We continue by discussing the implications of (25) for the density decay. From (19) and (21) one can view q as being a large sum of terms which are either ± 1 or zero, weighted with the corresponding G factors; thus, at long times the central limit theorem holds so that $q(\mathbf{x}, t)$ approaches a Gaussian distribution [3,14]. It follows that

$$q(t) = \langle |q(\mathbf{x}, t)| \rangle_{\mathbf{x}} = \left[\frac{2}{\pi} \langle q^2(t) \rangle \right]^{1/2} \ . \tag{26}$$

Furthermore, one expects from (16) that for very large κ one has $s^2(\mathbf{x}, t) \simeq q^2(\mathbf{x}, t)$, which means physically that the particles segregate in clusters [2,3]. Approximating hence $s(t) = \langle s(\mathbf{x}, t) \rangle_{\mathbf{x}}$ through $q(t)$ and using (25), it follows that

$$A(t) = \langle A(\mathbf{x}, t) \rangle_{\mathbf{x}} = \frac{1}{2} s(t) \geq \frac{1}{2} q(t) = [q_0 G(0, 2t)/2\pi]^{1/2} \ , \tag{27}$$

which, considering the power-law description for $G(0, t)$, leads to

$$A(t) \geq C_{\tilde{d}} A_0^{1/2} t^{-\tilde{d}/4} \ . \tag{28}$$

The constant is $C_d = \pi^{-1/2} (\tau/4\pi)^{d/4}$ for Euclidean lattices, where we introduced τ , the hopping time. We have now to establish just how far setting $q(t) \simeq s(t)$ is justified [3,14,23].

To settle the question, one has to center on (16), which, averaged over all lattice sites, gives

$$\langle \dot{s}(t) \rangle = \frac{\kappa}{2} \left[\langle s^2(t) \rangle - \langle q^2(t) \rangle \right] \ . \tag{29}$$

Here it is of interest to see if and how fast the ratio $\langle s^2(t) \rangle / \langle q^2(t) \rangle$ tends towards unity. Furthermore, we verify to what extent $s(\mathbf{x}, t)$ and $q(\mathbf{x}, t)$ are Gaussian distributed. A measure for this is how fast the ratios $\langle q^2(t) \rangle^{1/2} / \langle |q(t)| \rangle$ and $\langle s^2(t) \rangle^{1/2} / \langle s(t) \rangle$ reach the asymptotic value $(\pi/2)^{1/2}$.

One should note that for $\tilde{d} \geq 4$, (28) gives as lower bound $t^{-\tilde{d}/4}$, i.e., a decaying form faster than the kinetically expected t^{-1}, see (8). We are thus led to expect, as usual, the marginal dimension of the process to be $d = 4$. It then follows that the ratio $\langle s^2(t)\rangle/\langle q^2(t)\rangle$ should diverge for $d > 4$ [22,24]. On the other hand, convergence of this ratio means that an upper bound to $A(t)$ is also given by an expression akin (apart from the prefactor) to (28). We studied these points by solving (13) and (14) numerically. For the numerical treatment the ratio κ/τ^{-1} has to be fixed; for comparison to former approaches [23] we took κ as being equal to $2/\tau$.

In Fig. 2 various quantities are shown for $d = 1$. To demonstrate the region of long-time behavior clearly, the quantities were multiplied by their expected asymptotic forms, such that the asymptotic patterns appear as horizontal lines. A lattice of 4×10^5 sites was used. Plotted are $\langle|q(t)|\rangle t^{1/4}/(2C_1)$ and $\langle s(t)\rangle t^{1/4}/(2C_1)$. The displayed curves demonstrate that $\langle|q(t)|\rangle$ quickly reaches the asymptotic regime, whereas $\langle s(t)\rangle$ relaxes considerably more slowly. In the region of moderate times $\langle|q(t)|\rangle$ and $\langle s(t)\rangle$ differ significantly, and $\langle s(t)\rangle$, as presented in Fig. 2, shows a characteristic hump.

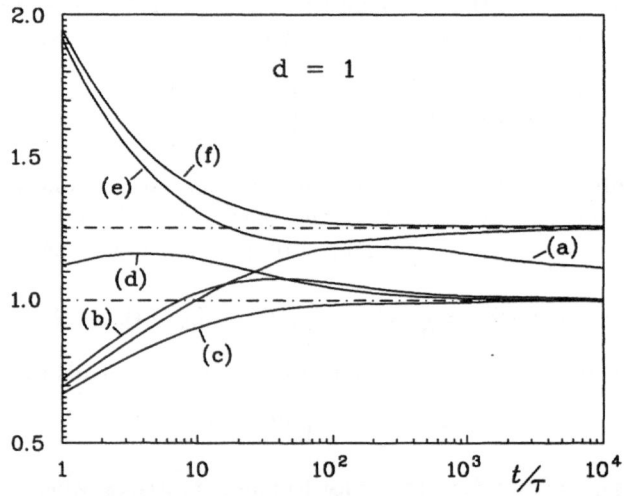

Fig. 2. Results in $d = 1$ from Monte Carlo (MC) simulations and from the deterministic approach [19] (a) MC result: $A_{\mathrm{MC}}(t)t^{1/4}C_1$. Deterministic results: (b) $\langle s(t)\rangle t^{1/4}/(2C_1)$, (c) $\langle|q(t)|\rangle t^{1/4}/(2C_1)$, (d) $\langle s^2(t)\rangle/\langle q^2(t)\rangle$, (e) $\langle s(t)\rangle/\langle s^2(t)\rangle^{1/2}$, (f) $\langle|q(t)|\rangle/\langle q^2(t)\rangle^{1/2}$. The initial concentrations are $A_0 = 0.05$ and $B_0 = 0.05$; the *dash-dotted* lines indicate the values 1 and $\sqrt{(\pi/2)}$

These patterns are compared with results taken from Monte Carlo (MC) calculations in which initially equal numbers of A and B particles were placed randomly on the lattice and typically 10^6 to 10^7 particles were used. Then a particle was picked randomly and was moved to a next neighbor position while simultaneously the time was incremented by the inverse of the number of particles still present in the sample. If one particle attempted to move into a site occupied by a particle of the opposite species, then both particles were removed from the lattice. The simulation results plotted as $A_{MC}(t)t^{1/4}/C_1$ [curve (a) in Fig. 2] show the same characteristic behavior as $\langle s(t)\rangle$ [curve (b)]; however, $A_{MC}(t)$ relaxes significantly more slowly to its asymptotic value than $\langle s(t)\rangle$. We view these differences between MC and deterministic data as resulting from the approximations introduced in the diffusion-reaction equations (13) and (14), which are thus limited in their ability to describe processes as complicated as particle annihilation. To complete the analysis, we also display in Fig. 2 the ratio $\langle s^2(t)\rangle/\langle q^2(t)\rangle$ [curve (d)], which shows a slow convergence to the value. Finally, also plotted are the two ratios $\langle q^2(t)\rangle^{1/2}/\langle|q(t)|\rangle$ and $\langle s^2(t)\rangle^{1/2}/\langle s(t)\rangle$. Both ratios converge to the asymptotic value of $(\pi/2)^{1/2}$, which is consistent with q and s being Gaussian distributed at long times. Again the sum variable relaxes more slowly than the difference variable to its limiting value.

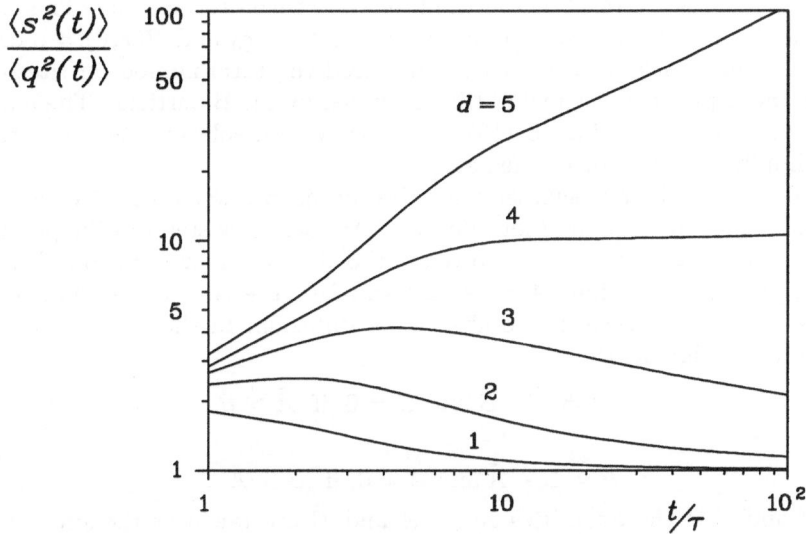

Fig. 3. The ratio $\langle s^2(t)\rangle/\langle q^2(t)\rangle$ for hypercubic lattices in one to five dimensions [19]

For a display of the situation for Sierpinski gaskets , see [19]. Here we close by showing in Fig. 3 the ratio $\langle s(t)\rangle/\langle q(t)\rangle$ for Euclidean lattices in 1, 2, 3, 4, and 5 dimensions. As is obvious from the figure, the behavior in $d = 4$ is marginal.

4   Reactions in Layered Systems

In this section and the next we display several results for reactions under stirring conditions, in which we highlight the interplay between simulation procedures and analytical approaches [25]. Stirring is of common occurrence both in chemical technology and also in everyday life. Laminar mixing, as found in viscous liquids, often occurs in polymer and glass processing and is the general feature in the physics of the Earth.

In this section we study reactions in premixed layered systems, and discuss the situations which evolve under mixing in the next section. In a premixed system the reaction is switched on after the mixing procedure has ended. As in Sect. 3 we focus on equal numbers of A and B particles, whose temporal evolution is governed by (13) and (14), restricted, because of the layered geometry, to one dimension.

This situation was first considered numerically by Muzzio and Ottino [11–13] and then analytically; our investigations [14,15] are analogous to the method reported below, whereas in [26] a perturbative approach was used.

We begin by discussing the numerical algorithms and the results of the simulations [25]. The starting point is a one-dimensional array of striations, which is a particular realization of a striation thickness distribution (STD). The initial concentrations of the reactants in the layers are taken to be equal. This means that initially one has at each site x either $A(x,0) = q_0$ and $B(x,0) = 0$ or vice versa. For the stoichiometrical case considered the total number of sites occupied by the A particles is equal to that occupied by the B particles. The reaction is then modeled according to (13) and (14), whose solution can be obtained through different numerical schemes.

Thus in [11–13] a two-substep numerical procedure was used. The method is applicable both for finite and for infinite κ. At each time step t of the procedure the diffusion equations were first solved in the absence of any reaction; this leads to some auxiliary functions $\tilde{A}(x, t + \Delta t)$ and $\tilde{B}(x, t + \Delta t)$. The second substep lets the reaction proceed in the absence of diffusion. In the case $\kappa \to \infty$ this leads to the assignment:

$$A = \tilde{A} - \tilde{B} \text{ and } B = 0 \text{ if } \tilde{A} > \tilde{B} \tag{30}$$

and to

$$B = \tilde{B} - \tilde{A} \text{ and } A = 0 \text{ if } \tilde{B} > \tilde{A} . \tag{31}$$

In (30) and (31) the quantities A, B, $\tilde{A}$ and $\tilde{B}$ are taken at the site x of the discretized system and at time $t + \Delta t$. In the case $\kappa < \infty$ the rules are more involved, and read:

$$A = \frac{\tilde{A} - \tilde{B}}{1 - \gamma \tilde{B}/\tilde{A}} \tag{32}$$

and

$$B = \frac{\tilde{B} - \tilde{A}}{(1 - \tilde{A}/\gamma \tilde{B})} \tag{33}$$

with $\gamma = \exp[\kappa \Delta t(\tilde{B} - \tilde{A}))]$. The main result of [11–13] is that both for finite and for infinite κ the long-time asymptotic behavior of the averaged concentration $A(t) = \langle A(x,t) \rangle = \langle B(x,t) \rangle = B(t)$ follows a power law

$$A(t) = B(t) \sim t^{-1/4} , \tag{34}$$

as is typical for $d = 1$ systems, see (12).

The numerical work of [11–13] and [26] also gives information on the spatial structures which emerge from the reaction. Thus, the total number N of striations decreases according to $N(t) \sim t^{-1/2}$. The numerical analysis also shows that the STD scales as a function of the dimensionless parameter $\eta = \sigma/L(t)$; here σ is the width of a particular striation and $L(t)$ is the mean width at time t.

In our own calculation we used a one-step algorithm for the diffusion-reaction problem [25]. In this case the time evolution of $A(x,t)$ and $B(x,t)$ is a discrete version $(x \to n)$ of (13) and (14). The algorithm reads [25]:

$$A(n, t + \Delta t) = A(n,t)+$$
$$+ \Delta t \left\{ D \left[A(n+1,t) + A(n-1,t) - 2A(n,t) \right] - \kappa A(n,t) B(n,t) \right\} \quad (35)$$

and

$$B(n, t + \Delta t) = B(n,t)+$$
$$+ \Delta t \left\{ D \left[B(n+1,t) + B(n-1,t) - 2B(n,t) \right] - \kappa A(n,t) B(n,t) \right\} \quad . \quad (36)$$

For a small enough Δt this algorithm is stable for small and medium values of κ. We discuss the results after considering the analytical approach to the problem, which, as before, is based on the sum and difference scheme.

The main change from Sect. 3 is that now the system is practically onedimensional and the initial distribution of reactants follows a STD, and not the (uncorrelated) prescription of (21). If the initial STD has a finite correlation length λ, the distribution of q at long-enough times will again be Gaussian. In a continuous picture one has namely

$$q(x,t) = \int\limits_{-\infty}^{\infty} q(\xi, 0) G(x - \xi, t) \mathrm{d}\xi \quad (37)$$

where $G(x,t) = (4\pi Dt)^{-1/2} \exp(-x^2/4Dt)$ is the (continuous) Green's function of the diffusion equation in 1D. The characteristic width of this bell-like function is $L_d \sim \sqrt{Dt}$. Therefore, for large enough t, when $L_d \gg \lambda$, the integral (37) can be viewed as being a sum of L_d/λ independent terms having zero mean and finite dispersion. Under the conditions of the central-limit theorem, the distribution of this sum will then be Gaussian. In a similar way, one can also make plausible that $q(x,t)$, viewed as a random function of the coordinate x at a fixed time t, corresponds to a Gaussian random process [27]. Such a process is fully characterized by its average value $\langle q(x,t) \rangle$ and by the correlation function $R_t(x_1, x_2) = \langle q(x_1,t) q(x_2,t) \rangle$. In our case we have as ensemble average $\langle q(x,t) \rangle = 0$ for all t and

$$R_t(x_1, x_2) = \int \int G(x_1 - \xi, t) \langle q(\xi, 0) q(\eta, 0) \rangle G(x_2 - \eta, t) \mathrm{d}\xi \mathrm{d}\eta \quad . \quad (38)$$

In the case when the two first moments $L = \langle x \rangle$ and $S = \langle x^2 \rangle$ of the STD exist and one also has $S \neq L^2$ the long-time (Gaussian) regime of (38) gives:

$$R_t(x) = \frac{\Gamma}{\sqrt{8\pi Dt}} \exp\left(-\frac{x^2}{8Dt}\right) \tag{39}$$

where the constant Γ equals $\Gamma = q_0^2(S - L^2)/L$ and we set $R_t(x) \equiv R_t(x,0)$. Furthermore, the concentration decay law in the long-time regime is:

$$A(t) = \frac{1}{2}\langle |q| \rangle = \frac{1}{2} \int |q| P(q) dq = \sqrt{\frac{\Gamma}{2\pi}} (8\pi Dt)^{-1/4} \tag{40}$$

as can be seen by inserting into (37) the explicit form of the q distribution, $P(q) = (2\pi)^{-1/2}\sigma^{-1}\exp[-(q - \mu)^2/\sigma^2]$ with $\mu = \langle q \rangle = 0$ and $\sigma^2 = R_t(0)$. One may note the emergence of the expected $t^{-1/4}$ behavior in (40).

Of interest for finite reaction rates κ is the behavior of $\langle s(t) \rangle$, see (29). For this we put $\langle s(t) \rangle = \langle s^2(t) \rangle^{1/2} = \hat{s}$ (see [16] for a discussion) and use the explicit expression for $\langle q^2 \rangle$ obtained with the help of $P(q)$. In [16] we found that for intermediate times $\hat{s}(t) \sim t^{-1}$ (classical behavior), while at long times $\hat{s}(t) \sim t^{-1/4}$.

Figure 4 shows the results of the numerical modeling of a reacting system with a finite κ. For comparison the analytical long-time behavior of $\langle |q(x,t)| \rangle$ is also given. One sees that at longer times the curves for $s(t)$ and $\langle |q(t)| \rangle$ merge and that they follow the analytical asymptotic form. Therefore the long-time kinetic behavior is determined by the fluctuations of the striation distribution.

The analytical approach used here also makes possible the investigation of the evolution of the clusters during the reaction, see [15]. Both the analytical and the numerical analysis show that for nonvanishing κ the system may be viewed as consisting of well-defined striations of A and B particles and that the reaction takes place within thin regions close to the clusters' boundaries. This confirms our picture that at longer times $s(x,t)$ tends to $|q(x,t)|$, or equivalently that the equations

$$A(x,t) = q(x,t)\Theta\left(q(x,t)\right) \tag{41}$$

and

$$B(x,t) = -q(x,t)\Theta\left(-q(x,t)\right) \tag{42}$$

hold asymptotically.

Here we view the clusters as being intervals between two subsequent simple roots $q(x,t) = 0$ of the difference variable q; hence the distribution of such clusters is equivalent to the zero-level crossing (ZLC) problem in the theory of random processes, see for example [27]. The solution of the ZLC problem gives the joint probability density $P(x_1, x_2)$ of finding a ZLC at the point x_1, provided there is a ZLC at the point x_2. The calculation proceeds by using the joint probability distribution $P(q_1, q_2, u_1, u_2)$ of the function q and of its derivative u at the two points x_1 and x_2. According to Rice's formula [28] for the density of ZLC $n = [-R_t''(0)/R_t(0)]^{1/2}/\pi$ one obtains $n = 1/\left(2\pi\sqrt{Dt}\right)$

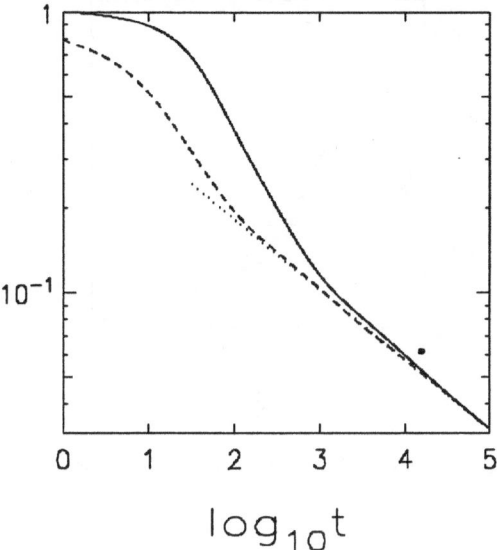

Fig. 4. Numerical solutions for $s(t)$ (*full line*) and $q(t)$ (*dashed line*) and the analytical asymptotic form (*dotted line*) for a system with a STD of width $W = 10$, and $D_A = 1/4 = D_B$, $\kappa = 0.05$, and $A(0) = 1.0$ [25]

and therefore $L(t) = n^{-1} \sim t^{1/2}$. The STD $\rho(x,t)$ at long times can also be expressed through $P(x_1, x_2)$, see [15]. From this expression the scaling form $\rho(x,t) = (8Dt)^{-1/2}\zeta(x/\sqrt{8Dt})$ follows, the function $\zeta(\xi)$ being universal. This scaling form is confirmed by direct computer modeling of the time evolution of the difference variable, see Fig. 5. Furthermore it follows that the function $g(\sigma,t) = \pi M \eta^2 \zeta \left(\pi\eta/\sqrt{2}\right)/\sqrt{2}$ (the occurrence frequency in a system of size M of lamellae of width σ, multiplied by σ^2) scales with the dimensionless variable $\eta = \sigma/L(t)$; it is precisely this fact which was discovered numerically in [11–13].

5   Reactions Under Mixing

Now we proceed by incorporating stirring aspects into the diffusion-reaction scheme. For this we focus on two basic models for mixing: on the one hand the Baker's transformation, which mixes strongly, and, on the other hand, shear-flow mixing, which is less effective. These rather simple-looking procedures are, however, related to industrially used mixing devices; for an overview one may consider [29] and [30].

Baker's transformation is one of the simplest theoretical models for mixing, see [31]. Each step of Baker's transformation (which requires, say, a time τ for completion) consists of three substeps: (a) squeezing the square to half its initial width and double height, (b) cutting the obtained object into two parts and

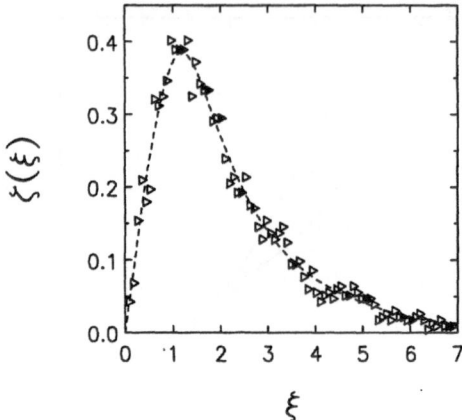

Fig. 5. The scaling function $\zeta(\xi)$ [25] obtained analytically (*dashed line*) and numerically (*triangles*)

(c) pasting the upper part to the right of the lower one. The second mixing model (shear flow) can be visualized as consisting of: (a) shearing a square into a rhombus with an acute angle of $\pi/4$, (b) cutting the rhombus into two equal rectangular triangles and (c) pasting the right triangle to the left of the left one.

The analytical procedure for mixing can be further simplified by using appropriate boundary conditions, which correspond to attaching to the initial system copies of itself along the x direction. In fact one may even go a step further and use statistical copies of the system, see [32,33]. Under such conditions one can dispense with the discontinuous step: the infinite domain now undergoes continuous squeezing for Baker's transformation and continuous shearing for shear-flow. In a liquid in motion both cases can be described by introducing position-dependent velocity fields.

For v we find the structure $v_i = \sum_j \alpha_{ij} r_j$, where i and j denote the coordinates and the α_{ij} are position and time independent. Moreover in both cases $\sum_i \alpha_{ii} = 0$ is obeyed, i.e. the liquid is incompressible, $\nabla \cdot v = 0$.

In two dimensions one finds for Baker's transformation that the matrix α_{ij} is diagonal with $\alpha_{xx} = -\alpha_{yy}$. For shear-flow only one element α_{xy} is nonzero. In three-dimensions the Baker's transformation has again a diagonal matrix with $\alpha_{xx} = -\alpha$ and $\alpha_{yy} = \alpha_{zz} = \alpha/2$ ($\alpha > 0$). For shear-flow the matrix α_{ij} again has only one nonzero element, namely α_{xy}.

As an extension of the former formalism a diffusion-controlled reaction in a moving, incompressible liquid is described by the following pair of differential equations, [11-13,16,17,32,33]:

$$\dot{A} + v \cdot \nabla A = D\Delta A - \kappa AB \ , \qquad (43)$$
$$\dot{B} + v \cdot \nabla B = D\Delta B - \kappa AB \ . \qquad (44)$$

Here A and B depend both on position and on time: (43) and (44) extend (13) and (14) through the inclusion of velocity-dependent (drift) terms. As before, we revert now to the difference q and the sum s of the local concentrations and obtain:

$$\dot{q} + v \cdot \nabla q = D \Delta q \tag{45}$$

and

$$\dot{s} + v \cdot \nabla s = D \Delta qs - \frac{1}{2}\kappa(s^2 - q^2) \ . \tag{46}$$

Thus we can proceed by paralleling the treatment of (17) and (18) above, the only difference being now that we need to have the Green's functions for diffusion under drift. Omitting the details (see [17] for the derivation) one finds for the Baker's transformation

$$G(\mathbf{r}, t; \mathbf{r}_0, 0) = (2\pi D)^{-d/2}$$

$$\times \prod_i \sqrt{\frac{\alpha_{ii}}{1 - \exp(-2\alpha_{ii}t)}} \exp\left(-\sum_i \frac{\alpha_{ii}\,(r_i - r_{0i}e^{\alpha_{ii}t})^2}{2D\left[\exp(2\alpha_{ii}t) - 1\right]}\right) \tag{47}$$

whereas shear-flow leads in 2D to [17]

$$G(\mathbf{r}, t; \mathbf{r}_0, 0) = \frac{\sqrt{3}}{2\pi Dt\sqrt{\alpha^2 t^2 + 12}}$$

$$\times \exp\left(-\frac{3\left[x - x_0 - \frac{\alpha t}{2}(y + y_0)\right]^2}{Dt(\alpha^2 t^2 + 12)} - \frac{(y - y_0)^2}{4Dt}\right) \tag{48}$$

and in 3D to a more complex form [17]

$$G(\mathbf{r}, t; \mathbf{r}_0, 0) = \left(\frac{3}{16\pi^3 D^3 t^3 (\alpha^2 t^2 + 12)}\right)^{1/2}$$

$$\times \exp\left(-\frac{3\left[x - x_0 - \frac{\alpha t}{2}(y + y_0)\right]^2}{Dt(\alpha^2 t^2 + 12)} - \frac{(y - y_0)^2}{4Dt} - \frac{(z - z_0)^2}{4Dt}\right) \ . \tag{49}$$

These forms can be used to analytically establish the decay forms, which can then be compared to simulation results. Analytically one finds for Baker's transformation at longer times,

$$q^2(t) \sim \exp\left(-|\alpha_{xx}|t\right) \ , \tag{50}$$

whereas for shear flow one obtains at longer times algebraic decays

$$q^2(t) \sim t^{-\beta} \ , \tag{51}$$

where $\beta = 2$ in 2D and $\beta = 5/2$ in 3D, see [17] for details. Taking also the quantity $s^2(t)$ into account allows one to show that in general the important initial stages of the decay (these are experimentally of main interest) are controlled by stirring. However, the duration of these stages varies, depending on the type of mixing considered and on the dimension of space in which the reactants move. On the other hand, one often recovers asymptotically, at long times, the classical kinetic behavior. Our calculations also allow us to conclude that the classical kinetic scheme is obeyed only in the limiting case of very effective mixing and very diluted solutions.

6   Reactions Controlled by Enhanced Diffusion

Much attention has been recently drawn to systems which display enhanced diffusion, where the mean-square displacement of a particle grows super-linearly in time [34–41]. Such enhancement has been experimentally observed, for instance, in a two-dimensional flow in a rotating annulus [42] and in self-diffusion studies in polymer-like breakable micelles [43]. In these cases, as well as in a broad range of numerical studies of dynamical systems , the enhancement has been attributed to Lévy walks which generalize the simple Brownian motion by extending the central-limit theorem [38,39,44,45].

We introduce Lévy statistics into reaction dynamics, a step which enables us to generalize the above-investigated reaction-diffusion schemes by including motional enhancement, and to demonstrate the continuous approach towards the mean-field results [40]. In this sense the Lévy-walk enhanced reactions present another model of simple mixing processes and broaden the scope of applicability of the above-mentioned reactions. We show that imposing the Lévy-walk aspect accelerates the reaction process, leads to different reaction patterns and lowers the critical dimension at which the mean-field behavior sets in.

As above we concentrate here on the transient A+B $\to$ 0 reaction with initially randomly placed A and B particles and with $A_0 = B_0$. The particles are considered to move at a constant velocity for time periods chosen randomly, according to a probability density $\psi(t)$. The density $\psi(t)$ is assumed to follow a power-law, $\psi(t) \sim t^{-\gamma-1}$. Here we restrict the range of the power-law exponents to $1 < \gamma < 2$. In this γ-regime the diffusion is enhanced, leading to a mean-squared displacement of a single particle that grows as $\langle r^2(t) \rangle \sim t^{2/\gamma}$.

We follow the analysis of the previous sections and study the time evolution of the particle densities in terms of the density-difference function, $q(\mathbf{x}, t) = A(\mathbf{x}, t) - B(\mathbf{x}, t)$; for its time evolution we write [40],

$$\dot{q}(\mathbf{x}, t) = \hat{L}q(\mathbf{x}, t) \ , \tag{52}$$

which is an extension of (15) to the case of enhanced diffusion. $\hat{L}$ is the operator which constitutes the Lévy process and which is defined in Fourier space ($\mathbf{x} \to \mathbf{k}$) as $\mathcal{F}\{\hat{L}f(\mathbf{x})\} = -c|\mathbf{k}|^\gamma \mathcal{F}\{f(\mathbf{x})\}$. The regular diffusion limit is recovered when $\gamma = 2$; in this case $\hat{L}$ is the Laplacian and c is the diffusion coefficient.

As in the previous sections we estimate the density of A and B particles from $A(t) = B(t) \gtrsim \langle |q(\mathbf{x}, t)| \rangle$. The latter quantity is obtained from the moment-generating function $\langle \exp[\phi q(\mathbf{x}=0, t)] \rangle$. At long times one has [3,46,48]

$$\langle \exp[\phi q(\mathbf{x}=0, t)] \rangle = \exp[\frac{1}{2}\phi^2 I(t)] \ , \tag{53}$$

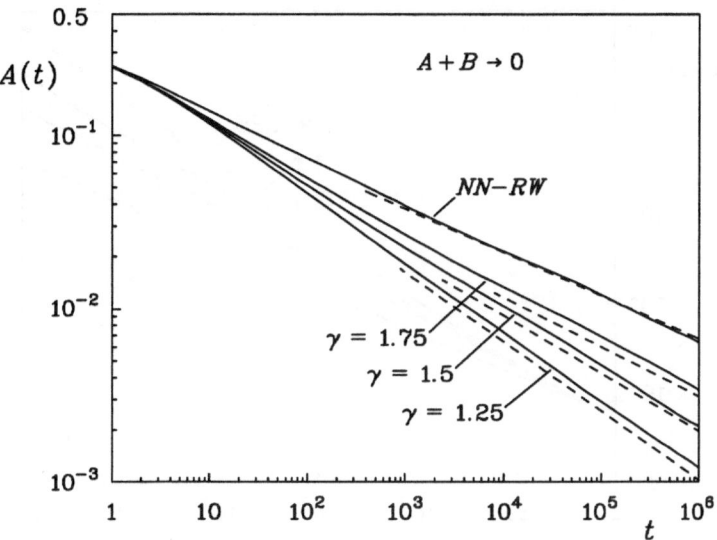

Fig. 6. The time evolution of the density $A(t)$ in the one-dimensional A+B $\to$ 0 reaction for nearest-neighbor random walks (NN-RW) and for enhanced diffusion, where the γ values are as indicated. Simulation results are given by *full lines*; the *dashed lines* are the predictions according to (54)

where the average is taken over all possible realizations of the initial conditions. This expression can be shown to hold equally well for regular and enhanced diffusion. It demonstrates that $q(\mathbf{x}=0,t)$ is Gaussian distributed; therefore asymptotically $\langle |q(\mathbf{x}=0,t)| \rangle = (2\langle q^2(\mathbf{x}=0,t)\rangle/\pi)^{1/2}$ also for enhanced diffusion. $I(t)$ was shown to be related to the Green's function [3,48], $I(t) = 2A_0 G(\mathbf{x}=0, 2t)$, where A_0 is the initial concentration of the A particles and $G(\mathbf{x},t)$ denotes the Green's function for enhanced diffusion. For Lévy walks one has $G(\mathbf{x}=0,t) \sim t^{-d/\gamma}$ in d dimensions; we thus obtain

$$A(t) \sim C(A_0/t^{d/\gamma})^{1/2} \ , \quad (d/2\gamma) \leq 1 \ , \tag{54}$$

which for $\gamma = 2$ reduces to the regular result of (12).

Segregation into A-rich and B-rich areas also occurs in reactions controlled by enhanced diffusion. The segregation is considered to take place on a scale $\Lambda(t)$. The temporal behavior of $\Lambda(t)$ can be discussed in terms of the position–position correlation function $\langle q(\mathbf{x},t)q(\mathbf{x}',t)\rangle$, that was shown to be related to the Green's function by [49]

$$\langle q(\mathbf{x},t)q(\mathbf{x}',t)\rangle \sim G(\mathbf{x}-\mathbf{x}', 2t) \ . \tag{55}$$

The behavior of the correlation length is thus equal to that of the characteristic length of the Green's function. From the scaling properties, $x \sim t^{1/\gamma}$, of the Green's function, we find

$$\Lambda(t) \sim t^{1/\gamma} \ . \tag{56}$$

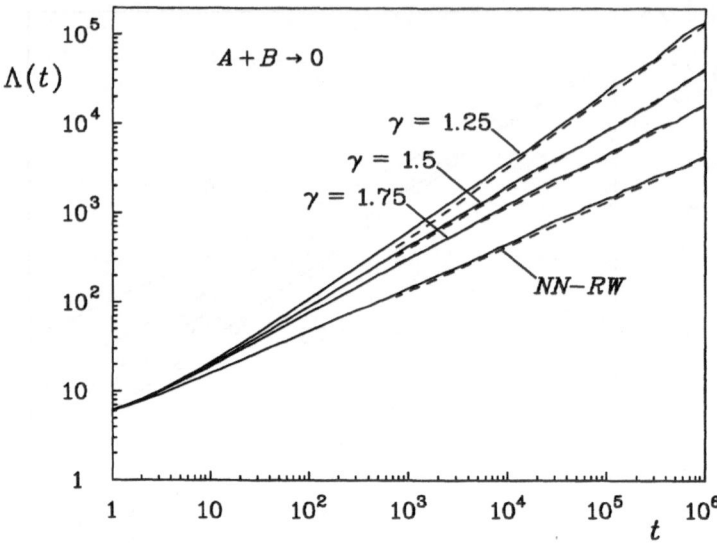

Fig. 7. The segregation length $\Lambda(t)$ as a function of time for the same diffusion controlled reactions considered for Fig. 6. *Full lines* give the simulation results and the *dashed lines* indicate the slopes according to $\Lambda(t) \sim t^{1/\gamma}$

In the simulations particles are initially dispersed randomly on the lattice and each particle is outfitted with a randomly chosen direction of motion and a randomly chosen duration time of constant velocity. After the duration time has elapsed a new direction and a new duration time are chosen at random. The particles are removed from the system at first encounter with an unlike species. Typically 10^6–10^7 particles are considered at the beginning of the reaction.

In Fig. 6 we show $A(t)$ both for regular nearest-neighbor random walks and also for various diffusional enhancements. The numerical results are compared with the predictions; a reasonable agreement is observed. From the above derivations the prefactors can also be derived, these prefactors are also considered for the presentation of the predictions in the figure. In Fig. 7 we show the segregation length $\Lambda(t)$ for the same set of parameters as in Fig. 6. Again the numerical results follow reasonably well the predicted slopes.

From this analysis we conclude that the enhanced diffusion, when introduced into diffusion-controlled reactions, manifests itself in a number of ways. As expected, the reaction is accelerated and the size of the clusters grows faster. Moreover, whereas in the simple diffusion case the critical dimension at which the classical rate-equation approach is applicable is $d_c = 4$ [2-6], here, due to the enhancement, the critical dimension is reduced to $d_c = 2\gamma$. Correspondingly, segregation, which slows down the reaction, is expected to also disappear for dimensions $2\gamma < d < 4$.

Acknowledgements

The support of the Deutsche Forschungsgemeinschaft (SFB 60), the Fonds der Chemischen Industrie, and a grant of the Rechenzentrum ETH-Zurich are gratefully acknowledged.

References

[1] A. Blumen, J. Klafter, and G. Zumofen, in *Optical Spectroscopy of Glasses*, ed. I. Zschokke, (Reidel, Dordrecht 1986), p. 199

[2] A.A. Ovchinnikov and Ya.B. Zeldovich, Chem. Phys. **28**, 215 (1978)

[3] D. Toussaint and F. Wilczek, J. Chem. Phys. **78**, 2642 (1983)

[4] K. Kang and S. Redner, Phys. Rev. Lett. **52**, 955 (1984)

[5] K. Kang and S Redner, Phys. Rev. A **32**, 435 (1985)

[6] G. Zumofen, A. Blumen, and J. Klafter, J. Chem. Phys. **82**, 3198 (1985)

[7] J. Laidler, *Chemical Kinetics*, (McGraw-Hill, New York 1950)

[8] T.R. Waite, Phys. Rev. **107**, 463 (1957), p. 471

[9] H. Eyring, S.H. Lin, and S.M. Lin, *Basic Chemical Kinetics*, (Wiley, New York 1980)

[10] J.M. Smith, *Chemical Engineering Kinetics*, (McGraw-Hill, Tokyo 1981)

[11] F.J. Muzzio and J.M. Ottino, Phys. Rev. Lett. **63**, 47 (1989)

[12] F.J. Muzzio and J.M. Ottino, Phys. Rev. A **40**, 7182 (1989)

[13] F.J. Muzzio and J.M. Ottino, Phys. Rev. A **42**, 5873 (1990)

[14] I.M. Sokolov and A. Blumen, Phys. Rev. A **43**, 2714 (1991)

[15] I.M. Sokolov and A. Blumen, Phys. Rev. A **43**, 6545 (1991)

[16] I.M. Sokolov and A. Blumen, Int. J. Mod. Phys. B **5**, 3127 (1991)

[17] I.M. Sokolov and A. Blumen, in *Synthesis, Characterization and Theory of Polymer Networks and Gels*, ed. S.M. Aharoni, (Plenum, New York 1992), p. 53

[18] V. Kuzovkov and E. Kotomin, Rep. Progr. Phys. **51**, 1479 (1988)

[19] G. Zumofen, J. Klafter, and A. Blumen, J. Stat. Phys. **65**, 1015 (1991)

[20] G. Zumofen, J. Klafter, and A. Blumen, Phys. Rev. A **44**, 8390 (1991)

[21] S. Alexander and R. Orbach, J. Phys. Lett. **43**, L625 (1982)

[22] S. Havlin and D. Ben-Avraham, Adv. Phys. **36**, 695 (1987)

[23] E. Clément, L.M. Sander, and R. Kopelman, Phys. Rev. A **39**, 6455 (1989)

[24] B.J. West, R. Kopelman, and K. Lindenberg, J. Stat. Phys. **54**, 1429 (1989)

[25] I.M. Sokolov and A. Blumen, Macromol. Chem. Macromol. Symp. **65**, 223 (1993)

[26] H. Taitelbaum, S. Havlin, J.E. Kiefer, B. Trus, and G.H. Weiss, J. Stat. Phys. **65**, 873 (1991)

[27] H. Cramér and M.R. Leadbetter, *Stationary and Related Stochastic Processes*, (Wiley, New York 1968)

[28] S.O. Rice, Bell Syst. Tech. J. **24**, 51 (1945)

[29] S. Middleman, *Fundamentals of Polymer Processing*, (McGraw-Hill, New York 1977)

[30] J.M. Ottino, *The Kinematics of Mixing: Stretching, Chaos and Transport*, (Cambridge University Press, Cambridge 1989)

[31] R.S. Spencer and R.M. Wiley, J. Coll. Sci. **6**, 133 (1951)

[32] I.M. Sokolov and A. Blumen, J. Phys. A **24**, 3687 (1991)

[33] I.M. Sokolov and A. Blumen, Phys. Rev. Lett. **66**, 1942 (1991)

[34] C.F.F. Karney, Physica D **8**, 360 (1983)
[35] B.V. Chirikov and D.L. Shepelyanski, Physica D **13**, 395 (1984)
[36] T. Geisel, J. Nierwetberg, and A. Zacherl, Phys. Rev. Lett. **54**, 616 (1985)
[37] J. Klafter, A. Blumen, and M.F. Shlesinger, Phys. Rev. A **35**, 3081 (1987)
[38] M.F. Shlesinger, B.J. West, and J. Klafter, Phys. Rev. Lett. **58**, 1100 (1987)
[39] G. Zumofen and J. Klafter, Phys. Rev. E **47**, 851 (1993)
[40] G. Zumofen and J. Klafter, Europhys. Lett. **26**, 565 (1994)
[41] M.F. Shlesinger, G.M Zaslavsky, and U. Frisch, eds. *Lévy Flights and Related Topics in Physics*, (Springer, Berlin 1995)
[42] T.H. Solomon, E.R. Weeks, and H.L. Swinney, Phys. Rev. Lett. **71**, 3975 (1993)
[43] A. Ott, J.P. Bouchaud, D. Lagevin, and W. Urbach, Phys. Rev. Lett. **65**, 2201 (1990)
[44] M.F. Shlesinger, G.M. Zaslavsky, and J. Klafter, Nature **263**, 31 (1993)
[45] G.M. Zaslavsky, D. Stevens, and H. Weitzner, Phys. Rev. E **48**, 1683 (1993)
[46] G. Zumofen and J. Klafter, Phys. Rev. B **50**, 5119 (1994)
[47] G. Zumofen, J. Klafter, and A. Blumen, Phys. Rev. A **45**, 8977 (1992)
[48] D.B. Abraham and P.J. Upton, Phys. Rev. B **39**, 736 (1989)
[49] K. Lindenberg, B.J. West, and R. Kopelman, Phys. Rev. Lett. **60**, 1777 (1988)

Random Walks on Fractals*

Armin Bunde[1], Julia Dräger[1,2], and Markus Porto[1]

[1] Institut für Theoretische Physik, Justus–Liebig–Universität Giessen,
D–35392 Giessen, Germany, e-mail: bunde@physik.uni-giessen.de
and Markus.Porto@physik.uni-giessen.de
[2] I. Institut für Theoretische Physik, Universität Hamburg,
D–20355 Hamburg, Germany

Abstract. In this paper we give a brief introduction into the fractal concept and discuss the way the laws of diffusion (mean square displacement as a function of time and spatial decay of the probability density) are modified on random fractal structures. We describe algorithms to generate random fractals and to simulate the diffusion process on these structures. We show how the theoretical predictions can be tested by computer simulations.

1 Introduction

The fractal concept is an important tool for characterizing irregular structures in nature that are self-similar on certain length scales [1, 2, 3, 4]. In this paper we study both analytically and numerically how the laws of diffusion are changed in these structures. Accordingly, the paper is divided into three parts. In the first part (Sects. 2–4), we discuss certain deterministic and random fractal structures that are widely used to mimic irregular structures in nature. In the second part (Sects. 5 and 6), we consider random walks on fractal structures and discuss how the laws of diffusion are changed compared with regular structures. In the third part (Sects. 7 and 8) we finally present the numerical methods for generating random fractals and simulating random walks on them and describe the programs used in the workshop.

Before starting with irregular structures, we would like to remind the reader of the concept of dimension in regular systems. It is well known that in regular systems (with uniform density) such as long wires, large thin plates, or large filled cubes, the dimension d characterizes how the mass $M(L)$ changes with the linear size L of the system. If we consider a smaller part of the system of linear size bL ($b < 1$), then $M(bL)$ is decreased by a factor of b^d, i.e.,

$$M(bL) = b^d M(L) \ . \tag{1}$$

The solution of the functional equation (1) is simply $M(L) = A L^d$. For the long wire the mass changes linearly with b, i.e., $d = 1$. For the thin plates we obtain $d = 2$, and for the cubes $d = 3$.

---

* Software included on the accompanying diskette.

Next we consider fractal objects. Here we distinguish between deterministic and random fractals. Deterministic fractals are generated iteratively in a deterministic way, whereas random fractals are generated using a stochastic process. Although fractal structures in nature are random, it is instructive to start with deterministic fractals where the fractal concept can be most easily introduced.

2   Deterministic Fractals

2.1   The Koch Curve

One of the most common deterministic fractals is the Koch curve [1, 2, 3]. Fig. 1 shows the first $n = 3$ iterations of this fractal curve. By each iteration the length of the curve is increased by a factor of 4/3. The mathematical fractal is defined in the limit of infinite iterations, $n \to \infty$, where the total length of the curve approaches infinity.

The dimension of the curve can be obtained just as for regular objects. From Fig. 1 we notice that, if we decrease the linear size by a factor of $b = 1/3$, the total length (mass) of the curve is decreased by a factor of 1/4, i.e.,

$$M\left(\frac{1}{3}L\right) = \frac{1}{4}M(L) \ . \tag{2}$$

This feature is very different from regular curves, where the length of the object decreases proportional to the linear scale. In order to satisfy (1) and (2) we are led to introduce a *noninteger* dimension d, satisfying $1/4 = (1/3)^d$, i.e., $d = \ln 4/\ln 3$. This noninteger dimension is smaller than the dimension of the embedding space, here $d = 2$, and is called the *fractal dimension*. In order to distinguish it from the space dimension d, we denote it by d_f. Structures described by a fractal dimension are called *fractals*. Thus, to include fractal structures, (1) is generalized by

$$M(bL) = b^{d_f} M(L) \ , \tag{3}$$

which is solved by $M(L) = AL^{d_f}$.

When generating the Koch curve and calculating d_f, we observe the striking property of fractals – the property of *self-similarity* – which is the basic feature of all deterministic and random fractals. If we take a part of a fractal and magnify it by the same magnification factor in all directions, the magnified picture cannot be distinguished from the original.

For the Koch curve as well as for all deterministic fractals generated iteratively, (3) is of course valid only for length scales L below the total linear size L_0 of the curve. If the number of iterations n is finite, then (3) is valid only above a lower cutoff length a_0, $a_0 = L_0/3^n$ for the Koch curve. Hence, for a finite number of iterations there exist two cutoff length scales in the system, an upper cutoff L_0 representing the total linear size of the fractal, and a lower cutoff a_0.

$n = 0$

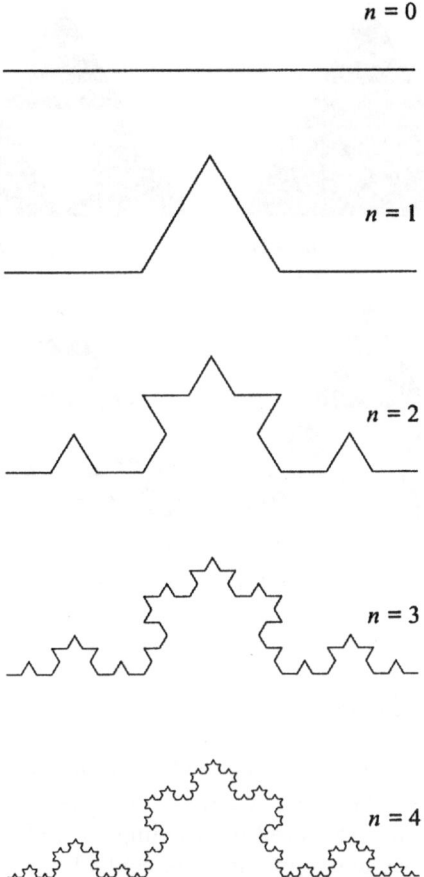

$n = 1$

$n = 2$

$n = 3$

$n = 4$

Fig. 1. The first iterations of the Koch curve. The fractal dimension of the Koch curve is $d_f = \ln 4/ \ln 3$

This feature of having two characteristic length scales is shared by all fractals in nature.

The Koch curve can be viewed as a mathematical model for coastlines. Similar to realistic coastlines such as the coast of Norway or the coast of Great Britain, the length of the Koch curve increases continuously when the length scale is decreased. Using as length scale a stick of length $\ell = (1/3)^m$ ($m = 0, 1, 2, \ldots$), one obtains for the length of the curve

$$L_C = (4/3)^m = [(1/3)^m]^{1-\ln 4/ \ln 3} = \ell^{1-d_f} \ ,$$

i.e., the fractal dimension determines the way L_C tends to infinity for ℓ approaching zero. Using good maps of Norway or Great Britain, it is not difficult to verify in this way that the fractal dimensions of the coastlines are $d_f \cong 1.5$ for Norway and $d_f \cong 1.3$ for Great Britain [1, 2].

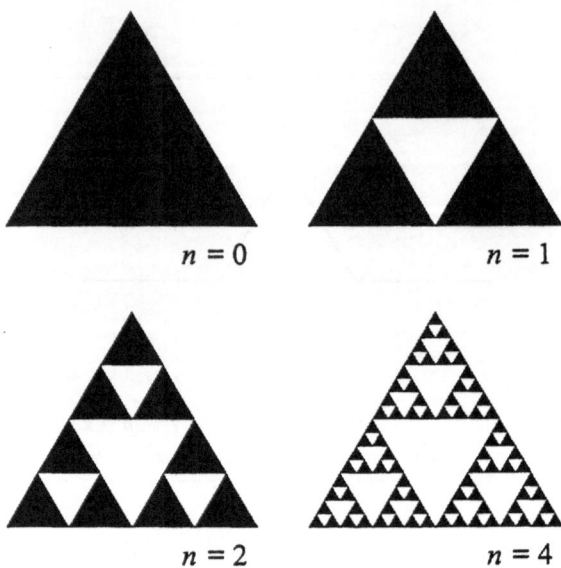

Fig. 2. The Sierpinski gasket. The fractal dimension of the Sierpinski gasket is $d_{\mathrm{f}} = \ln 3/\ln 2$

2.2   The Sierpinski Gasket

The Sierpinski gasket [1, 2, 3] is generated by dividing a full triangle into four smaller triangles and removing the central triangle (see Fig. 2). In the following iterations, this procedure is repeated by dividing each of the remaining triangles into four smaller triangles and removing the central ones.

To obtain the fractal dimension, we consider the mass of the gasket within a linear size L and compare it with the mass within $L/2$. Since $M(L/2) = M(L)/3$, we have $d_{\mathrm{f}} = \ln 3/\ln 2 \cong 1.585$.

Next we consider random fractal structures, which are more important since they occur in nature.

3   Random Fractals

3.1   The Random-Walk Trail

Imagine a random walker on a square lattice or a simple cubic lattice. In one unit of time the random walker advances one step of length a to a randomly chosen nearest neighbor site. Let us assume that the walker is unwinding a wire that he connects to each site along his way. The length (mass) M of the wire that connects the random walker with his starting point is proportional to the number of steps t performed by the walker (Fig. 3). After t steps, the actual

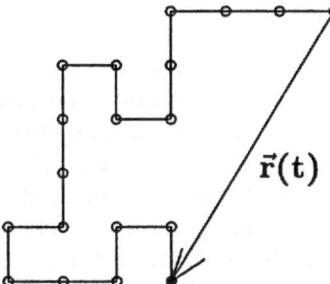

Fig. 3. A random walk in a square lattice. The lattice constant $a = 1$ is equal to the jump length of the random walker

position of the walker is described by the vector

$$\mathbf{r}(t) = a \sum_{\tau=1}^{t} \mathbf{e}_\tau \ , \tag{4}$$

where $\mathbf{e}_\tau$ denotes the unit vector pointing in the direction of the jump at the τth step and a is the lattice constant.

The mean distance the random walker has traveled after t steps is described by the root mean square displacement $R(t) \equiv \langle r^2(t) \rangle^{1/2}$, where the average $\langle \cdots \rangle$ is over all random walk configurations on the lattice. From (4) we obtain

$$R^2(t) \equiv \langle r^2(t) \rangle = a^2 \sum_{\tau,\tau'=1}^{t} \langle \mathbf{e}_\tau \cdot \mathbf{e}_{\tau'} \rangle = a^2 t + a^2 \sum_{\tau \neq \tau'} \langle \mathbf{e}_\tau \cdot \mathbf{e}_{\tau'} \rangle \ . \tag{5}$$

Since jumps at different steps τ and τ' are uncorrelated, we have $\langle \mathbf{e}_\tau \cdot \mathbf{e}_{\tau'} \rangle = \delta_{\tau\tau'}$, and therefore

$$R(t) = a\, t^{1/2} \ . \tag{6}$$

$R(t)$ characterizes the spatial extension of the curve generated by the random walker which we call the random-walk (RW) trail. According to (6), the length of the trail (which is proportional to the mass of the wire) increases as $R(t)^2$, and therefore the RW trail has the fractal dimension $d_f = 2$. Since $R^2(t) \sim t$ holds for all dimensions d, the fractal dimension of the RW trail does not depend on the embedding space dimension.

3.2 Self-Avoiding Walks

Self-avoiding walks (SAWs) are defined as the subset of all nonintersecting random walk configurations. As was found by Flory in 1944 [5], the end-to-end distance of SAWs scales with the number of steps t as

$$R(t) \sim t^\nu \ , \tag{7}$$

with $\nu = 3/(d+2)$ for $d < 4$ and $\nu = 1/2$ for $d \geq 4$. Since t is proportional to the mass of the RW trail, it follows from (7) that $d_f = 1/\nu$. Self-avoiding walks serve as a model for polymers in solution, see [6] and Chap. 6 in [3].

3.3   Percolation

Consider a square lattice, where each site is occupied randomly with probability p or empty with probability $1 - p$. For large lattices p is identical to the concentration of occupied sites. At low concentration p, the occupied sites are either isolated or form small clusters (Fig. 4a). Two occupied sites belong to the same cluster if they are connected by a path of nearest-neighbor occupied sites. When p is increased, the average size of the clusters increases. At a critical concentration p_c (also called the percolation threshold) a large cluster appears which connects opposite edges of the lattice (Fig. 4b). This cluster is called the *infinite* cluster, since its size diverges when the size of the lattice is increased to infinity. When p is increased further, the density of the infinite cluster increases, since more and more sites become part of the infinite cluster, and the average size of the *finite* clusters decreases (Fig. 4c).

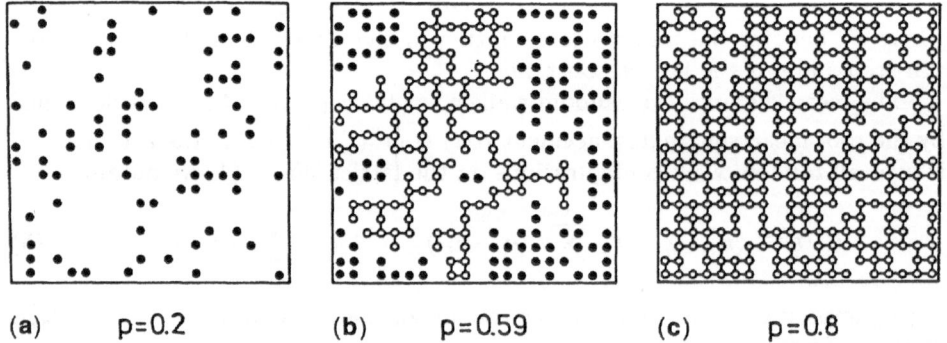

(a)   p=0.2   (b)   p=0.59   (c)   p=0.8

Fig. 4 a–c. Square lattice of size 20 × 20. Sites have been randomly occupied with probabilities p [$p = 0.20$ **(a)**, 0.59 **(b)**, 0.80 **(c)**]. Sites belonging to finite clusters are marked by *full circles*, whereas sites on the infinite cluster are marked by *open circles*

The percolation transition is characterized by the geometrical properties of the clusters near p_c. The probability P_∞ that a site belongs to the infinite cluster is zero below p_c and increases above p_c as

$$P_\infty \sim (p - p_c)^\beta \ . \tag{8}$$

The linear size of the *finite* clusters, below and above p_c, is characterized by the *correlation length* ξ. The correlation length is defined as the mean distance between two sites on the same finite cluster and represents the characteristic length scale in percolation. When p approaches p_c, ξ increases as

$$\xi \sim |p - p_c|^{-\nu} \ , \tag{9}$$

with the same exponent ν below and above the threshold. While p_c depends explicitly on the type of lattice (e.g., $p_c \cong 0.59277$ for the square lattice and

1/2 for the triangular lattice), the *critical exponents* β and ν are universal and depend only on the dimension d of the lattice, but not on the type of the lattice.

Near p_c, on length scales smaller than ξ, both the infinite cluster and the finite clusters are self-similar. Above p_c, on length scales larger than ξ, the infinite cluster can be regarded as an homogeneous system which is composed of many unit cells of size ξ. Mathematically, this can be summarized as

$$M(r) \sim \begin{cases} r^{d_f}, & r \ll \xi \\ \\ r^d, & r \gg \xi \end{cases} \tag{10}$$

The fractal dimension d_f can be related to β and ν:

$$d_f = d - \frac{\beta}{\nu} \ . \tag{11}$$

Since β and ν are universal exponents, d_f is also universal. One obtains $d_f = 91/48$ in $d = 2$ and $d_f \cong 2.5$ in $d = 3$ [4, 7].

The percolation model has found numerous applications in physics, chemistry, and biology, where occupied and empty sites may represent very different physical, chemical, or biological properties. Examples are the physics of two component systems (the random resistor, magnetic or superconducting networks), the polymerization process in chemistry, and the spreading of epidemics and forest fires, see [4, 7] and [8].

4 The "Chemical Distance" ℓ

The fractal dimension, however, is not sufficient to fully characterize a fractal. An important fractal substructure is the shortest path on the fractal between two distant fractal points (Fig. 5). The length ℓ of this path, also called the "chemical distance" , increases, on average, with the spatial distance r between both points as

$$\ell(r) \sim r^{d_{\min}} \ . \tag{12}$$

The average mass of a cluster within a chemical distance ℓ increases with ℓ as

$$M(\ell) \sim \ell^{d_\ell} \ , \tag{13}$$

which defines the "chemical dimension" d_ℓ. Since M scales with r as $M(r) \sim r^{d_f}$ it follows that $d_\ell = d_f/d_{\min}$. For linear fractal structures like coastlines or the RW trail one has $d_\ell = 1$ and thus $d_{\min} = d_f$. For percolation clusters one has $d_{\min} \cong 1.13$ $(d = 2)$ and $d_{\min} \cong 1.37$ $(d = 3)$ [4].

More information on the connectivity of a fractal is obtained from the probability $\Phi(\ell|r)$ of finding a site with a chemical distance ℓ at fixed distance r from a cluster site, and the related probability $\Phi(r|\ell)$ of finding a site with a spatial distance r at fixed chemical distance ℓ from a cluster site. Numerically, $\Phi(r|\ell)$ can be obtained as follows. First one chooses one cluster site as a center site and

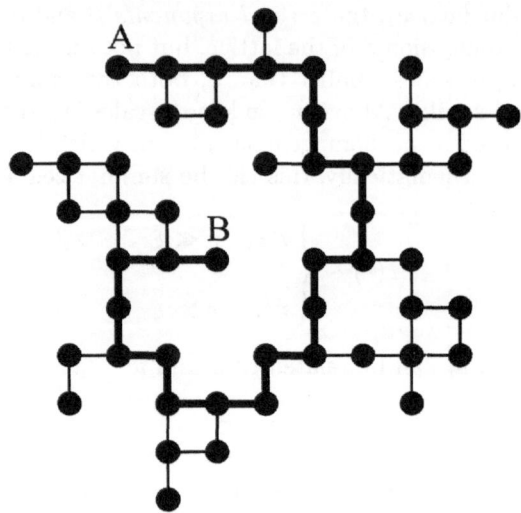

Fig. 5. The shortest path between two points A and B of the unfinite percolation cluster generated at the percolation threshold of the square lattice site problem. The chemical distance is the length of the shortest distance

counts the number $N(\ell)$ of all sites that are at a given chemical distance ℓ from this site. Among these $N(\ell)$ sites, there are $N(r, \ell)$ sites at a euclidean distance r from the center. The fraction $N(r, \ell)/N(\ell)$, averaged over many configurations, can be identified with $\Phi(r|\ell)$. The related probability $\Phi(\ell|r)$ is obtained in a similar way by calculating the number of sites $N(r)$ that are at euclidean distance r from the center, and averaging over the fraction $N(r, \ell)/N(r)$. Since by definition $\sum_r N(r, \ell) = N(\ell)$ and $\sum_\ell N(r, \ell) = N(r)$, the probability densities satisfy, in the continuum limit, the normalization condition $\int \Phi(r|\ell)\, dr = \int \Phi(\ell|r)\, d\ell = 1$. Since $N(\ell)$ and $N(r)$ scale as

$$N(\ell) \sim \ell^{d_\ell - 1} \quad \text{and} \quad N(r) \sim r^{d_f - 1} \ , \tag{14}$$

both probability densities are related by

$$\Phi(\ell|r) \sim \Phi(r|\ell) \frac{\ell^{d_\ell - 1}}{r^{d_f - 1}} \ . \tag{15}$$

For RW trails, $\Phi(\ell|r)$ can be determined analytically. By construction, the lenght ℓ of the trail is identical to the number of steps t performed by the random walker. Hence, for the RW trail, $\Phi(r|\ell)$ has the same form as the well-known probability $P(r, t)$ for finding the random walker at time step t a distance r from his starting point, i.e.

$$\Phi(r|\ell) \sim r^{d-1} \ell^{-d/2} \exp\left[-\frac{d\, r^2}{2\ell}\right] \ . \tag{16}$$

Thus we obtain with (15)

$$\Phi(\ell|r) \sim r^{d-2}\,\ell^{-d/2}\,\exp\left[-\frac{d\,r^2}{2\ell}\right] .\tag{17a}$$

It is convenient to rewrite (17a) as

$$\Phi(\ell|r) = \frac{C_1}{\ell}\left(\frac{r}{\ell^{1/d_{min}}}\right)^g\,\exp\left[C_2\left(\frac{r}{\ell^{1/d_{min}}}\right)^\delta\right]\tag{17b}$$

with $d_{min} = 2$, $\delta = d_{min}/(d_{min}-1)$, $g = d-2$ and $C_2 = d/2$. Equation (17b) holds also for percolation clusters with $g \cong 1.35$ $(d = 2)$ and $g \cong 1.5$ $(d = 3)$ [9, 10]. For fixed r, $\Phi(\ell|r)$ has a maximum at $\ell_{max} \cong r^{d_{min}}$. By definition we have $\Phi(\ell|r) \equiv 0$ below some cutoff length $\ell_{min}(r)$, which is the shortest chemical distance ℓ a site at the distance r from a central site can have when N configurations are considered. In contrast to ℓ_{max}, ℓ_{min} depends strongly on N [11]. This is shown explicitly in Fig. 6a, where for RW trails on the sc lattice ℓ_{min} is plotted as a function of r for N ranging from 1 to 10^4. The figure shows that

$$\ell_{min}(r, N) = \begin{cases} r, & r < r_c(N) , \\ \alpha_{min}(N)\,r^{d_{min}}, & r \gg r_c(N) , \end{cases}\tag{18}$$

with $d_{min} = 2$ for the RW trail.

To determine the crossover value $r_c(N)$ analytically, we note that in order to find one configuration with $\ell = r = r_c$ we have to generate about $N = z^{\ell-1} = z^{r_c-1}$ configurations, with the coordination number $z = 6$ for the sc lattice. This yields

$$r_c(N) = 1 + \frac{\ln N}{\ln z} .\tag{19}$$

To determine $\alpha_{min}(N)$ we assume scaling,

$$\ell_{min}(r, N) = r_c(N)\,g(r/r_c(N)) .\tag{20}$$

In order to satisfy (18), we must require $g(x) = x$ for $x < 1$ and $g(x) = x^{d_{min}}$ for $x \gg 1$. This yields

$$\alpha_{min}(N) = [r_c(N)]^{1-d_{min}} = \left(1 + \frac{\ln N}{\ln z}\right)^{1-d_{min}} .\tag{21}$$

Figure 6b shows $\ell_{min}/r_c(N)$ versus $r/r_c(N)$. The data collapse strongly supports the scaling assumption (20). Equations (18–21) also hold for percolation clusters at p_c when $r_c(N)$ is substituted by $r_c(N) = (\ln z + \ln N)/\ln(1/p_c)$ [11]. As discussed above we have $d_{min} \cong 1.13$ $(d = 2)$ and $d_{min} \cong 1.37$ $(d = 3)$.

The form of $\Phi(\ell|r)$ and the dependence of ℓ_{min} on N is characteristic for random fractals with $d_{min} > 1$.

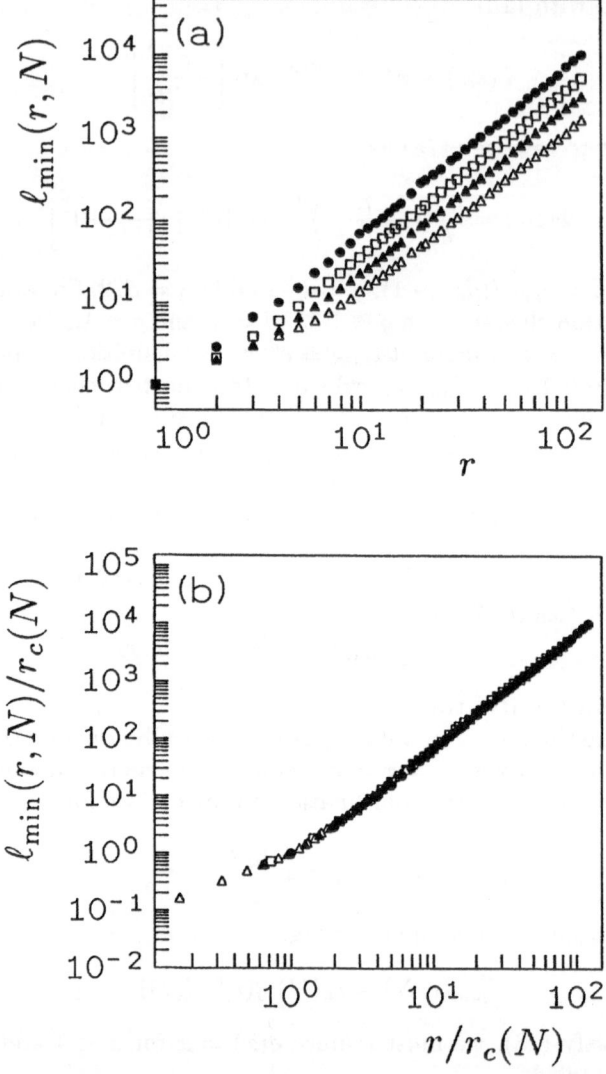

Fig. 6. (a) The minimum distance $\ell_{\min}(r, N)$ versus r for RW trails on the sc lattice, for $N = 1$ (*full circles*), 11 (*squares*), 128 (*full triangles*) and 10^4 (*triangles*). (b) Scale plot of $\ell_{\min}(r, N)/r_c(N)$ versus $r/r_c(N)$ for the same N values as above. For N below 10^4, averages have been performed over typically 100 sets of N configurations

5   Random Walks on Fractals

Next we discuss a random walker on a random fractal structure.

5.1   Root Mean Square Displacement $R(t)$

First we consider a random walker on the RW trail. We know that *along* the trail (in "ℓ-space") diffusion is one dimensional. After t time steps, the walker has traveled, on average, the chemical distance $\ell(t) = a\, t^{1/2}$ and thus has reached a distance $R(t) \sim [\ell(t)]^{1/d_{\min}} = [\ell(t)]^{1/2} \sim t^{1/4}$ from his starting point. Accordingly, instead of $R(t) \sim t^{1/2}$ we have now

$$R(t) \sim t^{1/d_{\mathrm{w}}} \tag{22}$$

with $d_{\mathrm{w}} = 2d_{\min} = 4$ the *fractal dimension of the random walk*. Equation (22) with $d_{\mathrm{w}} > 2$ is valid for all fractal structures. In general, however, one cannot express d_{w} rigorously by d_{f} or $d_{\min}$. For percolation clusters, one has approximately $d_{\mathrm{w}} \cong 3d_{\mathrm{f}}/2$ [12].

It is clear that the anomalous diffusion law (22) can only occur on length scales where the considered structure is self-similar. Let us consider, for example, the infinite percolation cluster above p_c that is fractal below the correlation length ξ and compact above ξ. According to (22), the random walker needs $t_\xi \sim \xi^{d_{\mathrm{w}}}$ time steps to travel a distance of the order ξ. For small times $t \ll t_\xi$, the random walker explores the fractal regime and (22) holds. For long times, $t \gg t_\xi$, on the other hand, he explores the whole compact cluster and diffusion is normal, $R(t) \sim t^{1/2}$.

The exponent d_{w} is experimentally accessible, for example, in chemical reactions on fractal structures. Equation (22) implies that the number of distinct sites visited scales as $S(t) \sim [R(t)]^{d_{\mathrm{f}}} \sim t^{d_{\mathrm{f}}/d_{\mathrm{w}}}$.

For annihilation reactions ($A + A \rightarrow \emptyset$), the reactants K of the diffusing A particles decreases as $dK/dt \sim t^{d_{\mathrm{f}}/d_{\mathrm{w}}-1}$. Hence we expect $K \sim t^{-1/3}$ on percolation clusters, which has been confirmed experimentally by [13]. For other ways of measuring d_{w} we refer to [4] and [8].

5.2   The Mean Probability Density

Let us again start with the simple RW trail. *Along* the trail, diffusion is one-dimensional and the probability of finding a random walker after t time steps on a site i at chemical distance ℓ from its starting point is given by

$$P(\ell, t) = P(0, t)\, \exp\left[-\left(\frac{\ell}{\xi_\ell(t)}\right)^2\right], \tag{23}$$

where $\xi_\ell(t)$ is proportional to the displacement $\ell(t) \sim t^{1/2}$ along the chain.

We now anticipate that the length ℓ of the shortest path connecting a site i with the origin is also the relevant physical length for diffusion in percolation clusters, such that the fluctuations of the probability $P_i(\ell, t)$ to find the random walker after t time steps on a site i at chemical distance ℓ from the origin are small (in contrast to the large fluctuation of the analogous quantity in r space). For simplicity, we assume that $P_i(\ell, t)$ only depends on ℓ and t for all sites i at fixed ℓ, and decays as

$$P_i(\ell, t) \cong P(\ell, t) = P(0, t) \exp\left[-\left(\frac{\ell}{\xi_\ell(t)}\right)^v\right] , \qquad (24)$$

where $\xi_\ell(t) \sim t^{d_{\min}/d_w}$ is proportional to the mean chemical distance traveled by the random walker and $v = d_w/(d_w - d_{\min})$ for $\ell \gg \xi_\ell(t)$ [14]. If we define by $P_i(r, t)$ the probability that the random walker is, after t time steps, on a site i at euclidean distance r, we obtain the mean probability $P(r, t)$ for a single configuration by averaging over all $N(r)$ sites i at fixed r,

$$P(r, t) = \frac{1}{N(r)} \sum_{i=1}^{N(r)} P_i(r, t) . \qquad (25)$$

Among the $N(r)$ sites at distance r, $N(\ell, r)$ sites are at chemical distance ℓ from the center. Therefore we can write (25) as [15]

$$P(r, t) = \frac{1}{N(r)} \sum_{\ell = \ell_{\min}(r, N)}^{\infty} N(\ell, r) P(\ell, t) . \qquad (26)$$

Averaging over $N \gg 1$ configurations and replacing the sum in (26) by an integral, yields

$$\langle P(r, t) \rangle_N = \int_{\ell_{\min}(r, N)}^{\infty} \Phi(\ell | r; N) P(\ell, t) \, d\ell , \qquad (27)$$

with $\Phi(\ell | r; N)$ from (17).

To evaluate the integral (27), we follow [15] and [16] and use the method of steepest descent. Using (17b) and (24) we obtain

$$\langle P(r, t) \rangle_N = P(0, t) \int_{\ell_{\min}(r, N)}^{\infty} \zeta(\ell) \exp[-\eta(\ell)] \, d\ell \qquad (28)$$

with $\eta(\ell) = C_2 \left(r/\ell^{1/d_{\min}}\right)^\delta + [\ell/\xi_\ell(t)]^v$ and $\zeta(\ell) = (C_1/\ell) \left(r/\ell^{1/d_{\min}}\right)^g$. The saddle $\ell^*(r)$ occurs at $d\eta/d\ell|_{\ell=\ell^*} = 0$:

$$\ell^*(r) = \xi_r^{d_{\min}} \left(\frac{r}{\xi_r}\right)^{u/v} , \qquad (29)$$

with

$$u = \frac{v d_{\min}}{1 + v(d_{\min} - 1)} \tag{30}$$

and

$$\xi_r^{d_{\min}} = \xi_\ell \left(\frac{C_2}{v(d_{\min} - 1)} \right)^{1/v} \tag{31}$$

where ξ_r is, up to a proportionality constant of order unity, the root mean square displacement $R(t)$ of the random walker. Figure 7a shows $P(\ell, t)$, $\Phi(r|\ell)$ and the product of both for diffusion on the RW trail, for $t = 2 \times 10^3$ and $r = 16$. The figure shows, that for the considered r value, the integrand of (28) is peaked strongly at ℓ^*.

Following the method of steepest descent, we can write approximately

$$\langle P(r,t) \rangle_N \cong P(0,t)\, \zeta(\ell^*(r))\, \exp\left[-\eta(\ell^*(r)) \right]$$

$$\int_{\ell_{\min}(r,N)}^{\infty} \exp\left\{ -\frac{1}{2}\eta''(\ell^*(r))\, [\ell - \ell^*(r)]^2 \right\}\, d\ell , \tag{32}$$

which yields

$$\ln \langle P(r,t) \rangle_N \sim -\eta(\ell^*) \sim -\left(\frac{r}{\xi_r} \right)^u . \tag{33}$$

By definition, (33) holds only for $\ell_{\min}(r, N) < \ell^*(r) < \ell_{\max}(r)$, and this restriction determines the r regime $r_1 < r < r_\times(N)$ where (33) is valid. We find with g and c from $\Phi(\ell|r)$ [11]

$$r_1 \cong \xi_r \left(\frac{g + d_{\min}}{C_2 \delta} \right)^{1/u} , \tag{34a}$$

$$r_\times(N) \cong \xi_r [r_c(N)]^{1/u} . \tag{34b}$$

Above $r_\times(N)$, the integrand in (27) is peaked sharply at $\ell = \ell_{\min}(r, N)$ (see Fig. 7b), and [11]

$$\ln \langle P(r,t) \rangle_N \sim -\left(\frac{\ell_{\min}(r, N)}{\xi_\ell} \right)^v$$

$$\sim -[r_c(N)]^{v(1-d_{\min})} \left(\frac{r}{\xi_r} \right)^{v d_{\min}} , \qquad r > r_\times(N) . \tag{35}$$

To determine the integral (27) in the short-distance regime ($r < r_1$), we note that $r < r_1$ implies $\ell_{\max}(r) < \xi_\ell$. Hence for $\ell < \xi_\ell$, the behavior of $\Phi(\ell|r)$ differs strongly from the behavior of $P(\ell, t)$. Whereas $\Phi(\ell|r)$ shows a steep maximum at ℓ_{max}, $P(\ell, t)$ is nearly constant (see Fig. 7c). Hence we can assume that to a very good approximation, the integrand of (27) can be written as [17]

$$\Phi(\ell|r)P(\ell, t) \cong \begin{cases} \Phi(\ell|r)\, P(0, t), & \ell \leq \xi_0 \\ \\ 0, & \ell > \xi_0 \end{cases} \tag{36}$$

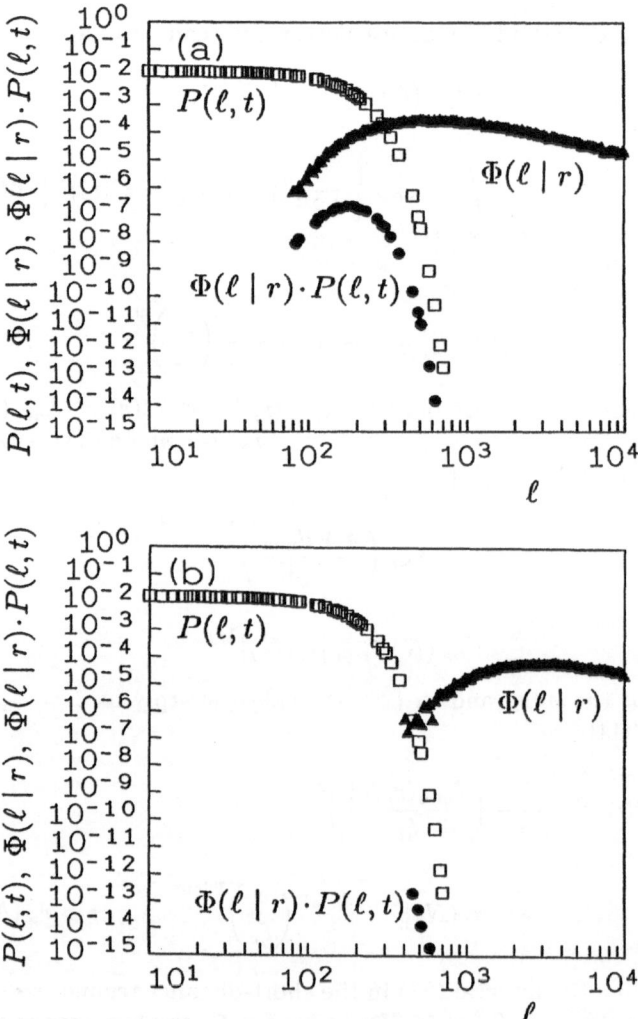

Fig. 7 a–c. (Continued on next page.)

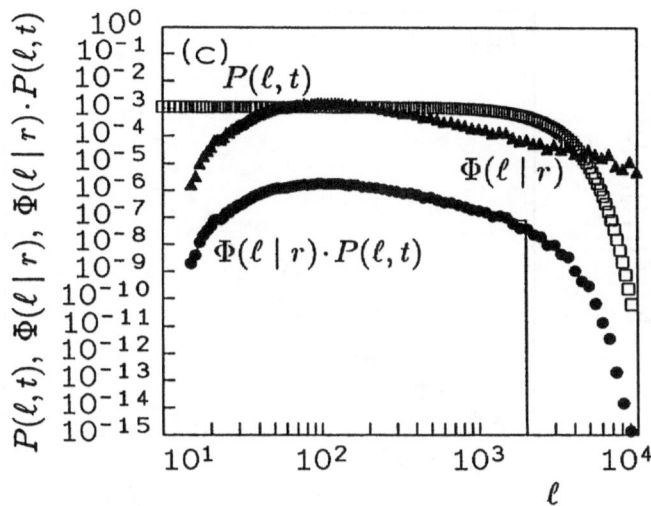

Fig. 7. The functions $P(\ell,t)$ (*squares*), $\Phi(\ell|r)$ (*full triangles*) and $P(\ell,t)\,\Phi(\ell|r)$ (*full circles*) for RW trails where (a) $t = 2 \times 10^3$ and $r = 16$ (b) $t = 2 \times 10^3$ and $r = 36$ and (c) $t = 5 \times 10^6$ and $r = 10$. In (a) the integrand $P(\ell,t)\,\Phi(\ell|r)$ of (27) shows a steep maximum at the saddle $\ell^*(r)$, while in (b) it shows a steep maximum at the cutoff value $\ell_{\min}(r,N)$, in (c) finally $\Phi(\ell|r)\,P(\ell,t)$ can be well approximated by the normalized step function of (36) (*full line*)

The cutoff length ξ_0 is determined by the normalization condition

$$\int\limits_0^\infty P(\ell,t)\,\ell^{d_\ell-1}\,\mathrm{d}\ell = \int\limits_0^{\xi_0} P(0,t)\,\ell^{d_\ell-1}\,\mathrm{d}\ell \ . \tag{37}$$

Inserting (24) into (37) we obtain $\xi_0 = \left[\Gamma((d_\ell/v)+1)\right]^{1/d_\ell}\xi_\ell$. Substituting (36) into (27) yields

$$\langle P(r,t)\rangle_N \cong P(0,t)\int\limits_0^{\xi_0} \Phi(\ell|r)\,\mathrm{d}\ell$$

$$= P(0,t)\left(\int\limits_0^\infty \Phi(\ell|r)\,\mathrm{d}\ell - \int\limits_{\xi_0}^\infty \Phi(\ell|r)\,\mathrm{d}\ell\right) \ . \tag{38}$$

Since $\int_0^\infty \Phi(\ell|r)\,\mathrm{d}\ell = 1$ and $\Phi(\ell|r) \cong (C_1/\ell)(r/\ell^{1/d_{\min}})^g$ for $\ell \gg \ell_{\max}$ we obtain finally in the short-distance regime

$$\frac{\langle P(r,t)\rangle_N}{P(0,t)} = 1 - a\left(\frac{r}{\xi_\ell^{1/d_{\min}}}\right)^g \ , \tag{39a}$$

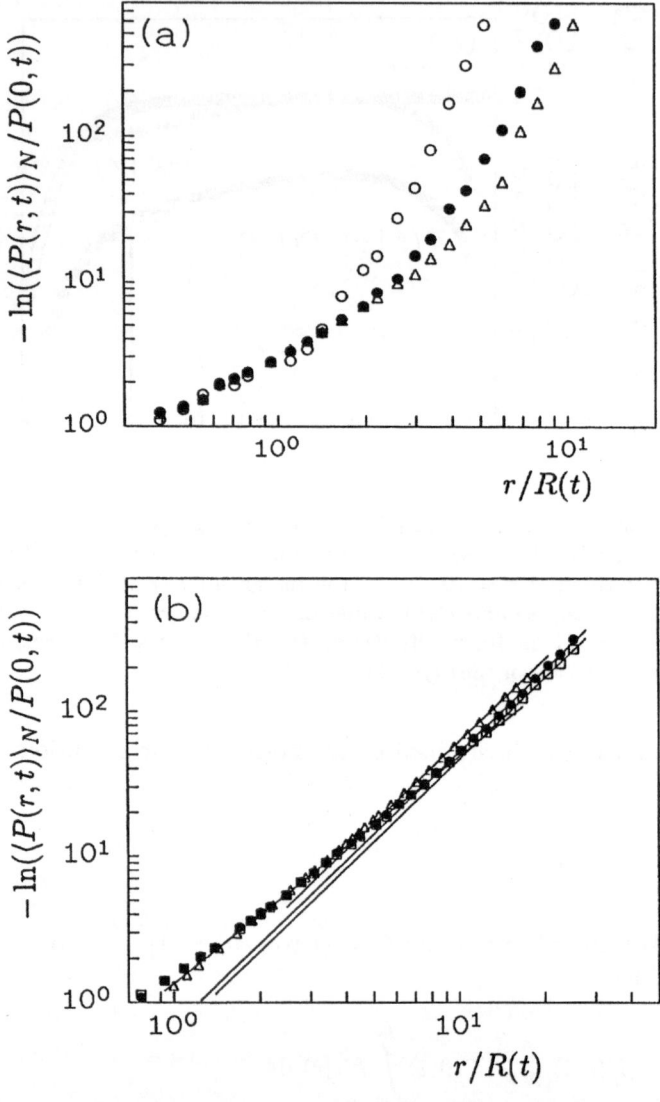

Fig. 8. Logarithm of the normalized mean probability density of random walks $-\ln[\langle P(r,t)\rangle_N /P(0,t)]$ versus $r/R(t)$ for **(a)** RW trails ($t = 10^4$, $N = 25$ (*full circles*) and 250 (*triangles*)) and **(b)** site percolation clusters on the square lattice ($t = 400$, $N = 5$ (*full circles*) and 50 (*triangles*)), compared with the typical probability density $-\ln[\langle P(r,t)_{\mathrm{typ}}\rangle /P(0,t)]$ (*circles*), which corresponds to the case $N = 1$ [11]; $R(t)$ is the root mean square displacement

with

$$a = \frac{C_1 d_{\min}}{g \left[\Gamma\left(\frac{d_\ell}{v}+1\right)\right]^{g/d_f}} \; , \tag{39b}$$

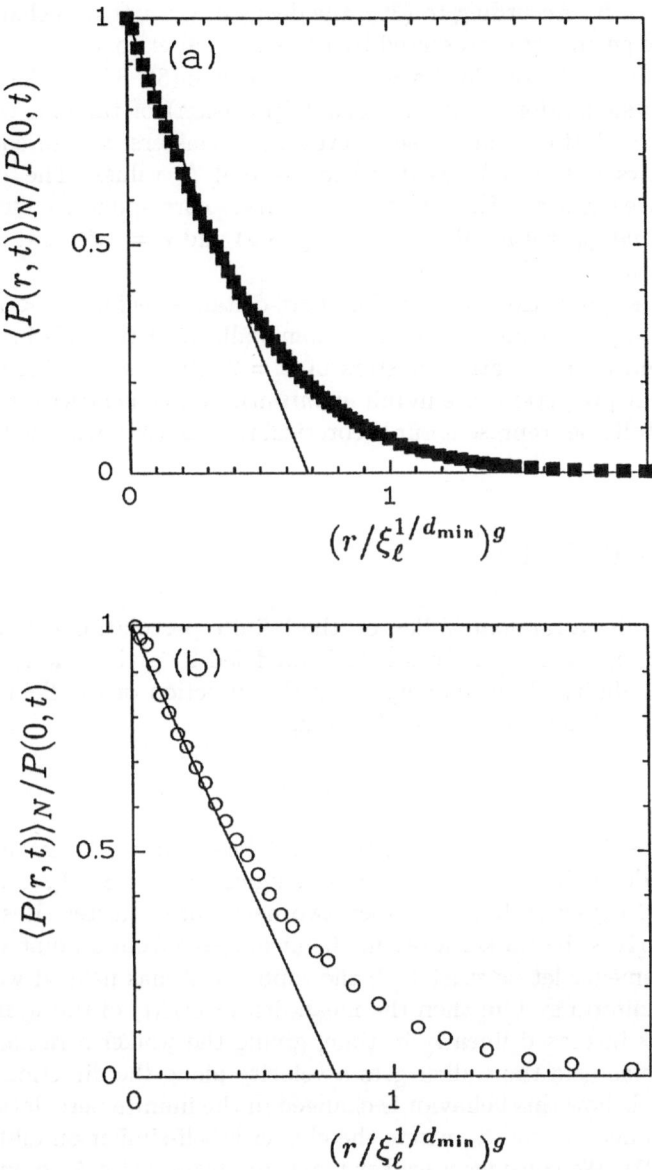

Fig. 9. The normalized mean probability density $\langle P(r,t)\rangle_N/P(0,t)$ of random walks versus $(r/\xi_\ell^{1/d_{\min}})^g$ for **(a)** RW trails on the s.c. lattice ($g = 1$, $t = 5 \cdot 10^6$, $\xi_\ell \cong 2200$, $N = 10^3$) and **(b)** site percolation clusters on the square lattice ($g \cong 1.35$, $t = 10^4$, $\xi_\ell \cong 40$, $N = 100$). The *full lines* in the plots represent the theoretical predictions of (39) without any fit parameters

independent of N. According to (39), the decay of $\langle P(r,t) \rangle_N$ is characterized by the substrate geometry represented by the structual exponent g.

To test our predictions in the asymptotic regime (33–35), we have performed Monte Carlo simulations (with quadruple precision) of random walks on RW trails in the sc lattice and on site percolation clusters on the square lattice. Figure 8 shows $\ln[\langle P(r,t) \rangle_N /P(0,t)]$ for several N values. The N-dependent crossover is clearly seen. The slopes of the curves correspond to our predictions, $u = 4/3$ and $vd_{\min} = 4$ for RW trails (Fig. 8 a) and $u = 1.53$ and $vd_{\min} = 1.86$ for percolation in $d = 2$ (Fig. 8 b).

To test the prediction (39) in the short-distance regime ($r < r_1$), we have performed computer simulations of random walks on RW trails in $d = 3$ where $a \cong 1.466$ and on percolation clusters in $d = 2$ where $a \cong 1.3 \pm 0.2$. Figure 9 shows that our predictions are in full *quantitative* agreement with the numerical results. The full lines represent our theoretical results, (39), with no fit parameter involved.

6   Biased Diffusion

Next we consider a random walker on the infinite percolation cluster under the influence of a bias field. The bias field is modeled by giving the random walker a higher probability P_+ of moving along the direction of the field and a lower probability P_- of moving against the field,

$$P_\pm \sim 1 \pm E \ , \qquad\qquad (40)$$

where $0 \le E \le 1$ is the strength of the field. The field can be either uniform in space ("euclidean" bias) or directed in topological space [18, 19]. In a topological bias (see Fig. 10) every bond between two neighbored cluster sites experiences a bias that drives the walker away in chemical space from a point source.

For convenience let us start with the topological bias field. If we apply such a field in a uniform system then the mean distance $R(t)$ of the walker from the "source" A is increased linearly in time, giving the walker a radial velocity. In a euclidean bias field the walker gets a velocity along the direction of the field. The question is how this behavior is changed in the infinite percolation cluster at the critical concentration p_c, where the cluster is self-similar on all length scales [18, 19, 20, 21]. We consider a walker travelling from a site A to another site B on the cluster. On his way the walker is driven into the loops and dangling ends that emanate from the shortest path between A and B. In a topological bias field the walker can get "stuck" in loops, as he can get stuck in dangling ends. Therefore, both loops and dangling ends act as random delays on the motion of the walker, and the percolation cluster can be imagined as a random comb where the teeth in the comb act as the random delays on the motion of the walker (see Fig. 10). The distribution of the length of the loops and dangling ends in the fractal structure determines the biased diffusion.

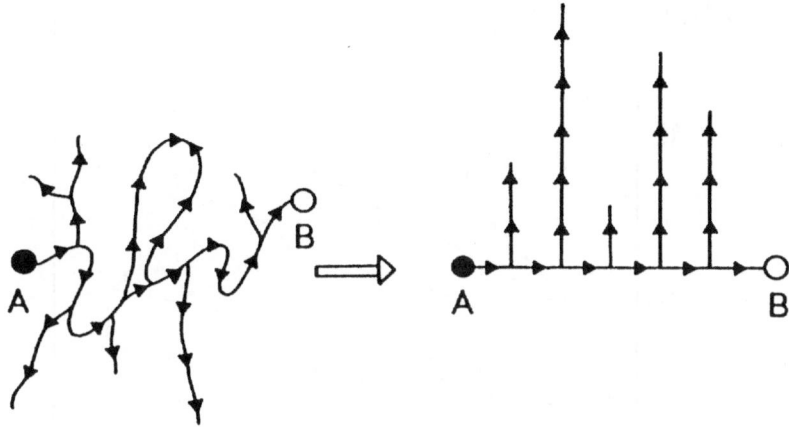

Fig. 10. Illustration of a percolation cluster under the influence of a topological bias field and its mapping to a random comb model (after [4])

At the critical concentration, due to self-similarity on all length scales, the lengths L of the teeth are expected to follow a power law distribution,

$$P(L) \sim L^{-(\alpha+1)}, \quad \alpha > 0 . \tag{41}$$

The time τ spent in a tooth increases exponentially with its length L, $\tau \sim [(1+E)/(1-E)]^L$ [18, 19]. Since the lengths of the teeth are distributed according to (41), it is easy to show that the waiting times τ follow the singular *waiting time distribution* [18, 19]

$$\Phi(\tau) \sim \left[\tau \, (\ln \tau)^{\alpha} \right]^{-1} , \tag{42}$$

and the system can be mapped onto a linear chain (the backbone of the comb) where each site i is assigned to a waiting time τ_i according to (42). A random walker has to wait on average τ_i time steps before he can jump from site i to one of the neighboring sites.

The singular waiting-time distribution changes the asymptotic laws of diffusion drastically, from the power law (22) to the *logarithmic* form [18, 19]

$$\ell \sim \left[\frac{\ln t}{A(E)} \right]^{\alpha} , \tag{43}$$

where

$$A(E) \sim \ln \left[\frac{1+E}{1-E} \right] , \tag{44}$$

and ℓ is the mean distance the walker has traveled along the backbone of the comb. Equation (43) is rigorous for the random comb, but is also in agreement with numerical data for the infinite percolation cluster at p_c, with $\alpha \cong 1$ [18, 19].

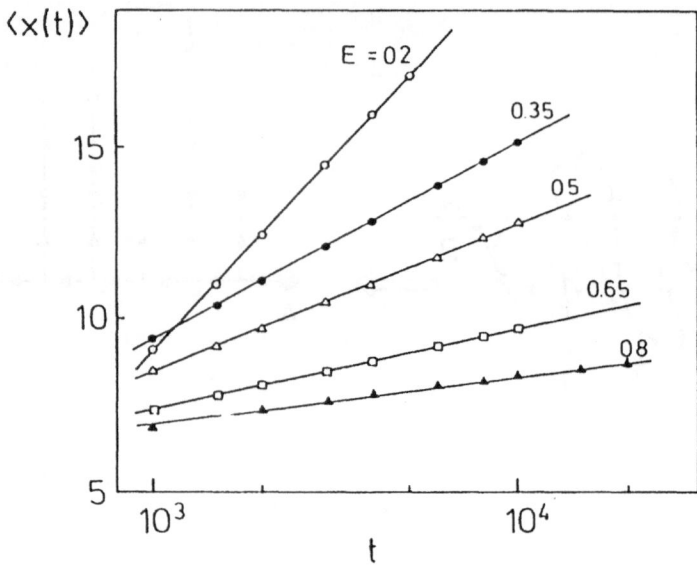

Fig. 11. Plot of $\langle x(t) \rangle$ versus $\ln t$ for different values of euclidean field strenghts E along the xy direction (after [20])

Accordingly, we have the paradoxical situation that on the fractal structure of the percolation cluster the motion of a random walker is dramatically *slowed down* by a bias field: the larger the bias field E the stronger the effect.

A logarithmic dependence of the mean displacement $\langle x \rangle$ along the direction of the field was also found for the euclidean bias (see Fig. 11),

$$\langle x \rangle \sim \frac{\ln t}{A(E)} \qquad (45)$$

with $A(E)$ from (44) for not too large values of E, $E \leq 0.6$ [20]. In a euclidean bias field, a random walker can also get stuck in backbends of the shortest path between two points, as also happens in linear fractal structures such as self-avoiding random walks or random walks. One can show rigorously [21] that in these structures (45) holds, and it is an open question how general the simple logarithmic behavior (45) and (44) is for diffusion in random fractals in the presence of an external bias field (see also [22]).

7   Numerical Approaches

In this section we discuss numerical methods for generating large percolation and for studying random walks on disordered structures. Our description of the algorithms follows closely [4].

7.1   Generation of Percolation Clusters

Percolation clusters can be generated either by the Leath method [23] or by the Hoshen–Kopelman algorithm [24], where all sites in the percolation system belonging to the same cluster are identified. We begin with the Leath Method.

Leath Method. In the Leath method [23] (see also [25, 26]), single percolation clusters are generated in the following way (see Fig. 12). In the first step the origin of an empty lattice is occupied, and its nearest neighbor sites become either occupied with probability p or blocked with probability $1 - p$. In the second step, the empty nearest neighbors of those sites occupied in the step before are occupied with probability p and blocked with probability $1 - p$. In each step, a new chemical shell is added to the cluster. The process continues until no sites are available for growth or the desired number of shells has been generated.

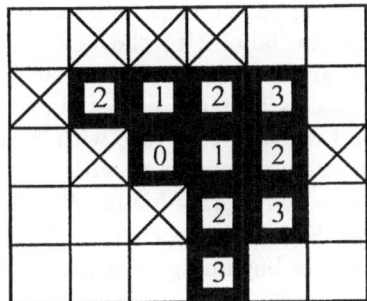

Fig. 12. The first four steps of the Leath cluster growth method

Since in the Leath method each site is labeled by its chemical distance from the origin, the method is particularly useful for studying structual and transport quantities related to the chemical space.

Hoshen–Kopelman Method. In the Hoshen–Kopelman algorithm [24], all sites in the percolation system are labeled in such a way that sites with the same label belong to the same cluster and different labels are assigned to different clusters. If the same label occurs at opposite sides of the system, an infinite cluster exists. In this way the critical concentration can be determined. By counting the number of clusters with s sites, we obtain the cluster distribution function.

The algorithm is quite tricky and we use a simple example to demonstrate it [4]. Consider the 5×5 percolation system in Fig. 13a which we want to analyze.

Beginning at the upper left corner and ending at the lower right corner, we assign cluster labels to the occupied sites. The first occupied site gets the label 1,

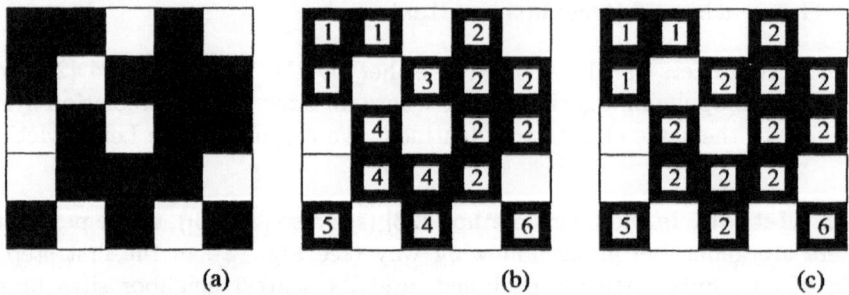

Fig. 13 a–c. Demonstration of the Hoshen–Kopelman algorithm

the neighboring site gets the same label because it belongs to the same cluster. The third site is empty and the fourth one is labeled 2. The fifth site is empty.

In the second line the first site is connected to its neighbor at the top and is therefore labeled 1. The second site is empty and the third one is labeled 3. The fourth site is now the neighbor of two sites, one labeled 2 and the other labeled 3. All three sites belong to the same cluster, which was first labeled 2. Accordingly, we also assign the label 2 to the new site, but we have to keep track that clusters 2 and 3 are connected. This is achieved by defining a new array $N_L(k)$: $N_L(3) = 2$ tells us that the correct label of cluster 3 is 2. If we continue the labeling we end up with Fig. 13b, with $N_L(3) = 2$ and $N_L(4) = 2$. Sites labeled 1, 2, 5, and 6 are not connected with sites with lower labels, and we define $N_L(1) = 1$, $N_L(2) = 2$, $N_L(5) = 5$ and, $N_L(6) = 6$.

In the second step (see Fig. 13c) we change the improper labels (where $N_L(k) < k$) into the proper ones beginning with the lowest improper label (here $k = 3$) and ending with the largest improper label (here $k = 4$).

The Hoshen–Kopelman algorithm is useful when investigating the distribution of cluster sizes as well as the largest cluster in *any* disordered system, not necessarily a percolation system. The method can also be used to determine p_c and to generate the infinite percolation cluster in percolation, but for this the Leath method described above is more efficient.

7.2 Simulation of Random Walks

Monte Carlo Method. In this method random walks on random fractals are simulated by the Monte Carlo technique, and the diffusion exponent d_w is determined from the mean square displacement. First the fractal structure of interest and one site is chosen randomly as the origin of the random walk. Then one of its z neighbor sites is chosen randomly and the random walker attempts to move to that site. If it does not belong to the fractal the move is rejected, otherwise the walker moves. In both cases, the time is enhanced by one unit. This procedure is repeated until the desired number of time steps has been performed. At each time step t, the square displacement r^2 of the random walker from the origin is recorded. Averages are performed over many starting points and a large number of fractals.

Exact Enumeration of Random Walks. In this technique [14, 27, 28, 29], the discretized version of the diffusion equation is solved numerically by iteration in time, using the fact that the probability of the random walker being at site **r** at time $t+1$ is determined by the probabilities of it being at its nearest neighbor sites at time t,

$$P(\mathbf{r}, t+1) = P(\mathbf{r}, t) + \sum_{\delta} w_{\mathbf{r}, \mathbf{r}+\delta} \left[P(\mathbf{r}+\delta, t) - P(\mathbf{r}, t) \right] . \qquad (46)$$

Here, the sum over δ is over the z nearest neighbor sites $\mathbf{r} + \delta$ of $\mathbf{r}$, and the transition probability $w_{\mathbf{r}, \mathbf{r}+\delta}$ is $1/z$ when $\mathbf{r} + \delta$ is a cluster site and 0 otherwise.

The iteration starts at time $t = 0$ when the random walker starts at the origin, i.e., $P(\mathbf{r}, 0) = 1$ for $\mathbf{r} = 0$ and $P(\mathbf{r}, 0) = 0$ otherwise, and continues until the desired number of time steps is reached. The method is illustrated in Fig. 14.

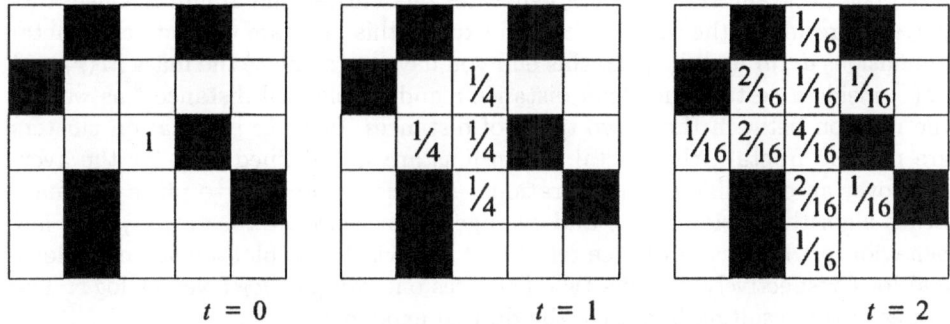

$t = 0$ $\qquad\qquad\qquad$ $t = 1$ $\qquad\qquad\qquad$ $t = 2$

Fig. 14. The evolution of the probability of a random walker for three successive time steps

To obtain the average probability density $\langle P(r, t) \rangle_N$ of Sect. 5.2 one has to average $P(\mathbf{r}, t)$ over all sites $\mathbf{r}$ of the clusters at fixed distance r and over N different cluster configurations. The mean square displacement is obtained from

$$\langle r^2(t) \rangle = \int \langle P(r, t) \rangle_N \; r^{d_f - 1} \; r^2 \; dr \sim t^{2/d_w} . \qquad (47)$$

8   Description of the Programs

The diskette enclosed with this book contains five FORTRAN programs dealing with topics discussed above. To compile the programs you need a graphics library called "Xflib". This library supplies a set of functions to handle the X window system and is freely available. To find the nearest FTP server you should use the archie command. If you want to use a different graphics library or compile the programs on a PC running MS-DOS you have to replace the library calls by appropriate ones. All used library functions are listed in the program headers and can be found and replaced easily. The function XINIT opens a window on the screen, XBACKSTORE tells the X window server to store the interior of the window and to restore it automaticaly if it was hidden by another window. The function XALLOCCOLOR allocates a color to be used by the program, XSETFGPIXVAL sets the foreground color, XFILLRECT draws a filled rectangle, XFLUSH flushes the previous draw commands to the X window server and XCLOSEDP frees the allocated colors and closes the window. The functions XBACKSTORE and XFLUSH do not have any appropriate counterparts on a PC.

The first program EXERCISE1.F draws a square lattice on the screen and marks sites with probability p. With this program you are able to study the geometrical phase transition in two-dimensional percolation by increasing p from 0 to 1. In addition one can change the system size and study its effect on the critical probability.

The four remaining programs use the Leath method to grow percolation clusters. The program EXERCISE2.F generates one cluster and simultaneously displays the result on the screen. You can expand this program to study the fractal dimensions d_f, d_ℓ and d_{min}. To this end you have to calculate the mass $M(r)$ and $M(\ell)$ inside a certain euclidean distance r and topological distance ℓ as well as the relation between these two types of distances. Because percolation clusters are random fractals, the fractal dimensions are well defined only for the averages over an ensemble of configurations. Therefore you have to place the main Leath algorithm inside a loop and average the measures. Expecting a power law behavior for the mass–distance relation it is advisable to plot $\log M$ versus $\log r$ and $\log \ell$ respectively. For the two distances one can plot $\log \ell$ versus $\log r$. The slopes of the resulting lines give the desired exponents.

The third program EXERCISE3.F generates a percolation cluster with the Leath algorithm and a random walk on the resulting cluster. The territory covered by the random walker is shown. The different colors indicate the number of time steps the random walker needed to reach the site for the first time. You can expand the program by numerically measuring the territory as a function of diffusion time and averaging over ensembles of random walks and clusters. It is interesting to increase the number of random walkers moving simultanously on one cluster and to study its effect on the territory covered by all random walkers as a function of time.

The two remaining programs EXERCISE4.F and EXERCISE5.F differ only slightly. Both generate a percolation cluster with the Leath method and calculate a diffusion process using the method of exact enumeration. By this method you can directly study the probability of a random walker to be at a certain distance after a number of time steps. The first program enumerates normal unbiased diffusion in contrast to the second one, which enumerates topologically biased diffusion. The strength of the bias is determined by the parameter EPSILON. For EPSILON equal to 0 one gets the unbiased case, the maximum bias is for EPSILON equal to 1 or -1 depending on its direction. Both programs display the probability distribution for a random walker for fixed time. The different colors indicate the probability in a logarithmic scale. In addition, both programs store the probability distribution $P(r, t)$ for fixed time t as a function of the distance r in files PROBDIS.EX4 and PROBDIS.EX5 respectively and the root mean square displacement for the set of times defined in the header of both programs in files RMS.EX4 and RMS.EX5 respectively. You can change the parameter EPSILON from 0 to 1 and study its effect on the probability distribution and on the root mean square displacement. You can also exchange the topological bias with a euclidean one by simply modifing the transition rates. In addition, for both types of biased diffusion as well as for the normal random walk, you can modify the programs to study also the probability distribution in topological space $\langle P(\ell, t) \rangle$ and the topological mean square displacement $\langle \ell^2(t) \rangle$.

Acknowledgements

We would like to thank S. Havlin, A. Ordemann, and H.E. Roman for valuable discussions. This work has been supported by the Deutsche Forschungsgemeinschaft.

References

[1] B.B. Mandelbrot, *The Fractal Geometry of Nature* (Freeman, San Francisco 1982)

[2] J. Feder, *Fractals* (Plenum, New York 1988)

[3] A. Bunde and S. Havlin, eds., *Fractals in Science* (Springer, Heidelberg 1994)

[4] A. Bunde and S. Havlin, eds., *Fractals and Disordered Systems*, 2nd ed. (Springer, Heidelberg 1996)

[5] P.J. Flory, *Principles of Polymer Chemistry* (Cornell University Press, New York 1971)

[6] P.G. de Gennes, *Scaling Concepts in Polymer Physics* (Cornell University Press, Ithaca 1979)

[7] D. Stauffer and A. Aharony, *Introduction to Percolation Theory* (Taylor and Francis, London 1985)

[8] M. Sahimi, *Applications of Percolation Theory* (Taylor and Francis, London 1994)

[9] U.A. Neumann and S. Havlin, J. Stat. Phys. **52**, 203 (1988)

[10] A. Bunde and J. Dräger, Phil. Mag. B **71**, 721 (1995)

[11] A. Bunde and J. Dräger, Phys. Rev. E **52**, 53 (1995)

[12] S. Alexander and R. Orbach, J. Phys. Lett. **43**, L625 (1982)

[13] R. Kopelman, Science **241**, 1620 (1988)

[14] S. Havlin and D. Ben-Avraham, Adv. in Phys. **36**, 695 (1987)

[15] A. Bunde, S. Havlin, and H. E. Roman, Phys. Rev. A **42**, 6274 (1990)

[16] A. Bunde, H.E. Roman, St. Russ, A. Aharony, and A.B. Harris, Phys. Rev. Lett. **69**, 3189 (1992)

[17] J. Dräger, St. Russ, and A. Bunde, Europhys. Lett. **31**, 425 (1995)

[18] A. Bunde, S. Havlin, H.E. Stanley, B. Trus, and G.H. Weiss, Phys. Rev. B **34**, 8129 (1986)

[19] S. Havlin, A. Bunde, H.E. Stanley, and D. Movshovitz, J. Phys. A **19** L693 (1986)

[20] A. Bunde, H. Harder, S. Havlin, and H.E. Roman, J. Phys. A **20**, L865 (1987)

[21] H.E. Roman, M. Schwartz, A. Bunde, and S. Havlin, Europhys. Lett. **7**, 389 (1988)

[22] J.P. Bouchaud and A. Georges, Phys. Rep. **195**, 127 (1990)

[23] P.L. Leath, Phys. Rev. B **14**, 5046 (1976)

[24] J. Hoshen and R. Kopelman, Phys. Rev. B **14**, 3428 (1976)

[25] Z. Alexandrowicz, Phys. Lett. A **80**, 284 (1980)

[26] J.M. Hammersley, Meth. in Comp. Phys. **1**, 281 (1963)

[27] D. Ben-Avraham and S. Havlin, J. Phys. A **15**, L691 (1982)

[28] S. Havlin, D. Ben-Avraham, and H. Sompolinsky, Phys. Rev. A **27**, 1730 (1983)

[29] I. Majid, D. Ben-Avraham, S. Havlin, and H.E. Stanley, Phys. Rev. B **30**, 1626 (1984)

Multifractal Characteristics of Electronic Wave Functions in Disordered Systems*

Michael Schreiber

Institut für Physik, Technische Universität, D-09107 Chemnitz, Germany,
e-mail: schreiber@physik.tu-chemnitz.de

Abstract. We investigate the Anderson model of localization in order to describe electronic states in disordered materials. The eigenfunctions of the Anderson Hamiltonian, which are obtained from direct diagonalization by means of the Lanczos algorithm, are analyzed with respect to their spatial distribution. It is demonstrated that the wave functions display multifractal behavior up to length scales which are typically beyond the sizes of numerically accessible systems. On the enclosed diskette, wave functions are provided for different strengths of disorder and the reader is guided to write small computer programs in order to derive the characteristic multifractal properties such as the generalized dimensions, the mass exponents, and the singularity spectra. Special emphasis is laid on programming tricks which either save computer time or increase the accuracy. Finally, some results of large-scale numerical investigations by such a multifractal analysis are presented in order to demonstrate how this statistical method is applied in current research on the electronic properties of disordered materials.

1 Electronic States in Disordered Systems

In the beginning of condensed matter physics, the investigation of electronic (and other) properties concentrated on perfect crystals, because the translational invariance simplifies the calculations significantly and often allows us to successfully employ analytical methods. The electronic eigenstates are given by Bloch functions which are extended over the whole crystal. They appear homogeneous if one neglects fluctuations on the length scale of the lattice constant. Small deviations from the lattice structure of perfect crystals may be treated by perturbation theory. In this way it is, for example, possible to deal with a single impurity.

For a sufficiently (energetically) deep impurity exponentionally localized states are found. These wave functions are thus strongly localized and the electron is bound at the impurity. If one adds more impurities to the crystal, the exponential tails of the wave functions overlap, tunneling is enabled between the impurities, and consequently the electrons become less localized, which can be understood as a hybridization of energetically close impurity levels or as an orthogonality constraint on the wave functions of different impurities. If in the extreme case all lattice sites of the crystal are replaced by impurities, the resulting fluctuations of the wave functions may be so strong that the exponential decay is completely masked.

---

* Software included on the accompanying diskette.

On the other hand, the extended wave functions of the crystal are already disturbed by a single shallow impurity. If there are more impurities, interference effects become more significant, and in this case the resulting fluctuations of the wave functions may completely mask the extended character of the eigenstates. Nevertheless, even in the extreme case of replacing all lattice sites by impurities, the wave function may still be extended over the entire system. In three-dimensional (3D) samples this is the case for sufficiently weak disorder, i.e., for sufficiently shallow impurities.

It is the purpose of this chapter to present a possible method for describing the spatial fluctuations of the wave functions in a quantitative way. For this purpose, the probability density of the electronic eigenstates will be used as a measure. This measure is demonstrated to show fractal behavior. Different moments of the measure display different fractal behaviors, this leads to the concept of multifractality.

In Sect. 2 the Anderson model of localization is introduced on which the subsequent numerical investigations are based. The diagonalization of the Hamiltonian matrix is discussed and examples of eigenstates are presented. In the following sections the multifractal analysis of the wave functions is presented. Finally it is shown, how the multifractal analysis can be used to determine the metal–insulator transition which separates extended and localized wave functions and thus distinguishes metallic and insulating behavior. In this way the statistical analysis of the fluctuations of the wave functions will be shown to yield a means of determining the metal–insulator transition which occurs with an increase in the disorder or the energy in 3D disordered samples.

2   The Anderson Model of Localization

To investigate the electronic states of disordered materials we use a lattice model with random potential energies as discussed in the introduction. Of course, this is a severe simplification. Amorphous materials and alloys do not show a regular lattice structure. However, the Anderson model [1] described below has become a paradigm and has been widely used in the study of localization properties.

Starting from a general Hamiltonian $H = p^2/2m + V$ and expressing the second derivative in the kinetic energy by finite differences, one obtains the following stationary Schrödinger equation in 1D systems:

$$-\frac{\hbar^2}{2m} \frac{\psi(x + \Delta) - 2\psi(x) + \psi(x - \Delta)}{\Delta^2} + V(x)\,\psi(x) = E\,\psi(x)\ . \qquad (1)$$

In the lattice model we are not interested in the behavior of the wave function on length scales smaller than a lattice constant. Thus the smallest distance Δ which can be used in (1) is the lattice constant a. For the wave function ψ_n at the position $x = na$, i.e., at the nth site, one obtains the Schrödinger equation in site representation

$$-\frac{\hbar^2}{2ma^2}\,(\psi_{n+1} - 2\psi_n + \psi_{n-1}) + V_n\,\psi_n = E\,\psi_n\ , \qquad (2)$$

which can be written in a more comprehensive way as

$$t \, \psi_{n+1} + t \, \psi_{n-1} + \epsilon_n \, \psi_n = E \, \psi_n \ . \tag{3}$$

It should be noted that due to the underlying regular lattice the parameters t are independent of the lattice site. They are called transfer matrix elements because the respective matrix elements reflect the transfer of an electron from the nth site to a neighbouring site. As usual we express all energies in units of this parameter, i.e., we choose $t = 1$.

Besides a constant term, the parameters ϵ_n contain the potential energies V_n. To simulate the random potential of the impurities these parameters are chosen independently from a box distribution

$$P(\epsilon_n) = W^{-1} \, \Theta(W/2 - |\epsilon_n|) \ . \tag{4}$$

Here Θ denotes the step function. The width W of this distribution reflects the strength of the disorder. Choosing a zero mean value of the distribution as in (4) fixes the origin of the energy scale. Of course, the box distribution means a severe simplification of the random potential. A binary distribution, e.g., would be more realistic for describing binary alloys. On the other hand if one takes into account that there may be several different features which contribute independently to the disorder, then the central limit theorem would suggest a Gaussian distribution. Due to its numerical simplicity, however, the box distribution has been used in most investigations of the Anderson model.

The Schrödinger equation (3) can be solved as a recursion equation as in the chapter by Kramer et al. in this book. Here we solve the corresponding eigenvalue problem which can be written in matrix representation for a finite system with N sites as

$$\mathbf{H}^{(1)} \psi_i = E_i \, \psi_i \ . \tag{5}$$

The diagonalization of the respective Hamiltonian matrix $\mathbf{H}^{(1)}$ yields the eigenvalues E_i and corresponding eigenvectors $\psi_i = (\psi_{i1}, \psi_{i2}, ..., \psi_{in}, ..., \psi_{iN})^{\mathrm{T}}$. It is easy to construct the matrix $\mathbf{H}^{(1)}$ from (3). For a chain of five sites it reads, e.g.,

$$\mathbf{H}^{(1)} = \begin{pmatrix} \epsilon_1 & 1 & 0 & 0 & 1 \\ 1 & \epsilon_2 & 1 & 0 & 0 \\ 0 & 1 & \epsilon_3 & 1 & 0 \\ 0 & 0 & 1 & \epsilon_4 & 1 \\ 1 & 0 & 0 & 1 & \epsilon_5 \end{pmatrix} \tag{6}$$

where $t = 1$ was used as discussed above. The nonzero matrix elements $H_{15}^{(1)}$ and $H_{51}^{(1)}$ are due to the employed periodic boundary conditions, i.e., the chain is closed to a ring so that the first and last sites are neighbors.

It is straightforward to generalize the above derivation for higher dimensions. In 2D samples the kinetic energy contains a second derivative with respect to the y direction. The respective finite differences lead to another two terms in (2) and (3) which reflect the transfer to the neighboring sites in the y direction. As an

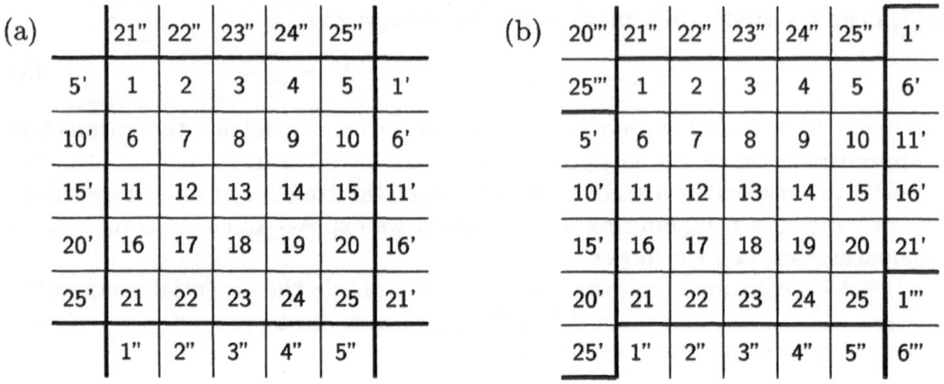

(a)

	21"	22"	23"	24"	25"	
5'	1	2	3	4	5	1'
10'	6	7	8	9	10	6'
15'	11	12	13	14	15	11'
20'	16	17	18	19	20	16'
25'	21	22	23	24	25	21'
	1"	2"	3"	4"	5"	

(b)

	20'''	21"	22"	23"	24"	25"	1'
25'''		1	2	3	4	5	6'
5'		6	7	8	9	10	11'
10'		11	12	13	14	15	16'
15'		16	17	18	19	20	21'
20'		21	22	23	24	25	1'''
25'		1"	2"	3"	4"	5"	6'''

Fig. 1a,b. Sites of a 5 × 5 square lattice numbered as appropriate for (7). The boundaries of the unit cells are indicated by *thick lines*. The *dashed* and *double-dashed* sites reflect the copies of the system in the x and in the y direction, respectively, due to periodic boundary conditions in (a) and due to helical boundary conditions in (b)

example we consider a 5 × 5 sample and count the sites as in Fig. 1a. Each row of this system can be represented by a 5 × 5 matrix $\mathbf{H}^{(1)}$, automatically including the transfer due to the periodic boundary conditions in the x direction. These matrices yield the diagonal of the Hamiltonian matrix for the 2D sample which can be written in block from as

$$
\mathbf{H}^{(2)} =
\begin{pmatrix}
\mathbf{H}^{(1)} & 0 & 0 & 0 & 0 \\
0 & \mathbf{H}^{(1)} & 0 & 0 & 0 \\
0 & 0 & \mathbf{H}^{(1)} & 0 & 0 \\
0 & 0 & 0 & \mathbf{H}^{(1)} & 0 \\
0 & 0 & 0 & 0 & \mathbf{H}^{(1)}
\end{pmatrix}
+
\begin{pmatrix}
0 & 1 & 0 & 0 & 1 \\
1 & 0 & 1 & 0 & 0 \\
0 & 1 & 0 & 1 & 0 \\
0 & 0 & 1 & 0 & 1 \\
1 & 0 & 0 & 1 & 0
\end{pmatrix}.
$$

$$
=
\begin{pmatrix}
\mathbf{H}^{(1)} & 1 & 0 & 0 & 1 \\
1 & \mathbf{H}^{(1)} & 1 & 0 & 0 \\
0 & 1 & \mathbf{H}^{(1)} & 1 & 0 \\
0 & 0 & 1 & \mathbf{H}^{(1)} & 1 \\
1 & 0 & 0 & 1 & \mathbf{H}^{(1)}
\end{pmatrix}.
$$

(7)

Here each $\mathbf{0}$ represents a 5 × 5 zero matrix and each $\mathbf{1}$ represents a 5 × 5 unity matrix. The latter describe the transfer to neighboring sites in the y direction as can be easily verified by inspection of Fig. 1a. The periodic boundary conditions in the y direction are reflected by the entries $\mathbf{1}$ in the lower left and upper right corner of the matrix $\mathbf{H}^{(2)}$ in analogy to the above discussion of the respective matrix elements of $\mathbf{H}^{(1)}$.

The generalization to 3D samples can be performed in the same way putting five 25 × 25 matrices of the form (7) as blocks onto the diagonal of $\mathbf{H}^{(3)}$ and adding **0** and **1** matrices yielding

$$
\mathbf{H}^{(3)} = \begin{pmatrix}
\mathbf{H}^{(2)} & \mathbf{1} & \mathbf{0} & \mathbf{0} & \mathbf{1} \\
\mathbf{1} & \mathbf{H}^{(2)} & \mathbf{1} & \mathbf{0} & \mathbf{0} \\
\mathbf{0} & \mathbf{1} & \mathbf{H}^{(2)} & \mathbf{1} & \mathbf{0} \\
\mathbf{0} & \mathbf{0} & \mathbf{1} & \mathbf{H}^{(2)} & \mathbf{1} \\
\mathbf{1} & \mathbf{0} & \mathbf{0} & \mathbf{1} & \mathbf{H}^{(2)}
\end{pmatrix} . \tag{8}
$$

Now, however, these zero and unity matrices are of size 25×25. Altogether we ended with the 125×125 matrix (8) which is a sparse matrix, because only seven matrix elements in each row and column differ from zero.

3   Calculation of the Eigenvectors

The sparseness of the Hamiltonian matrix makes it ideally suitable for the application of the Lanczos algorithm [2]. This algorithm has been widely used in recent years and shall not be described in detail here. It is sufficient to note that it is an iterative procedure which requires the repeated multiplication of the Hamiltonian matrix with some initial vector. These matrix–vector multiplications consume most of the computer time. Usually such a step would require N^2 multiplications and additions. Due to the particular shape of the matrix (8) it is easy to avoid all multiplications with 0 or 1 and all summations of zeros. Thus in each step only N multiplications and $6N$ additions are required. This makes the Lanczos algorithm very effective in the present case, not only as usual for extreme eigenvalues but also for eigenvalues in the middle of the spectrum, although in this case a large number of Lanczos iterations is necessary.

In order to increase the efficiency of the matrix–vector multiplication even further, especially on vectorizing supercomputers, it has been most useful to slightly alter the boundary conditions. The periodic boundary conditions as discussed above have to be programmed separately for each row of the 2D sample and for each 2D slice of the 3D system, requiring either a relatively large number of loops or conditional statements in the program. This can be avoided by employing so-called helical boundary conditions. In 2D samples these are achieved by shifting the unit cell, which is periodically repeated in the x direction, by one lattice constant into the y direction. This is illustrated in Fig. 1b. Now the 5th site is not a neighbor of site 1 but of site 6. This leads to a much simpler structure of the first matrix in (7). Instead of five 5×5 matrices it can now be written as one 25×25 matrix of the form (6). The second matrix in (7) remains unaltered. In order to achieve the same advantage in 3D systems, one has to shift the unit cells, which are periodically repeated in the y direction, by one lattice constant into the z direction. The resulting matrix has a very simple structure which cannot only be programmed most easily in the matrix–vector multiplication but which becomes also most efficient on vectorizing computers as mentioned above. Physically the change of boundary conditions is of no relevance, because it only means replacing the originally cubic unit cell by an affinely distorted cell.

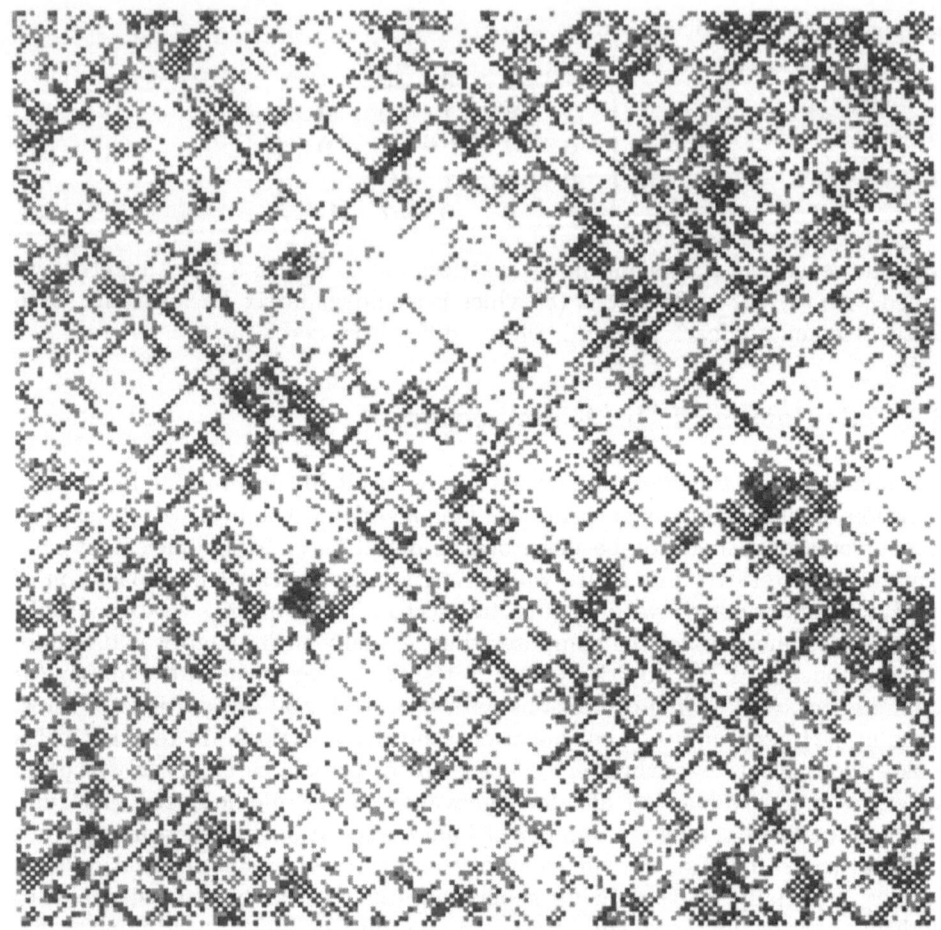

Fig. 2. Probability amplitude of eigenstates of the Anderson Hamiltonian on a square lattice of $N = 200 \times 200$ sites for energy $E = 0$ and disorder parameter $W = 1$. Every site with a probability larger than average is shown, i.e., $|\psi_{in}| > N^{-1/2}$. Four different grey levels ($j = 0, 1, 2, 3$) distinguish whether $\psi_{in}^2 > 2^j/N$

Employing the helical boundary conditions in this way it has been possible [3] to determine eigenstates of 3D samples with up to $N = 74^3$ sites in the middle of the spectrum, i.e., for eigenenergies close to $E = 0$. We note that the number of nonzero elements in each row and column of the Hamiltonian matrix remains seven in 3D and five in 2D also for the helical boundary conditions. Typical wave functions for 2D systems are illustrated in Figs. 2 and 3. These plots demonstrate the fragmentation of the probability density of the wave function. It is a characteristic feature that the speckles which reflect lumps of high probability occur on all possible scales and intensities and are separated by openings

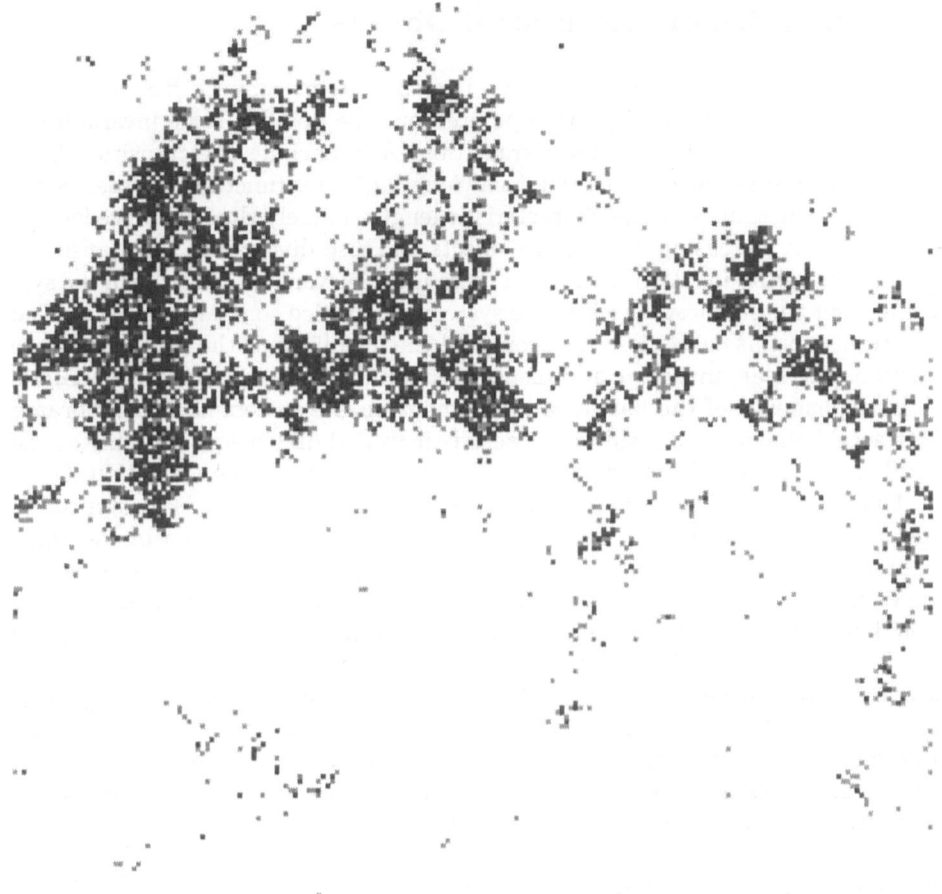

Fig. 3. Same as Fig. 2, but for disorder $W = 5$

or gaps which likewise appear with all sizes. For larger disorder the probability is concentrated in certain areas. Mandelbrod [4] has coined the illustrative phrase *curdling* for this behavior. Strong curdling of the electronic wave function occurs when the electron is put into a system with too much disorder just in the same way as old milk curdles when it is poured into coffee which is too hot. As demonstrated in Figs. 2 and 3, the characteristics of the curdling depend on the disorder parameter.[1] In the following sections a quantitative way of describing this behavior shall be presented.

---

[1] The data for these wave functions are included on the enclosed diskette together with programs for plotting these functions in the same way as in Figs. 2 and 3. In the World Wide Web (http://www.tu-chemnitz.de/home/schreiber/springer) data of more wave functions for other parameters can be found.

4   Description of Multifractal Objects

The notion that the eigenstates of disordered systems may be fractal objects was first suggested by Aoki [5] who pointed out the possible scale invariance of eigenstates at the metal–insulator transition. Approaching the transition from the insulating side, the exponential decay length of wave function diverges. Similarly, on the metallic side the characteristic length, namely the coherence length, increases with increasing interference effects and also diverges at the transition. Thus at the metal–insulator transition there is no relevant length scale in the system (except the smallest length in the system, the lattice constant, which is of no importance in this context). This prompted the idea [5] that the respective wave functions are scale invariant and display self-similarity. A quantitative analysis of the dependence of the participation number on the system size corroborated the fractal behavior and yielded characteristic fractal dimensions not only at the metal–insulator transition but also in the neighborhood of the transition [6].

The fractality of the wave functions around the transition is an appealing concept, because it enables us to assume a continuous transition of the characteristics of the wave function at the metal–insulator transition. Moreover, it allows us to accommodate two seemingly contradictory notions. A localized state should occupy only an infinitesimal fraction of space even arbitrarily close to the metal–insulator transition, on the other hand an extended state is expected to be spread over the entire system. A filamentary structure over the whole sample like a mesh with openings on all scales or a curdled structure with lumps of all sizes could represent such an effectively extended state which nevertheless does not fill any finite fraction of the volume. This is exactly what happens in our disordered samples as illustrated in Figs. 2 and 3.

However, it has become clear in recent years that the simple fractal description cannot be sufficient to characterize the electronic states in disordered systems. Rather one has to take into account the spatial distribution of the probability density of the wave function over the entire system in a quantitative way. This shall be explained in the following.

The fractal picture as it is, e.g., used in Bunde's chapter of this book, is based on the analysis of the spatial distribution of a set of points. A prominent example is given by the set of lattice sites in the Sierpinski gasket. Covering the set of points with squares of size δ^2 in 2D or cubes of size δ^3 in 3D systems requires $\mathcal{M}(\delta)$ boxes. Introducing generalized cubes of size δ^d as test boxes one can define a measure $M_d = \sum_k \delta^d = \mathcal{M}(\delta)\,\delta^d$ by summing over all boxes k. This quantity allows us to measure the "mass" of the fractal set. The behavior of the measure

$$M_d = \mathcal{M}(\delta)\,\delta^d \overset{\delta\to 0}{\longrightarrow} \begin{cases} 0 & , \quad d > D_0 \,, \\ \infty & , \quad d < D_0 \,, \end{cases} \tag{9}$$

defines the fractal dimension D_0 of the set of points or lattice sites. In order to measure the finest details of the set, the boxes should become as small as possible, which in our case means that the ratio δ of lattice constant and system size has to become small. We note that the investigated set may be rigorously defined by a simple construction law like the Sierpinski gasket, which is exactly self-similar on all scales. The set of lattice points may also be defined in a statistical way as in the case of random walk trails or percolation clusters (cf. Bunde's chapter of this book) yielding a stochastic fractal in which self-similarity is achieved only in a statistical sense.

However, counting in our case all boxes in which the wave function is nonzero would mean counting all sites, because due to the random potential the wave function does not vanish at any site, although it may become exponentially small. Consequently the simple measure (9) includes all sites and yields the Euclidean dimension $D_0 = 2$ for the square lattice and $D_0 = 3$ for the cubic system. Thus D_0 is not a particularly useful quantity in our case. For a more interesting evaluation of the fractal properties we have to take the "contents" of each box into account. This means that we attribute an appropriate weight to each box. We consider the so-called box probability, i.e., the probability of finding an electron in a box of linear size La

$$\mu_k(L) = \sum_{n(k)} |\psi_{in}|^2 \ , \qquad k = 1, ..., N_L \ . \tag{10}$$

Here the summation is performed over the L^2 or L^3 sites in the kth box, i.e., $N_L = N/L^2$ or $N_L = N/L^3$. This quantity, which is normalized because the wave function is normalized, thus measures the spatial distribution of the probability density of the wave function. Obviously, it can be a fractal object only in a statistical sense.

The scaling of this measure and its qth moments in the limit of small box sizes δ

$$M_d(q) = \sum_k \mu_k^q \, \delta^d \xrightarrow{\delta \to 0} \begin{cases} 0 \ , & d > \tau(q) \\ \infty \ , & d < \tau(q) \end{cases} \tag{11}$$

defines the so-called mass exponents $\tau(q)$. It is sometimes more illustrative to discuss the generalized dimensions

$$D(q) = \tau(q)/(1-q) \ , \tag{12}$$

because for a homogeneous distribution of the mass, or in our case a homogeneous distribution of the probability density of the wave function, all generalized dimensions equal the Euclidian dimension. If the measured quantity is distributed homogeneously over a fractal set, then all generalized dimensions equal the fractal dimension. But for a nonhomogeneous distribution, different fractal dimensions $D(q)$ occur and we have a multifractal rather than a fractal object. It should be noted that for $q = 0$ we always have $D(0) = \tau(0)$, and in this case the definitions (9) and (11) coincide yielding the fractal dimension D_0 of the underlying lattice which supports the wave function. Therefore D_0 is called the dimension of the support or the similarity dimension, because it reflects the self-similarity of the support.

The physical meaning of $\tau(q)$ and $D(q)$ is that the different moments of the mass distinguish intertwined regions of the wave functions which scale in different ways according to the mass exponents $\tau(q)$. The generalized dimensions characterize the different fractality of the different subsets of the measure corresponding to the mentioned intertwined regions of the wave function. Accordingly the multifractal wave function is not really self-similar, but rather self-affine, again in a statistical sense.

Another quantity which is often used for a description of multifractal objects is the Lipschitz–Hölder exponent α which reflects the strength of the singularity of the box probability in the kth box:

$$\mu_k(\delta) \sim \delta^\alpha . \tag{13}$$

It describes how the mass in a particular box changes when the size of the box is reduced. Of course, for a homogeneous quantity one obtains $\alpha = D_0$. The set containing all boxes in which a particular value α of the singularity strength is observed is a fractal itself. Its fractal dimension f describes the scaling of the number of respective boxes

$$\mathcal{N}(\alpha) \sim \delta^{-f(\alpha)} . \tag{14}$$

For a homogeneous quantity we obtain only one value $f(D_0) = D_0$. For a non-homogeneous quantity the so-called singularity spectrum $f(\alpha)$ results. Characterizing multifractal objects by the singularity spectrum $f(\alpha)$ is equivalent to the determination of the mass exponents $\tau(q)$. In fact both quantities are connected by a Légendre transformation [7]. In particular, one can determine the singularity strength,

$$\alpha(q) = -\frac{\mathrm{d}\tau(q)}{\mathrm{d}q} , \tag{15}$$

and the singularity spectrum,

$$f(\alpha(q)) = q\alpha(q) + \tau(q) , \tag{16}$$

in a parametrized way from the mass exponents.

5   Multifractal Analysis of the Wave Functions

For simplicity the following discussion is formulated for 2D systems. The generalization to other dimensions is straightforward, but in 2D systems the proceedings can be illustrated more easily.

For the multifractal analysis we cover the system with boxes of linear size δ. Here the size is taken relative to the extent of the system, i. e., $\delta = L/N^{1/2}$. In our case, in which the wave functions do not vanish exactly at any site, this means we cover the *entire* system with squares of size $L \times L$, so that the number of boxes which are required is fixed as $N_L = N/L^2$. For the so-called box-counting method we only have to count the number of boxes (cf. (9)). Of course

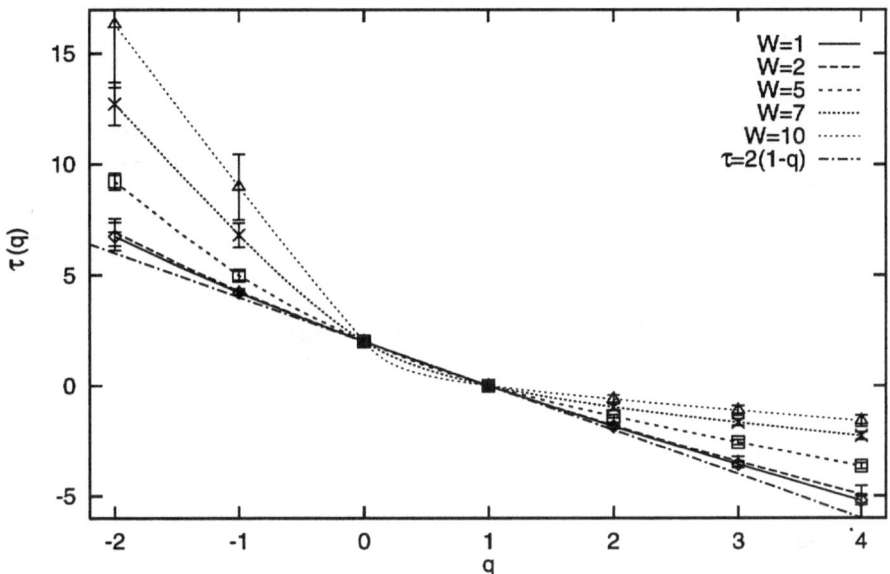

Fig. 4. Mass exponents $\tau(q)$ of the measure $\mu_k(L)$ for various wave functions in a 2D sample of $N = 200 \times 200$ sites at different values of the disorder W. The *error bars* reflect the accuracy of the linear regression. The *dash-dotted line* shows the mass exponents for a homogeneously extended wave function

in our multifractal case we have to take the appropriate weights into account. For each box the probability of finding the electron in the box is calculated according to (10). According to (11) the sum of the qth moments over all boxes should be proportional to $\delta^{-\tau(q)}$. Thus the mass exponents can be determined by plotting $\sum_k \mu_k^q$ on a doubly logarithmic scale and measuring the steepness of the curve for small δ. This can be easily performed by fitting a straight line to the logarithmic data using a standard procedure for linear regression [8]. Of course, such a routine will always yield some result. Whether this is a meaningful result has to be checked by ensuring that the data follow a linear behavior. In other words, it is necessary to check whether the qth moments of the box probability obey a scaling law at all. This test may seem obvious, but experience shows that it is sometimes forgotten. Obviously, deviations for the smallest possible size (the lattice constant, i.e., $L = 1$) and the largest possible size (the system size $L = N^{1/2}$) cannot be avoided. This makes the determination of the mass exponents a numerically delicate task. It is particularly difficult for large negative values of q for which the moments μ_k^q are dominated by the small amplitudes of the wave function which are most susceptible to numerical inaccuracies. This is the reason why we do not include data for $q < -2$ in the following.

Typical examples for the mass exponents are shown in Fig. 4 for different values of the strength of the disorder. With increasing disorder the deviation

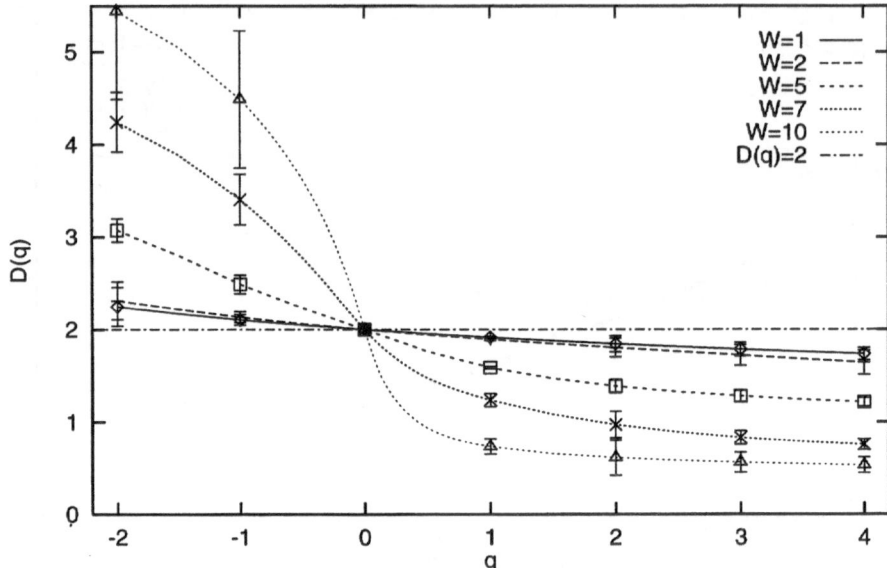

Fig. 5. Generalized dimensions $D(q)$ of the measure $\mu_k(L)$ for various wave functions in a 2D sample of $N = 200 \times 200$ sites at different values of the disorder W. The *error bars* reflect the accuracy of the linear regression. The *dash-dotted line* shows the dimension of a homogeneously extended wave function

from the straight line $\tau(q) = 2\,(1-q)$, which describes a homogeneous object in 2D systems, become more pronounced. The linear behavior for large positive and negative values of the parameter q is a typical feature of multifractal objects. The values $\tau(q = 0) = 2$ and $\tau(q = 1) = 0$ coincide for all disorder values due to the 2D support and due to the normalization of the wave functions, respectively.

The generalized dimensions can be easily calculated according to (12) from the mass exponents. Results are shown in Fig. 5. A homogeneous object would yield $D(q) = 2$ in 2D systems. With increasing disorder the curves in Fig. 5 show stronger deviations from this constant. The overall shape of the curves is again typical for multifractal objects. The asymptotically constant behavior for $q \to \pm\infty$ corresponds to the linear behavior of the mass exponents. For large positive and negative q the steepness of $\tau(q)$ equals the minimal and maximal generalized dimensions and, according to (15), the minimal and maximal values of the singularity strength α, i.e., $\alpha_{\min} \leq D(q) \leq \alpha_{\max}$. These values correspond to the subset where the measure and thus the wave function is most concentrated or rarified, respectively.

In principle, the singularity spectrum can be computed analogously. Replacing the exponent α in (13) by the coarse-grained Hölder exponent

$$\alpha_k = \log \mu_k(\delta) / \log \delta \,, \tag{17}$$

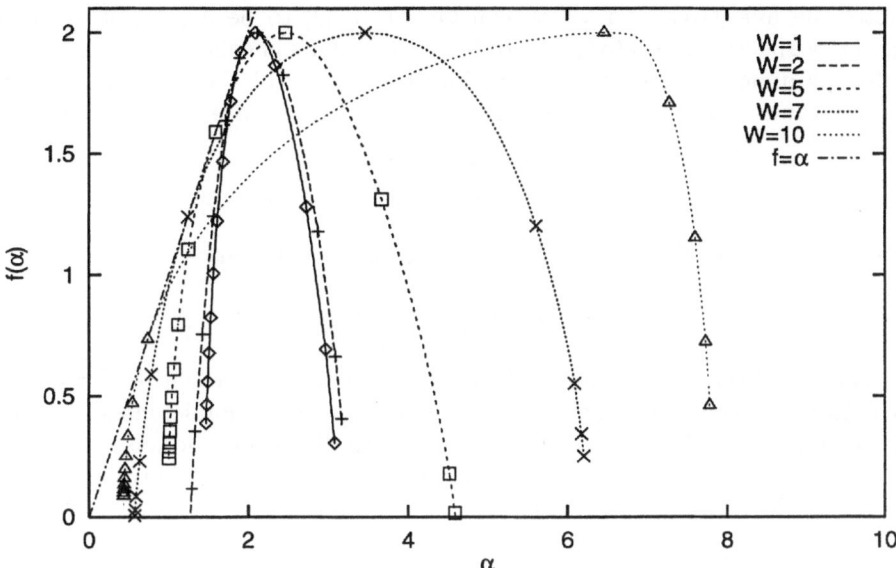

Fig. 6. Singularity spectrum $f(\alpha)$ of the measure $\mu_k(L)$ for various wave functions in a 2D sample of $N = 200 \times 200$ sites at different values of the disorder W. Integer values of the implicit parameter q are indicated by *symbols* for $-4 \leq q \leq 10$. The *dash-dotted line* is given by $f(\alpha) = \alpha$, which is a tangent of the singularity spectrum for every wave function

which is equivalent to (13) in the limit $\delta \to 0$, one attributes the singularity strength α_k to each box. Then one counts the number $\mathcal{N}(\alpha)$ of boxes· which occur for each singularity strength or rather which yield a value of α within a small interval. The resulting histogram (hence the name histogram method) is used to determine the singularity spectrum according to (14) as

$$f(\alpha) = -\log \mathcal{N}(\alpha)/\log \delta \qquad (18)$$

in the limit $\delta \to 0$. Typical results are presented in Fig. 6. The displayed shape is again characteristic for a multifractal object. It is a convex curve with its maximum given by the fractal dimension of the support of the measure, i.e., $\max f(\alpha) = D_0 = 2$. Around the maximum each curve can be approximated by a parabola. In the limit of large positive or negative q the singularity spectrum $f(\alpha)$ should approach zero with infinite slope. Due to numerical difficulties the curves do not reach the α axis. But the corresponding minimal and maximal values of α have already been derived above as the extreme values of the generalized dimensions. Accordingly the width of the singularity spectrum increases with increasing disorder as can be seen in Fig. 6. We also note in Fig. 6 that the maximum of the singularity spectrum moves towards larger values of α with increasing disorder. The dash-dotted line in Fig. 6 is of interest in so far as it denotes the line $f(\alpha) = \alpha$ which can be shown [9] to be a tangent of every singularity spectrum at $f(\alpha(1)) = \alpha(1) = D(1)$. This is called the information or entropy dimension because it describes the scaling of the entropy S of the measure, which is defined as usual in statistical physics from the box probabilities as

$$S = -\sum_k \mu_k \ln \mu_k \ . \tag{19}$$

It can be seen from (12) by expanding μ_k^q around $q = 1$ that

$$D(1) = \lim_{\delta \to 0} -S/\ln \delta \ . \tag{20}$$

One can show that in the limit of infinite system size the entire measure is concentrated into a fractal set with this dimension. For completeness, we note that the dimension $D(2)$ is called the correlation dimension, because it reflects the scaling of the density–density correlation function. Higher correlations are characterized by larger values of q.

6   Computation of the Multifractal Characteristics

The above-described method of computing the multifractal characteristics can be easily applied and can be programmed in a straightforward way. Two wave functions are supplied on the enclosed diskette. The reader is encouraged to calculate $\tau(q), D(q)$, and $f(\alpha)$ as a simple exercise. However, the convergence is usually not very good and it is advisable to try to improve the accuracy. One way in which this can be done is by averaging over different choices of the division of the system into the N_L boxes. Let us consider a 6×6 unit cell as an example. There is a natural way of dividing it into four boxes of size 3×3. However, due to the periodic boundary conditions there are eight equally allowed divisions which one obtains by shifting the "natural" boxes one or two sites into the x direction and/or one or two sites into the y direction. This means that we can use 36 values for μ_k instead of four values, thus improving the statistics significantly. Of course we have to take into account that every site is included nine times so that an appropriate normalization is necessary.

Another important aspect of this consideration is that it also allows us to use box sizes for which N_L is not an integer, i.e., boxes which cannot be used in a straightforward way to cover the whole system without overlap or holes. Let us consider again as an example the 6×6 system which cannot be covered directly with 5×5 boxes. However, if we take into account all possible 36 nonequivalent positions for placing a 5×5 box in the periodically repeated system (shifting the box up to five sites in the x and/or in the y direction) we include every lattice site exactly 25 times. This yields the normalization in analogy to the previous example of 3×3 boxes. There is another way of visualizing this average. If one periodically repeats the original 6×6 system four times in the x direction and the resulting 30×6 strip four times in the y direction, the resulting 30×30

system can be divided into 5×5 boxes in a natural way. The 36 boxes thus obtained are equivalent to the 36 shifted boxes. In this way many more data points can be computed for the linear regression for the evaluation of (11) as well as for the calculation of (17) and (18), because all integer values of L are now allowed.

If the number of data points is still not sufficient for a statistically reasonable result another improvement can be made by allowing slightly rectangular boxes, e.g., of size 2×3 and 3×2. In this way two more data points for an effective size $L = 6^{1/2}$ are obtained.

The reader is encouraged to use these tricks in order to improve the statistics of the simple exercise program. It is also interesting to note that the derivation of the singularity spectrum from the mass exponents according to (15) and (16) is not very accurate due to the necessary numerical derivative. This can be seen by comparing the numerically obtained results of (15) and (16) with the results of (18).

For a more accurate computation of the singularity spectrum it turned out to be better to use the qth moment of the measure in the separate boxes which, properly normalized,

$$
\mu_k(q, L) = \mu_k^q(L) \, / \sum_{k'=1}^{N_L} \mu_{k'}^q(L) \, , \tag{21}
$$

constitutes a measure itself. In the method of moments this measure is used to directly calculate the Lipschitz-Hölder exponents

$$
\alpha(q) = \lim_{\delta \to 0} \sum_k \mu_k(q, L) \log \mu_k(1, L) / \log \delta \tag{22}
$$

and the corresponding value of the singularity spectrum

$$
f(q) = \lim_{\delta \to 0} \sum_k \mu_k(q, L) \log \mu_k(q, L) / \log \delta \tag{23}
$$

in a parametric representation. These formulae can be evaluated by a standard linear regression, determining the steepness of $\mu \log \mu$ versus $\log \delta$. Of course, the above-explained tricks to improve the accuracy by averaging over different nonequivalent divisions of the system into boxes, by considering box sizes which do not naturally fit into the system size, and by taking rectangular boxes into account can all be applied for the evaluation of (22) and (23), too.

In fact, the data in Fig. 6 have been obtained in this way. The mass exponents can then be easily calculated from (16) and the generalized dimensions from (12), yielding Figs. 4 and 5. These data are also available on the enclosed diskette, together with plot programs to display the data in different combinations. We have also added data for different system sizes and for different values of the random potential energies. Comparing these data in the plots, the finite-size effects and the dependence on the actual values of the random numbers can be estimated.

In actual applications it is necessary to average over several realizations of the random potential. It is useful to note one more hint in this case: it is advantageous to average the sums over $\mu \log \mu$ in (22) and (23) and to perform the linear regression afterwards for the averaged data instead of averaging the results of the linear regression for different realizations of the random potential. Usually one also performs an average over several eigenstates which are close in energy. This average is numerically less expensive, because the eigenstates can be computed in one run of the Lanczos procedure. Nevertheless, to obtain sufficiently accurate values for reasonably large systems an unpleasantly large amount of computer time is necessary.

7   Topical Results of the Multifractal Analysis

Following Aoki's suggestion [5] that the eigenstates of disordered systems should be scale invariant at the metal–insulator transition, the fractal character of the electronic wave functions of the Anderson Hamiltonian has been demonstrated by showing the power-law dependence of the participation number on the system size [6,10]. The respective fractal dimension coincides with the correlation dimension $D(2)$, which has also been derived directly from an analysis of the density-density correlation of the wave function [11]. Other generalized dimensions have been computed at the metal–insulator transition in 3D systems [12,13] as well as for the critical states in the lowest Landau band of a 2D system under the influence of a strong magnetic field [14] and in 2D systems with spin-orbit coupling [15]. These systems display a metal–insulator transition in 2D, too, and can be investigated in terms of the Schrödinger equation (3) with appropriately chosen transfer integral t.

When it became clear that the simple fractal picture was not sufficient, it was found that the multifractal character of the electronic states at the metal–insulator transition could be established for the 3D Anderson model [16] as well as for the 2D model with strong magnetic field [17]. It is interesting to note that the resulting singularity spectrum for the 2D tight-binding system coincides with that for the continuum model of 2D samples with magnetic field [18]. Likewise, the singularity spectrum of the 3D model (3) with Gaussian distribution of the random potential of the impurities was found [19] to be the same as for the model with box distribution (4) of the potential energies when both models are evaluated at the metal–insulator transition which is known [20] in the band centre, i.e., for $E = 0$, from respective studies by means of the transfer-matrix method (cf. the chapter of Kramer et al. in this book) to occur at $W_c = 16.5\ (21.0)$ for the box (Gaussian) distribution. These observations stimulated the conclusion [19] that a characteristic singularity spectrum is universal for the metal–insulator transition regardless of the specific parameters of the model (except the dimensionality). Around its maximum the characteristic singularity spectrum in 3D was also shown to agree with the parabolic approximation

$$f(\alpha) = 3 - (4 - \alpha)^2 / 4 \,, \tag{24}$$

which was obtained from the $2 + \epsilon$ expansion within the nonlinear σ model [21].

Taking into account the pronounced dependence of the singularity spectrum on the disorder that can be seen in Fig. 6, the comparison with the characteristic singularity spectrum for the transition can then be used as a tool [22] to distinguish localized and extended states. As it is impractical to compare the entire singularity spectrum, two specific points of the spectrum have been used for this purpose, namely $\alpha(0) = \max f(\alpha)$ and $\alpha(1) = f(\alpha(1)) = D(1)$. Both show a pronounced dependence on the system parameters: $\alpha(0)$ increases, $\alpha(1)$ decreases with increasing disorder and/or energy. Taking only the transfer-matrix-method value of the critical disorder ($W_c = 16.5$) in the band centre ($E = 0$) as input parameter, the critical values at the metal–insulator transition have been determined as $\alpha_c(0) = 4$ and $\alpha_c(1) = 2$. These values were then employed to distinguish localized and extended wave functions for various parameter combinations of energy and disorder. As a result the phase diagram in the energy-disorder parameter space could be determined for the box, the Gaussian, and the binary distribution of random potential energies [20]. For all distributions the entire trajectory of the metal–insulator transition was in good agreement with respective results from the transfer-matrix method. Already for very small systems with $20^3 = 8000$ sites the accuracy of the multifractal analysis is sufficient to obtain the phase boundary, whereas the system size for the mentioned calculations by means of the transfer-matrix method was three orders of magnitude larger.

For the relatively small systems it is not surprising that multifractal characteristics can also be derived for wave functions not exactly at the metal–insulator transition but in its neighborhood. In this case, however, the wave functions should display scale invariance not on all scales but only up to the correlation length or the localization length. Close to the transition, these lengths are much larger than the size of the investigated systems so that the multifractal analysis is feasible. As a consequence, however, the singularity spectrum shows a specific dependence on the size N of the system. With increasing system size the singularity spectrum for the extended wave functions becomes narrower and narrower in agreement with the expectation that for infinite systems the singularity spectrum $f(\alpha)$ of an extended state would consist only of one point $f(D_0) = D_0$. On the other hand, for localized states the singularity spectrum would consist of two points: $f(0) = 0$ would reflect the localization centre, whereas $f(\infty) = D_0$ would represent the large part of the system in which the wave function is effectively zero. Accordingly, in the regime of localized states the singularity spectrum becomes broader and broader with increasing system size. Only exactly at the metal–insulator transition does the singularity spectrum not change upon altering the system size. Thus the analysis of the dependence of the singularity spectrum on the system size enables us to distinguish localized and extended states. Again it is impractical to analyze the entire singularity spectrum. Instead we have again evaluated the data for $\alpha(0)$ and $\alpha(1)$. The results agree with the above-described evaluation of the disorder and/or energy dependence of the singularity spectrum [20]. However, to obtain sufficiently accurate data for the system-size dependence of the singularity spectrum a much larger numerical expense is necessary. Therefore this method is impractical for a determination of the complete phase boundary. But it is conceptually satisfying, because it yields a possibility of determining the characteristic singularity spectrum for the metal–insulator transition without relying on input data from the transfer-matrix method.

The discussed investigations have been concerned with the multifractal properties of single eigenstates. Transport, however, is usually not connected with single eigenstates, but is rather concerned with wave packets traveling through the system. Such a wave packet can be constructed by linear combination from a large number of eigenstates within a given energy range. A priori, it is by no means clear that such a linear combination maintains the multifractal properties. Looking at Figs. 2 and 3, one can easily imagine that the sum of several wave functions, the curdling of which occurs at different positions in the system, yields a less fragmented picture. (It is of course a significant difference whether one linearly combines the wave functions before calculating a box probability in (10) or whether one averages the results of the singularity spectrum over several different wave functions as discussed in Sect. 6.) A detailed analysis, however, has shown [23] that a linear combination of a large number of wave functions does not give a homogeneous distribution of the probability amplitude, but rather yields the same singularity spectrum and consequently the same set of generalized dimension as the original wave functions. Therefore the characteristics of a single eigenstate are sufficient to analyze the influence of multifractal properties on the transport properties of electrons in disordered systems [23].

In conclusion, we have shown that the multifractal analysis of the wave functions yields a straightforward and independent way of determining the trajectory of the metal–insulator transition, i.e., the phase boundary in the energy/disorder diagram. The large-scale numerical investigation was concentrated on two points $\alpha(0)$ and $\alpha(1)$ of the singularity spectrum. In principle one can apply the method to any other point $\alpha(q)$ or $f(q)$ but the numerical accuracy will be more difficult to achieve for large q, especially for large negative values of q.

Conceptually the multifractal method is an appealing concept, because it exploits profitably the strong fluctuations of the wave functions which hitherto were considered a nuisance in the numerical investigations. From a physical point of view it is interesting that these fluctuations, which are evaluated on relatively short scales, can be successfully used to determine the metal–insulator transition which reflects the long-range behavior of the wave functions. The reason for this unexpected success is of course the scale invariance of the wave functions at the metal–insulator transition, where the characteristic fluctuations are independent of the system size and therefore show up in small systems in a characteristic way.

Acknowledgement

I thank Thomas Rieth for bringing the enclosed programs into such a form that they can be used without further advice by a specialist.

References

[1] P.W. Anderson, Phys. Rev. **109**, 1492 (1958)
[2] J.K. Cullum and R.A. Willoughby, *Lanczos Algorithms for Large Symmetric Eigenvalues Computations, Vol.1, Theory* (Birkhäuser, Basel 1985)
[3] H. Grussbach, Doctor thesis: *Fraktale und Multifraktale im Anderson-Modell der Lokalisierung*, Universität Mainz (1995)
[4] B.B. Mandelbrot, *The Fractal Geometry of Nature* (Freemann, New York 1982)
[5] H. Aoki, J. Phys. C **16**, L205 (1983)
[6] M. Schreiber, Phys. Rev. B **31**, 6146 (1985)
[7] T.C. Halsey, M.H. Jensen, L.P. Kadanoff, I. Procaccia, and B.I. Shraiman, Phys. Rev. A **33**, 1141 (1986)
[8] W.H. Press, B.P. Flannery, S.A. Teukolski, and M.T. Vetterling, *Numerical Recipes* (Cambridge University Press, Cambridge 1989)
[9] J. Feder, *Fractals* (Plenum, New York 1988)
[10] M. Schreiber, Physica A **167**, 188 (1990)
[11] C.M. Soukoulis and E.N. Economou, Phys. Rev. Lett. **52**, 565 (1984)
[12] T.M. Chang, J. Bauer, and J.L. Skinner, J. Chem. Phys. **93**, 8973 (1990)
[13] S. Evangelou, J. Phys. A **23**, L317 (1990)
[14] Y. Ono, T. Ohtsuki, and B. Kramer, in *High Magnetic Fields in Semiconductor Physics III*, ed. G. Landwehr, Springer Series in Solid-State Sciences 101, (Springer, Berlin 1992), p. 60
[15] S. Evangelou, Physica A **167**, 199 (1990)
[16] M. Schreiber and H. Grussbach, Phys. Rev. Lett. **67**, 607 (1991)
[17] B. Huckestein and L. Schweitzer, in *High Magnetic Fields in Semiconductor Physics III*, ed. G. Landwehr, Springer Series in Solid-State Sciences 101, (Springer, Berlin 1992), p. 84
[18] W. Pook and M. Janssen, Z. Phys. B **82**, 295 (1991)
[19] M. Schreiber and H. Grussbach, Mod. Phys. Lett. B **6**, 851 (1992)
[20] H. Grussbach and M. Schreiber, Phys. Rev. B **51**, 663 (1995)
[21] F. Wegner, Nucl. Phys. B **316**, 663 (1989)
[22] M. Schreiber and H. Grussbach, J. Fractals **1**, 1037 (1993)
[23] H. Grussbach and M. Schreiber, Phys. Rev. B **48**, 6650 (1993)

Transfer-Matrix Methods and Finite-Size Scaling for Disordered Systems*

Bernhard Kramer[1] and Michael Schreiber[2]

[1] I. Institut für Theoretische Physik, Universität Hamburg, Jungiusstrasse 9,
D-20355 Hamburg, Germany, e-mail: `kramer@physnet.uni-hamburg.de`
[2] Institut für Physik, Technische Universität, D-09107 Chemnitz, Germany,
e-mail: `schreiber@physik.tu-chemnitz.de`

Abstract. In this chapter we introduce recursive scaling methods for the calculation of electronic properties of disordered systems. First we introduce in some detail the basic physical concept by considering the one-dimensional limit. Then, we describe the recursive procedure and demonstrate the finite-size scaling method for analyzing the data. Finally, we briefly describe the status of the results for the disorder-induced metal–insulator transition.

1 Introduction

Real solids, such as metals, alloys, glasses, and doped semiconductors, always contain a certain degree of disorder induced by impurities, defects, and/or dislocations. The arrangement of the atomic constituents is far from being ideally ordered. Their physical properties are therefore to a considerable extent determined by "randomness". The understanding of the latter is of immense practical importance. Detailed knowledge of material properties in general, the properties of thin films, especially electrical and magnetic, is an unavoidable ingredient of everyday technology in practically all areas of our modern civilization. Its most recent achievement, communication technology, is hard to imagine without understanding the transport behavior of electrons in the disordered potential landscape of doped semiconductors.

Methods that are borrowed from the theory of the random spectra of heavy atomic nuclei can also be used to treat certain questions of a fundamental nature in connection with the more or less random electronic spectra of heavy atoms and large molecules, and of noninteracting disordered systems. As all of these are basically dominated by the laws of quantum physics, the *motion of a particle in a random potential* can be considered as one of the paradigmatic quantum problems which is both of outstanding practical *and* fundamental importance. The most striking example is an electron moving in a random potential. The understanding of its quantum properties yields not only the key to the understanding of the metallic and insulating nature of matter. It is also a crucial ingredient for modern electronic devices. In this chapter we concentrate on this example. We

---

* Software included on the accompanying diskette.

will especially deal with the phenomenon of *localization* which is closely related to the electronic transport properties [1].

Essentially, we have to solve the stationary Schrödinger equation for a single particle,

$$-\frac{1}{2m}\nabla^2\psi(x) + V(x)\,\psi(x) = E\,\psi(x) \ . \tag{1}$$

Here, $V(x)$ is the random potential energy, m the (effective) mass of the particle, and E the energy of the stationary wave function $\psi(x)$. For numerical purposes, the lattice equivalent of (1) is more convenient,

$$\sum_{n'} t_{nn'}\,\psi_{n'} + \varepsilon_n\,\psi_n = E\,\psi_n \ . \tag{2}$$

As discussed in this volume in the chapter by Schreiber, the first term reflects the kinetic energy in (1), the second comprises the potential energy. The matrix elements $t_{nn'}$ describe the transitions of the particle between the lattice sites, ε_n are the site energies. For the sake of simplicity, we assume in the following that randomness is incorporated only in the latter. They are chosen independently at random according to some distribution $p(\varepsilon)$ with a vanishing mean and a variance σ^2.

Qualitatively, the properties of such a model can easily be understood. When the variance of the potential vanishes the Schrödinger equation describes an ordered lattice. The energy eigenvalues form bands of finite widths, the corresponding eigenstates are Bloch states. They extend throughout the whole infinite lattice. An electron in such a state – having an energy E and a momentum k – is spread across the infinite system. Such an electron will be highly mobile, and contribute to the transport of charge. For nonzero fluctuations of the potential, the bands of energy eigenvalues are broadened. Near the band edges, states are formed which are spatially localized within finite regions of accidental potential wells. Electrons in these localized states are spatially confined. They "cannot move" and do not contribute to charge transport. One of the questions to ask in this context is what the precise conditions are under which localized and extended states occur in a random system. Another one is how physical quantities behave near the localization–delocalization transition. The *theory of localization* deals with these questions. The recursive numerical scaling methods to be described below can provide quantitative answers.

In the next section we treat the example of a one-dimensional (1D) disordered lattice model in some detail [2, 3, 4]. The results provide the basic ingredients for the numerical procedures.

2 One-Dimensional Systems

The quantum mechanics of an electron in a 1D random potential constitutes the paradigm of localization theory. It can be shown in a mathematically rigorous way that all eigenstates of the Hamiltonian are exponentially localized in the thermodynamic limit.

2.1   The Transfer Matrix

For simplicity, we assume that the kinetic energy term in (2) contains only hopping processes between nearest neighbors. We choose $t = 1$, fixing the energy scale. Then the Schrödinger equation may be rewritten as

$$\psi_{n+1} = (E - \varepsilon_n)\psi_n - \psi_{n-1} \ . \tag{3}$$

Together with $\psi_n = \psi_n$ this can be cast into the matrix notation

$$\begin{pmatrix} \psi_{n+1} \\ \psi_n \end{pmatrix} = \begin{pmatrix} E - \varepsilon_n & -1 \\ 1 & 0 \end{pmatrix} \begin{pmatrix} \psi_n \\ \psi_{n-1} \end{pmatrix} \equiv \mathbf{T}_n \begin{pmatrix} \psi_n \\ \psi_{n-1} \end{pmatrix} \ . \tag{4}$$

The random matrices $\mathbf{T}_n$ *transfer* amplitudes between the sites $n, n-1$ and $n+1, n$. Equation (4) may be solved recursively for arbitrary initial conditions ψ_1 and ψ_0 . The amplitudes at $N+1, N$ are

$$\begin{pmatrix} \psi_{N+1} \\ \psi_N \end{pmatrix} = \mathbf{T}_N \mathbf{T}_{N-1} \cdots \mathbf{T}_1 \begin{pmatrix} \psi_1 \\ \psi_0 \end{pmatrix} \equiv \mathcal{T}_N \begin{pmatrix} \psi_1 \\ \psi_0 \end{pmatrix} \ . \tag{5}$$

The calculation of the amplitudes is thus equivalent to the calculation of the product of random matrices,

$$\mathcal{T}_N = \prod_{n=1}^{N} \mathbf{T}_n \ . \tag{6}$$

The important point now is that there exists a mathematical theorem due to Oseledec [5] which guarantees the existence of the product in the thermodynamic limit,

$$\Gamma \equiv \lim_{N \to \infty} \left(\mathcal{T}_N \mathcal{T}_N^\dagger \right)^{1/2N} \ , \tag{7}$$

with eigenvalues $e^{\gamma_m} < \infty$ $(m = 0, 1)$ where γ_m are the Lyapunov exponents. Physically, the theorem of Oseledec means that when starting from one end of the system, the amplitudes can be written as a superposition of exponentially increasing and decreasing functions, since $\gamma_0 = -\gamma_1$ due to the symplecticity[1] of the matrix (6). For definiteness we select the label so that $\gamma_0 < 0 < \gamma_1$. For large distances from the origin, the behavior of the amplitude will be dominated by the exponentially increasing component.

We will now argue that γ_1 determines the exponential decay of the *eigenstates* such that the localization lengths may be defined by

$$\lambda \equiv -\gamma_0^{-1} \equiv \gamma_1^{-1} \ . \tag{8}$$

The exponentially increasing states obtained from the transfer-matrix equation for an arbitrary energy are in general not eigenstates. The latter may in principle

---

[1] A matrix $\mathbf{A}$ is called symplectic if $\mathbf{A}\mathbf{J}\mathbf{A}^T = \mathbf{J}$ with $\mathbf{J} = \begin{pmatrix} 0 & -1 \\ 1 & 0 \end{pmatrix}$, where $\mathbf{0}$ and $\mathbf{1}$ are zero and unity matrices. Then the eigenvalues of $\mathbf{A}\mathbf{A}^T$ occur in pairs a_m and a_m^{-1}.

be constructed by starting the iteration (4) from both ends of the system, and matching the wave function and its derivative continuously at some site within the bulk. The eigenvalues E_i of the Schrödinger equation are given by those energies for which such a continuous matching is successful. It follows that the localization lengths of the eigenfunctions $\psi_i = (\psi_{i1}, \psi_{i2}, ..., \psi_{in}, ..., \psi_{iN})^{\mathrm{T}}$ are given by $\lambda(E_i)$.

In the limit of infinite system length the eigenvalues are densely distributed. Their mean distance decreases as N^{-1}. Therefore, it does not make any sense to consider individual eigenstates. Eventually, it will be the *average* of the localization lengths within a certain energy interval ΔE which determines the transport properties,

$$\lambda(E) \equiv \frac{1}{N_{\Delta E}} \sum_i \lambda(E_i) \ . \tag{9}$$

It is one of the key assumptions of localization theory that this average can be calculated by averaging $\gamma_1^{-1}(E)$ over an ensemble of macroscopically equivalent systems. In 1D, this *hypothesis of ergodicity* for the localization length can be proven.

2.2   The Ordered Limit

In order to demonstrate how the transfer-matrix equation works, we will now consider a trivial example, namely the limit of an ordered system ($\sigma = 0$) with periodic boundary condition, $\psi_{N+1} = \psi_1$. Unfortunately,

$$\mathbf{T}_n(E) = \begin{pmatrix} E & -1 \\ 1 & 0 \end{pmatrix} \equiv \mathbf{T} \tag{10}$$

is not diagonal. Diagonalization of the transfer matrix $\mathbf{T}$ yields the eigenvalues

$$\tau_{1,2} = \frac{E}{2} \pm \sqrt{\frac{E^2}{4} - 1} \ , \tag{11}$$

with some unitary matrix $\mathbf{U}$ which contains the corresponding eigenvectors. Then

$$\mathcal{T}_N = \mathbf{U} \begin{pmatrix} \tau_1^N & 0 \\ 0 & \tau_2^N \end{pmatrix} \mathbf{U}^{-1} \ , \tag{12}$$

and after multiplication with $\mathbf{U}^{-1}$ from the left one obtains for the amplitudes

$$\begin{pmatrix} a_{N+1} \\ a_N \end{pmatrix} = \mathbf{U}^{-1} \mathbf{T}^N \mathbf{U} \begin{pmatrix} a_1 \\ a_0 \end{pmatrix} = \begin{pmatrix} \tau_1^N & 0 \\ 0 & \tau_2^N \end{pmatrix} \begin{pmatrix} a_1 \\ a_0 \end{pmatrix} \ , \tag{13}$$

with

$$\begin{pmatrix} a_{N+1} \\ a_N \end{pmatrix} \equiv \mathbf{U}^{-1} \begin{pmatrix} \psi_{N+1} \\ \psi_N \end{pmatrix} \ . \tag{14}$$

Therefore

$$a_{N+1} = \tau_1^N a_1 \equiv a_1 e^{N\gamma} \ , \tag{15}$$

with $\gamma \equiv \log\left(E/2 + \sqrt{E^2/4 - 1}\right)$.

For $|E| > 2$ there is no solution possible which fulfills the above periodic boundary condition. When $|E| \leq 2$, then γ becomes complex, with $\mathrm{Re}\gamma = 0$ and $\mathrm{Im}\gamma = \mathrm{arctg}(\sqrt{4 - E^2}/E)$. Then

$$a_{N+1} = a_1 \exp\left(iN\mathrm{arctg}(\sqrt{4 - E^2}/E)\right) \equiv a_1 e^{ikN} \ . \tag{16}$$

This defines the energy eigenvalues

$$E(k) = 2\cos k \ , \tag{17}$$

which correspond to the Bloch states $\psi_n(k) \propto e^{ikn}$. On the other hand, because of the boundary condition, $ikN = 2\pi i\kappa$, such that $k = 2\pi\kappa/N$ with $\kappa = 0, \pm 1, \pm 2, \dots$. The "allowed" range of k values is therefore given by the 1D Brillouin zone $-\pi < k \leq \pi$.

2.3 The Localization Length

Now, we consider the case of a disordered chain, i.e. $\sigma \neq 0$. We cannot, obviously, apply the same method as in the above example, since the eigenvalues

$$\tau_{1,2}^{(n)} = \frac{E - \varepsilon_n}{2} \pm \sqrt{\frac{(E - \varepsilon_n)^2}{4} - 1} \tag{18}$$

depend on the lattice site, as do also the unitary matrices $\mathbf{U} = \mathbf{U}_n$. How can we proceed in this case?

Since we are interested in the average exponential increase (or decrease) of the wave function, it is natural to ask for the dependence of $\overline{|\psi_N|^2}$ on N. Here, $\overline{\cdots}$ denotes the average with respect to the realizations of the disorder,

$$\overline{A} \equiv \prod_{n=1}^{N} \int d\varepsilon_n \, p(\varepsilon_n) \, A(\varepsilon_1 \cdots \varepsilon_N) \ , \tag{19}$$

where we make use of the assumption that the distributions $p(\varepsilon_n)$ of the site energies do not depened on the other sites as mentioned in the introduction.

For a localized state, the modulus of the amplitude should decrease exponentially, or equivalently – if ψ is not an eigenstate – increase exponentially. Formally, it is necessary to consider

$$|\psi_{N+1}|^2 + |\psi_N|^2 = (\psi_1^* \psi_0^*) \, T_N^\dagger T_N \begin{pmatrix} \psi_1 \\ \psi_0 \end{pmatrix} \ , \tag{20}$$

in order to avoid the effect of accidental vanishing of the amplitude on two successive sites. We use the initial conditions $\psi_1 = 1$, $\psi_0 = 0$. Then

$$|\psi_{N+1}|^2 + |\psi_N|^2 = \left(T_N^\dagger T_N\right)_{11} \ . \tag{21}$$

The task is now to compute this matrix element as a function of N. First, we note that

$$T_N^\dagger T_N = \mathbf{T}_1^\dagger \tilde{T}_{N-1}^\dagger \tilde{T}_{N-1} \mathbf{T}_1 \ . \tag{22}$$

For convenience we have slightly changed the notation here, and have abbreviated $\mathbf{T}_N \mathbf{T}_{N-1} \cdots \mathbf{T}_2 = \tilde{T}_{N-1}$. This yields a recursive equation for the matrix elements $T_{\nu\mu}(N)$ of $T_N^\dagger T_N$. For $E = 0$,

$$T_{11}(N) = \overline{\varepsilon_1^2} T_{11}(N-1) - \overline{\varepsilon_1}[T_{12}(N-1) + T_{12}(N-1)] + T_{22}(N-1) \ ,$$

$$T_{22}(N) = T_{11}(N-1) \ ,$$

$$T_{12}(N) = \overline{\varepsilon_1} T_{11}(N-1) - T_{21}(N-1) \ , \tag{23}$$

$$T_{21}(N) = \overline{\varepsilon_1} T_{11}(N-1) - T_{12}(N-1) \ ,$$

since the random matrix elements $T_{\nu\mu}(N-1)$ are statistically independent of ε_1. For $T_{11}(N)$, this yields

$$T_{11}(N) = \overline{\varepsilon_1^2} T_{11}(N-1) + T_{11}(N-2) \ , \tag{24}$$

because the average of ε_1 vanishes. One obtains eventually, by rearranging suitably,

$$\frac{\overline{\varepsilon_1^2}}{2} = \frac{T_{11}(N) - T_{11}(N-2)}{2T_{11}(N-1)} \equiv \Delta\big(\log T_{11}(N-1)\big) \ . \tag{25}$$

When the disorder is small, such that the localization length is large ($\lambda \gg 1$), we can replace the finite difference Δ by the total differential

$$d\log T_{11}(N) \approx \frac{\sigma^2}{2} dN \ , \qquad (\sigma \to 0) \ , \tag{26}$$

where $\sigma^2 = \overline{\varepsilon_1^2}$. By integration and exponentiation, this yields

$$T_{11}(N) = T_{11}(0) e^{\sigma^2 N/2} \equiv e^{2N/\lambda} \ . \tag{27}$$

On average, the wave function increases exponentially with N.

For a box distribution of the random site energies with width W we have $\sigma^2 = W^2/12$, such that the localization length

$$\lambda \equiv \frac{48}{W^2} \tag{28}$$

is finite for any finite disorder, and diverges at $W = 0$ with a *critical exponent* $\nu = 2$. The prefactor in this expression, when calculated more carefully, turns out to be 105 [8], instead of 48. This is a consequence of pecularities in the statistics of the localization properties [9], and a singularity at $E = 0$ of the model [10] that have to be treated more carefully than the somewhat hand-waving argument used above.

2.4   Resolvent Method

We will now discuss a recursive method which is slightly more general than the above transfer-matrix method. It allows us not only to calculate the localization length, but also other electronic properties, i.e., the density of energy levels and the conductivity.

The solutions of the Schrödinger equation, E_i, $|i\rangle$, are contained in the resolvent operator $G(z)$ [11] defined by

$$G(z)(z - H) = 1 , \tag{29}$$

where $z \equiv E + i\eta$ is the complex energy variable. The spectral representation of the resolvent is

$$G(z) = \sum_i \frac{|i\rangle\langle i|}{z - E_i} . \tag{30}$$

In the representation of the site states,

$$G_{jk} \equiv \langle j|G(z)|k\rangle = \sum_i \frac{\langle j|i\rangle\langle i|k\rangle}{z - E_i} \tag{31}$$

is the (time-integrated) *probability amplitude* for an electron with energy z to hop from the site j to the site k.

The energy eigenvalues of H correspond to the poles of G while the residues are given by the projection operators onto the basis states. By decomposing $H = H_0 + V$ and expanding it in powers of $V/(z - H_0)$, we obtain the resolvent equation

$$G(z) = G_0(z) + G_0(z)VG(z) , \tag{32}$$

with the resolvent of the *unperturbed* system, $G_0(z) = (z - H_0)^{-1}$.

If all of the wave functions with energies near E are localized, say, $\langle j|i\rangle = f_i(j)\exp\left(-|j - j_i|/\lambda(E_i)\right)$ with a random phase factor $f_i(j)$ and localization centre j_i, then the transition probability between site 1 and site N of the system will be exponentially decreasing,

$$|G_{1N}|^2 \propto e^{-2N/\lambda(E)} . \tag{33}$$

The localization length may thus be obtained from the limiting behavior of G_{1N}.

In order to calculate G_{1N} we decompose the Hamiltonian for the system consisting of N sites,

$$H^N = H^{N-1} + \varepsilon_N|N\rangle\langle N| + V|N\rangle\langle N - 1| + V^\dagger|N - 1\rangle\langle N| \tag{34}$$

$$\equiv H_0 + H_1 ,$$

with $H_1 \equiv V|N\rangle\langle N - 1| + V^\dagger|N - 1\rangle\langle N|)$ comprising the coupling between the $(N-1)$th and the Nth site. The corresponding resolvent equation

$$G_{jk}^N = G_{0jk}^N + \sum_{j'k'} G_{0jj'}^N V_{j'k'}^N G_{k'k}^N , \tag{35}$$

with

$$G_0^N \equiv G^{N-1} + \frac{|N\rangle\langle N|}{z - \varepsilon_N} \; , \tag{36}$$

may be cast into a set of recursive equations

$$G_{jk}^N = G_{jk}^{N-1} + G_{jN-1}^{N-1} V G_{Nk}^N \; ,$$

$$G_{jN}^N = G_{jN-1}^{N-1} V G_{NN}^N \; ,$$

$$G_{Nk}^N = (z - \varepsilon_N)^{-1} V^\dagger G_{Nk}^N \; , \tag{37}$$

$$G_{NN}^N = (z - \varepsilon_N)^{-1}(1 + V^\dagger G_{N-1N}^N) \; .$$

This may be solved for $g_N \equiv G_{NN}^N$,

$$g_N = \frac{1}{z - \varepsilon_N - V^\dagger g_{N-1} V} \; . \tag{38}$$

This is the key equation for the recursive calculation of electronic properties. We demonstrate this for two examples. It is easy to see from (37) that

$$G_{1N}^N = G_{1N-1}^{N-1} V g_N = g_1 \prod_{j=2}^{N} V g_j \; , \tag{39}$$

such that we obtain for the localization length

$$\lambda(E) \equiv \lim_{N \to \infty} \frac{1}{\gamma^{(N)}} \tag{40}$$

the recursive equation

$$\gamma^{(N)} = \frac{N-2}{N-1} \gamma^{(N-1)} - \frac{1}{N-1} \log |g_N| \tag{41}$$

or

$$\frac{1}{\lambda(E)} = - \lim_{N \to \infty} \frac{1}{N-1} \log |G_{1N}^N(E)| \; . \tag{42}$$

This is completely equivalent to our earlier definition,[2] cf. (7,20). Although the above recursive procedure will lead to a numerical problem, since after relatively few iterations G_{1N} becomes exponentially small,[3] it can nevertheless be successfully applied even to very large systems $N = \mathcal{O}(10^{10})$, since the theorem of Oseledec guarantees convergence, and can even be used to construct a prescription of how to estimate the statistical accuracy of the result as a function of the length of the system [7].

---

[2] As an exercise, show that $a_N^{-1} = G_{1N}^N$. For a solution, see [7].
[3] How can the numerical instability related with the exponential decrease of G_{1N} be avoided? For a solution, see [7].

A similar, though considerably more complicated procedure may be constructed for the average linear conductivity as given by the Kubo formula [12],

$$\sigma = \lim_{\eta \to 0} \lim_{N \to \infty} \frac{2e^2\eta^2}{\pi\hbar N^2} \sum_{jk} (k-j)^2 |G_{jk}^N(E)|^2 \ , \tag{43}$$

which can be cast into the form [1]

$$\sigma_N(E,W,\eta) = -\frac{4e^2}{\pi\hbar N} \left(\eta^2 \sum_{jk} jk G_{jk}^N (G_{kj}^N)^* + \eta \sum_j j^2 \mathrm{Im} G_{jj}^N \right) \tag{44}$$

by using homogeneity of the system, i.e., that the configurational average of the absolute square of the Green's function at the right hand side of (43) depends only on the distance between j and k.

Using (37) we obtain the nonlinear set of recursive equations for the conductivity

$$\sigma_N \equiv \frac{4e^2}{\pi\hbar N} s_N \ , \tag{45}$$

with

$$s_N = s_{N-1} + \mathrm{Re}\left(g_N \left[b_{N-1} - g_N^* |d_{N-1}|^2 - 2\eta g_N^* d_{N-1} N + (i\eta - \eta^2 g_N^*) N^2 \right] \right) \ ,$$

$$d_N = |g_N|^2 (d_{N-1} + \eta N) \ , \tag{46}$$

$$b_N = g_N^2 \left[b_{N-1} - 2g_N^* |d_{N-1}|^2 + (i\eta - 2\eta^2 g_N^*) N^2 - 4\eta d_{N-1} g_N N \right] \ .$$

This can be used for a numerical determination of the conductivity. The results for $E = 0$ are consistent with a scaling law [12]

$$\sigma(E = 0, W, \eta) = \frac{1}{W^2} \sigma\left(0, \frac{\eta}{W^2}, 1 \right) \ , \tag{47}$$

which can be used to extrapolate the conductivity to its value for $\eta \to 0$ from the numerically obtained data,

$$\lim_{\eta \to 0} \sigma(0, W, \eta) \propto \lim_{\eta \to 0} \eta = 0 \ . \tag{48}$$

If we identify $W^{-2} \propto \lambda^{-1}$, $\eta \propto \tau_\varphi \propto \ell_\varphi$, where τ_φ and ℓ_φ are a "phase-coherence time" and the corresponding "phase-coherence length", respectively, we obtain

$$\sigma(0, \lambda, \ell_\varphi) = \lambda \, \sigma\left(0, 1, \frac{\lambda}{\ell_\varphi} \right) \ . \tag{49}$$

When the temperature goes to zero, the phase-coherence length will extrapolate to infinity, and the conductivity to zero, since phase-breaking processes, such as electron–phonon scattering, will be frozen out.

3   Finite-Size Scaling

Both of the above methods can be generalized to dimensions $d > 1$, by considering quasi-1D systems of cross-sectional area M^{d-1} and length $N(\gg M)$. One has to replace ε_N, V, and g_N by matrices of dimension M^{d-1}. The transfer-matrix equation (4) becomes a set of $2M^{d-1}$ equations. The set (37) becomes then a set of recursive matrix equations of the dimension M^{d-1}. As a result, one obtains M^{d-1} localization lengths which correspond to the Lyapunov exponents.[4] The smallest of them defines the localization length

$$\lambda_M \equiv \frac{1}{\min_{m=1\ldots M} |\gamma_m|} , \tag{50}$$

since it eventually determines the exponential decay of the wave functions.

Being interested in the d-dimensional infinite system we have to extrapolate the localization length for $M \to \infty$, a task which is not so easy to perform, given the usual computational restrictions. Using the above procedures, one can, for instance, determine data for M smaller than, say, 20 in 3D, at best.

It is therefore imperative to search for scaling laws to be fulfilled by the data, in order to make reliable extrapolations possible. We have already encountered an example of such a scaling law in the previous section.

As another example, let us consider the classical conductance of a metallic cube of the size L^d at zero temperature. The residual conductivity σ is in this case a finite material parameter that characterizes the electrical transport behavior. The conductance is

$$\mathcal{G}(L, W) = \sigma L^{d-2} , \tag{51}$$

where σ depends on the disorder, $\sigma \propto W^{-2}$. Depending on the dimensionality, the conductance behaves therefore according to a *one-parameter scaling law*

$$\mathcal{G}(L, \sigma) = f_\infty(L/\xi) , \tag{52}$$

with $\xi = \sigma^{2-d}$. For insulators,

$$\mathcal{G}(L, W) \propto |G_{1N}|^2 \propto e^{-2L/\lambda(W)} = f_0(L/\lambda) . \tag{53}$$

In the following we want to generalize this one-parameter scaling idea. Let $\mathbf{a}(L) \equiv (a_1(L), a_2(L), \cdots, a_M(L))$ be a vector with components that describe certain properties of the system. The dimension M of this vector can also be infinite, and instead of the vector we can also consider a (continuous) function $a(\mu, L)$. How does $\mathbf{a}$ behave under scale transformations

$$L' = bL , \tag{54}$$

---

[4] Since the transfer matrices are symplectic, the Lyapunov exponents come in pairs, γ_m and $-\gamma_m$. Therefore, only M^{d-1} of the $2M^{d-1}$ exponents are significant. They depend on the cross-sectional diameter of the system, and can always be arranged such that $0 < \gamma_m < \gamma_{m+1}$.

with b a real number? Then we can certainly assume the general relation

$$\mathbf{a}(bL) \equiv F\big(\mathbf{a}(L), b\big) \ . \tag{55}$$

More restrictive is the following assumption,

$$\frac{d\mathbf{a}}{d\log L} \equiv \mathbf{f}\big(\mathbf{a}(L)\big) \equiv \lim_{\Delta L \to 0} \frac{F\big(\mathbf{a}(L), 1 + \Delta L/L\big) - F\big(\mathbf{a}(L), 1\big)}{\Delta L/L} \ . \tag{56}$$

Let us further assume that there exists a *fixpoint* $\mathbf{f}(\mathbf{a}^*) = 0$, and $\mathbf{a} \approx \mathbf{a}^*$. By expanding into powers of $\mathbf{a} - \mathbf{a}^*$, one obtains a first-order linear differential equation for $\mathbf{a}(L)$ which can be solved,

$$\mathbf{a}(L) = \mathbf{a}^* + \sum_m \left(\frac{L}{\xi_m}\right)^{f'_m} \varphi_m \ . \tag{57}$$

Here, $\mathbf{f}' \cdot \varphi_m = f'_m \varphi_m$, with the Jacobian matrix $\mathbf{f}'$ of the partial derivatives of $\mathbf{f}$, and its eigenvalues f'_m and eigenvectors φ_m. On the other hand, when $\mathbf{a}$ is a function of an additional parameter, say $\mathbf{a}(L) \equiv \mathbf{a}(L, W)$, one can expand

$$\mathbf{a}(L, W) = \mathbf{a}^* + \left.\frac{d\mathbf{a}(L, W)}{dW}\right|_{W^*} (W - W^*) \ . \tag{58}$$

By comparing (57) and (58), one obtains

$$\xi_m(W) = \frac{1}{(W - W^*)^{1/f'_m}} \ , \tag{59}$$

such that

$$\mathbf{a}(L, W) = \mathbf{a}^* + \sum_m L^{f'_m} \varphi_m (W - W^*) \ . \tag{60}$$

For large L the largest positive f'_m dominates the behavior near the fixpoint. The components of $\mathbf{a}$ that correspond to positive (negative) f'_m are denoted *relevant (irrelevant) scaling variables*. The largest f'_m defines the *critical exponent*, $\nu \equiv 1/\max f'_m$.

Let us consider two instructive examples. For the above-mentioned conductance we can define the β *function*

$$\frac{d\log \mathcal{G}}{d\log L} \equiv \beta(\log \mathcal{G}) \ . \tag{61}$$

In the metallic limit $\beta = d - 2$, whereas in the insulating region $\beta = \log \mathcal{G}$. Assuming that β is continuous and monotonically increasing, and recalling that in the metallic regime the *maximally crossed* diagrams yield a negative quantum correction to the conductance [13], one arrives at the conclusion that a disorder-induced metal–insulator transition (the so-called Anderson transition) can only occur in 3D. For $d \leq 2$ no genuine metallic behavior can exist if interaction effects are ignored. The critical exponent is given by the inverse of the derivative of β at the critical point defined by $\beta(\log \mathcal{G}^*) = 0$.

As mentioned above, the set of M^{d-1} Lyapunov exponents $\gamma_m(M, W)$ determines the behavior of the states of a quasi-1D system. We define renormalized dimensionless Lyapunov exponents by $\tilde{\gamma}_m \equiv M\gamma_m$. If $\tilde{\gamma}_1 \overset{M\to\infty}{\longrightarrow} \infty$, the system is insulating, if $\tilde{\gamma}_1 \overset{M\to\infty}{\longrightarrow} 0$ then it is metallic. The critical point W^* is given by $\lim \tilde{\gamma}_1 > 0$.

4 Numerical Evaluation of the Anderson Transition

4.1 Localization Length of Quasi-1D Systems

As an example for the application of the methods described in the previous section we consider a strip of $M \times N$ sites. If $N \gg M$, this a quasi-1D system. The length N will be increased in the calculation until a preset accuracy is achieved. The width M has to be relatively small due to the available computer power. But as mentioned in the previous section an extrapolation towards infinite width will be possible by finite-size scaling so that 2D systems can be described.

As in the 1D case the Schrödinger equation (2) can be rewritten as a recursion. In analogy to (4) it reads

$$\begin{pmatrix} \boldsymbol{\psi}_{n+1} \\ \boldsymbol{\psi}_n \end{pmatrix} = \begin{pmatrix} E\mathbf{1} - \mathbf{H}_n & -\mathbf{1} \\ \mathbf{1} & \mathbf{0} \end{pmatrix} \begin{pmatrix} \boldsymbol{\psi}_n \\ \boldsymbol{\psi}_{n-1} \end{pmatrix} = \mathbf{T}_n \begin{pmatrix} \boldsymbol{\psi}_n \\ \boldsymbol{\psi}_{n-1} \end{pmatrix} . \tag{62}$$

But in this case the transfer matrices $\mathbf{T}_n$ transfer the amplitudes of the wave function between all the M sites of the cross-sectional chains $n, n-1$ and $n+1, n$. There are M linearly independent ways to distribute the amplitude of the initial state on the M sites of the first chain. These possibilities can be simultaneously treated if the respective M vectors of expansion coefficients are comprised in an amplitude matrix $\boldsymbol{\psi}$ of size $M \times M$. The initial condition can then be expressed by $\boldsymbol{\psi}_1 = \mathbf{1}$ and $\boldsymbol{\psi}_0 = \mathbf{0}$, where $\mathbf{1}$ and $\mathbf{0}$ are $M \times M$ unity and zero matrices, respectively.

The $2M \times 2M$ transfer matrix $\mathbf{T}_n$ in (62) is the respective generalization of the transfer matrix in (4): the unity matrices in $\mathbf{T}_n$ transfer the particle (or rather amplitude of its wave function) between neighboring chains. The Hamiltonian matrix $\mathbf{H}_n$ comprises the random site energies ε_{nm} of the sites $m = 1, ..., M$ in the nth chain as well as the matrix elements $t(= 1)$ describing the transition of the particle within the chain. $\mathbf{H}_n$ is given by

$$\mathbf{H}_n = \begin{pmatrix} \varepsilon_{n1} & 1 & 0 & 0 & 1 \\ 1 & \varepsilon_{n2} & 1 & 0 & 0 \\ 0 & 1 & \varepsilon_{n3} & 1 & 0 \\ 0 & 0 & 1 & \varepsilon_{n4} & 1 \\ 1 & 0 & 0 & 1 & \varepsilon_{n5} \end{pmatrix} \tag{63}$$

for a chain of $M = 5$ sites. Here the nonzero matrix elements in the upper right and lower left corner of the matrix reflect the periodic boundary condition. We note that (63) is exactly equivalent to the Hamiltonian matrix for a 1D system as constructed in Schreiber's chapter of this book.

As in the 1D case, the product matrix $\mathcal{T}_N$ of the transfer matrices satisfies Oseledec's theorem, i.e., the limiting matrix Γ as given by (7) exists. Since $\mathbf{T}_n$ and thus $\mathcal{T}_N$ are symplectic matrices the eigenvalues of Γ occur pairwise as $e^{-\gamma_m}$ and $e^{+\gamma_m}$, where $\gamma_m (> 0)$ and $-\gamma_m$ are the Lyapunov exponents describing the exponentially increasing and decreasing amplitudes along the quasi-1D system. These Lyapunov exponents characterize how the amplitudes of the initial states which are described by the expansion coefficients in ψ_1 "drift apart" exponentially. The inverse values reflect the different characteristic length scales. The largest length (i.e., the inverse of the smallest positive Lyapunov exponent) describes the weakest possible decay of the transmission probability along the quasi-1D system for a state at a given energy. This length is commonly associated with the localization length, implicitly assuming that the electronic states are exponentially localized. We shall denote this length by $\lambda_M = \gamma_1^{-1}(M)$.

In principle the eigenvalues can be determined by the repeated multiplication of the transfer matrices onto an arbitrary initial vector. This product converges towards the largest eigenvalue of $\mathcal{T}_N$ times its eigenvector. That eigenvalue, however, reflects the fastest exponential increase. But as the localization length λ_M is given by the slowest exponential increase, we have to determine that eigenvalue of $\mathcal{T}_N$ which is closest to unity, or, equivalently, the smallest (positive) Lyapunov exponent $\gamma_1(M)$. This can be achieved by the repeated multiplication of the transfer matrices $\mathbf{T}_n$ onto M orthogonal initial amplitude vectors, which will eventually provide all the Lyapunov exponents if the orthogonalization of the amplitude vectors is maintained.

This determination of the Lyapunov exponents from the asymptotic behavior of the recursion (4) is conceptually slightly different from the diagonalization of the product matrix (6). However, due to the initial values $\psi_1 = \mathbf{1}$ and $\psi_0 = \mathbf{0}$ the left-hand side of the 2D analogon of (5) is identical with the first M columns of the product matrix. Due to the symplecticity of the product matrix the knowledge of these columns is sufficient for the determination of all Lyapunov exponents.

But the numerical problem is that the ratio of the smallest to the largest eigenvalue of $\mathcal{T}_N$ becomes comparable with the machine accuracy after very few multiplications. This means that the smallest eigenvalue would be lost very soon. These convergence problems are similar to those mentioned in Sect. 2.4 with respect to the calculation of the localization length from the resolvent according to (42). To circumvent this problem one has to reorthonormalize the left-hand side of (62) regularly during the procedure. To be specific, we orthogonalize each of the M columns of the $M \times 2M$ vector onto the previous columns, i.e., we perform the standard Gram-Schmidt orthogonalization procedure. Then the first column converges towards the eigenvector corresponding to the largest eigenvalue, the second column to the second largest, and so on. The normalization of the eigenvectors yields asymptotically the respective eigenvalues. In practice, it is sufficient to perform the reorthonormalization after every 10 steps of the iteration (62). The Mth normalization constants of all these Gram-Schmidt procedures have to be multiplied to determine the overall normalization of the Mth eigenvector and thus the eigenvalue closest to unity. In practice, the logarithms are summed yielding the smallest Lyapunov exponent and thus the inverse localization length. This sum can also be interpreted as an average over the respective Lyapunov exponents of many short strips of length 10. Accordingly, the fluctuations of these data can be used to determine the statistical error of the result in a straightforward way by computing the variance.

The FORTRAN program loc2d1.for on the enclosed diskette[5] performs the recursion (62) for a quasi-1D strip. Due to the very simple structure of the transfer matrices, the matrix–matrix multiplication in (62) can be programmed in a very efficient way, avoiding all multiplications with 1 and all summations of 0. The recursion (62) is performed until the requested relative accuracy is achieved unless the preset maximum number of allowed recursion steps is reached earlier. The data for the localization length and its variance are written into the file lambda_N.dat and can be visualized with the gnuplot program lambda_N.gnu. A typical example is presented in Fig. 1, demonstrating the large fluctuations and the slow convergence of the transfer-matrix procedure.

4.2 Dependence of the Localization Length on the Cross Section

A second version of the program, loc2d2.for, is provided for calculating the converged values of the localization length for various input parameters. For each combination of energy and disorder the results are stored in a separate file lambda##.dat, where ## denotes a consecutive number. These data and their dependence on the width M of the quasi-1D strip can then be visualized by means of the gnuplot program lambda_M.gnu. In this way the dependence of the localization length on the various parameters can be investigated. The reader is encouraged to try different combinations, but should be aware that the localization length for small values of the disorder may become very large so that a reasonable accuracy can be achieved only for very long strips. For zero disorder the energy eigenvalues of the Schrödinger equation are limited to $|E| \leq 4$, which follows from the 2D version of (17). Therefore reasonable values for the energy parameter are given by $|E| \leq 4 + W/2$. The width M of the strip should not be chosen too small to avoid strong finite-size effects. The available computer power will limit the possible values for the width. The arrays in the program allow values up to $M = 64$. Of course, this limit as well as the preset accuracy of 1% can easily be changed. We have estimated that the necessary computation time increases proportional to the inverse square of the relative accuracy and the fifth power of the strip width in 2D systems.

---

[5] For copyright reasons we have not included a random-number generator with the program. Any generator will be sufficient, e. g. the very simple one described in Stauffer's chapter of this book.

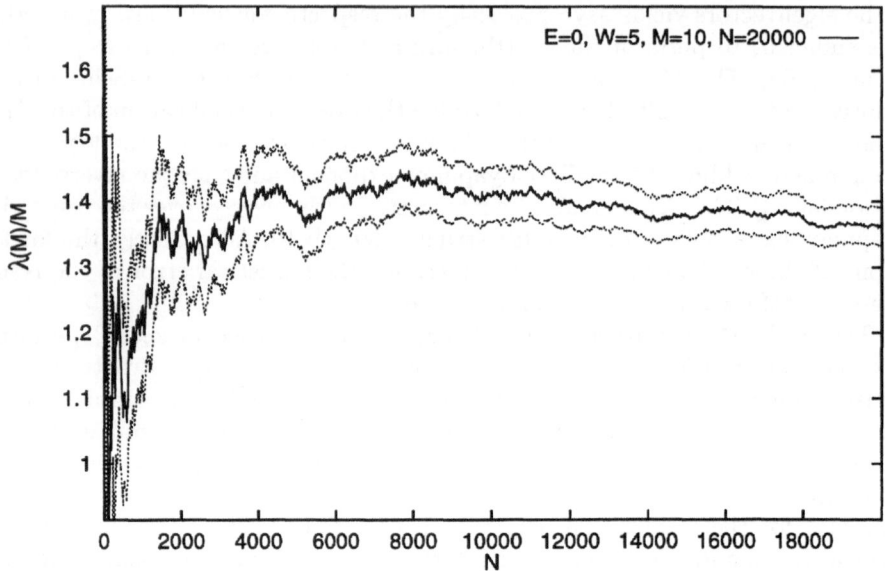

Fig. 1. Dependence of the localization length λ_M of a quasi-1D strip on the length N of the system for energy $E = 0$ and disorder $W = 5$. The cross section is a linear chain of $M = 10$ sites with periodic boundary condition. The *thick line* reflects the localization length, the *thin lines* indicate the accuracy of the data determined from the variance as discussed in the text. At the end of this calculation ($N = 20000$) the achieved accuracy was 2.6%

It is also instructive to vary the number of recursion steps which are performed between subsequent reorthonormalizations. If it is chosen smaller, the computer time increases. If it is chosen larger, wrong results are calculated. This can be easily verified by comparing various runs of the program `loc2d1.for`.

As discussed in the previous section no genuine metallic behavior can exist in 2D systems if interaction effects or magnetic fields are ignored. In 3D samples metallic behavior is expected for small energy and small disorder, while insulating behavior is expected for large energy and/or large disorder. Of course, it is most interesting to investigate the metal–insulator transition which separates these two regimes. For this purpose, it is straightforward to generalize (62) to 3D systems. Now the cross section will be a slice consisting of a square lattice of $M \times M$ sites. There are M^2 linearly independent possibilities for distributing the amplitudes on the first slice so that ψ_n becomes a $M^2 \times M^2$ matrix. The respective Hamiltonian matrix $\mathbf{H}_n$ for the cross section is the 2D analogon of (63). It is explicitly constructed in Schreiber's chapter in this book.

However, the calculation of a sufficiently large data set of localization lengths for different parameter values in 3D is too time consuming for simple exercises. Therefore we have included respective data sets on the diskette for 24 different

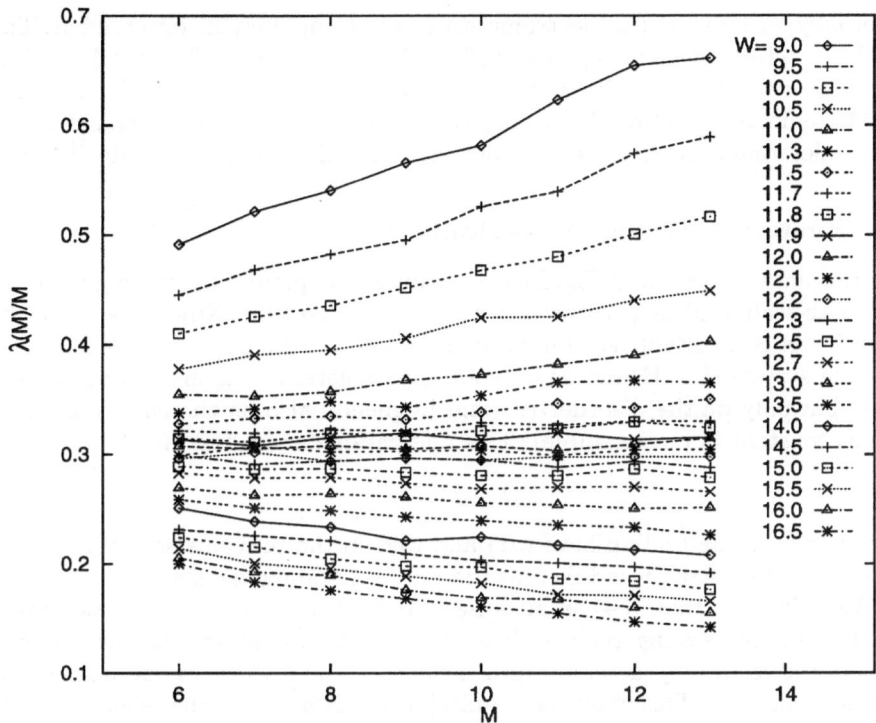

Fig. 2. Localization length λ_M of anisotropic quasi-1D bars with cross section of $M \times M$ sites for different disorder W and energy $E = 0$. The length of the bar was increased in each case until the accuracy of 1% was achieved for the data points. Connecting lines are for guiding the eye only

disorders W (keeping the energy $E = 0$ constant) in the files `raw##.dat`. Each data set contains the localization lengths for 8 different values of M. The data can be plotted with the gnuplot program `raw.gnu` in the same way as the 2D data of the above described exercise.

In order not to duplicate previously published data we have determined these data for a slightly modified system, choosing different transition matrix elements between nearest neighbors along and across the quasi-1D bar. Specifically we have used $t = 0.4$ along the bar and $t = 1$ within the slices. The data are presented in Fig. 2.

These raw data can already be evaluated for the determination of the metal–insulator transition as indicated at the end of the previous chapter with regard to the renormalized Lyapunov exponents $\tilde{\gamma}_1$. For large disorder, $\tilde{\gamma}_1^{-1} = \lambda_M/M$ increases with increasing M. This means that λ_M grows faster than the extension of the system. Consequently, the localization length will become infinite in an infinitely wide system. This corresponds to metallic behavior. On the other hand, for large disorder λ_M/M decreases with increasing M, which means that eventually for large M the electronic states will completely fit into the bar. They will be localized, this is a signature of insulating behavior. The metal–insulator transition occurs at the fixpoint, i.e., for that value of the disorder for which λ_M/M remains constant. Due to the fluctuations of the data in Fig. 2 it is difficult to determine this critical disorder. But one can already estimate $W^* \approx 12$.

4.3   Finite-Size Scaling Numerically

The raw data presented in Fig. 2 can be fruitfully exploited by making use of the finite-size-scaling ideas presented in the previous section. Now we assume that $\lambda_M/M \equiv \Lambda$ is a suitable scaling variable, which can be expressed as a function f of the system size M and some scaling parameter ξ, which does not depend on M but only on the disorder W (and in general also on the energy E, which is kept constant in our example). This one-parameter scaling law

$$\Lambda = f(\xi/M) \tag{64}$$

corresponds to the scaling laws (51) and (52) in the previous section, in which the conductance itself was taken as the relevant scaling variable.

We point out that (64) is an ansatz and whether or not the raw data in Fig. 2 fulfill this scaling relation has to be verified. This can be performed in a quantitative way by attempting a mean least squares fit of the data onto a common curve f. This requires a suitable adjustment of the scale of M (or equivalently of $1/M$) by means of the scaling parameter $\xi(W)$ for each disorder W. We have performed such a mean least squares fit successfully, the results for $\xi(W)$ are supplied in the file xi.dat. The program scaling.for performs nothing but a multiplication of the sets of raw data with the corresponding scaling parameter. The scaled data can then be plotted using scacurve.gnu. The result is presented in Fig. 3. This figure demonstrates that the attempt to find a common functional relationship $f(\xi/M)$ has been successful within the accuracy of our raw data. Thus the assumption of one-parameter scaling has been numerically corroborated.

We note that there is an ambiguity in the scaling curve in Fig. 3, because the entire curve can be shifted by multiplying all scaling parameters ξ with a common factor. But for small Λ the cross-sectional diameter of the investigated bar is already much larger than the localization length λ_M. It is therefore a reasonable approximation that in this case λ_M is already close to the 3D localization length. This means that $f(x) = x$. Applying this relation to the data set for the largest disorder in Fig. 2 resolves the mentioned ambiguity.

The most prominent feature of the scaling curve in Fig. 3 is the existence of two branches. These correspond of course to the two qualitatively different behaviors of the localization length in Fig. 2 increasing faster or slower than the lateral system size M, as discussed in the previous subsection. Accordingly, the upper branch of the scaling curve in Fig. 3 corresponds to the metallic regime, the lower branch to the insulating regime. The branches touch at the metal–insulator transition at which the scaling parameter ξ should diverge. Of course

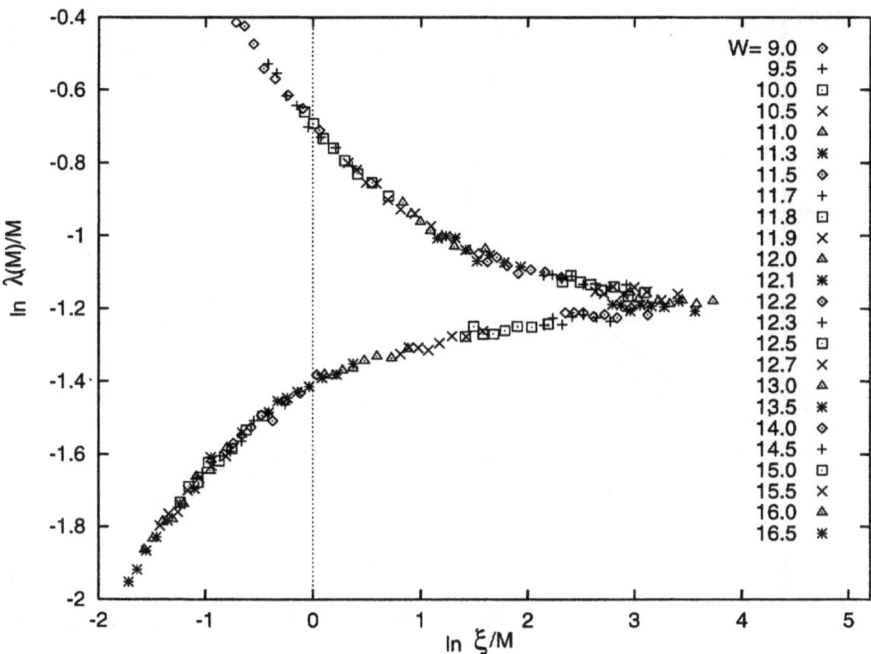

Fig. 3. Scaling function for the anisotropic 3D system. All raw data of Fig. 2 are scaled onto a common curve f by changing the scale of M^{-1} via fitting parameters $\xi(W)$

such a divergence cannot be expected in the numerical analysis due to rounding errors. A close inspection of this regime in Fig. 3 shows that several data sets overlap in this area. This is not surprising, if one takes into account that the raw data have been determined with a statistical accuracy of 1%. By the mean least squares fit procedure we minimize the vertical deviations of the data from the scaling curve in Fig. 3. However, a large horizontal shift of a data set in this regime produces only a small difference in the vertical agreement of the data set with the scaling curve. Therefore the fit cannot be accurate in this regime.

This is a common problem in the numerical analysis of phase transitions. The divergence of any curve is rounded. The usual impression is that it should be easily possible to improve the fit by hand. The reader is encouraged to do this and to perform the entire scaling procedure by hand. For this purpose we have supplied a second file of scaling parameters xi.own in which all values are set to unity in the beginning. Running the scaling program and plotting the scaling curve with these parameters reproduces Fig. 2, but in a doubly logarithmic plot. The reason for the logarithmic Λ axis is only a better resolution. But the reason for the logarithmic M axis is more important for the following exercise. Now changing the scale by a factor ξ means adding a constant $\log \xi$ to $\log M$ so that the change of scale in the plot only means a horizontal shift of the data set for each disorder W. The reader is encouraged to perform this shift consecutively for

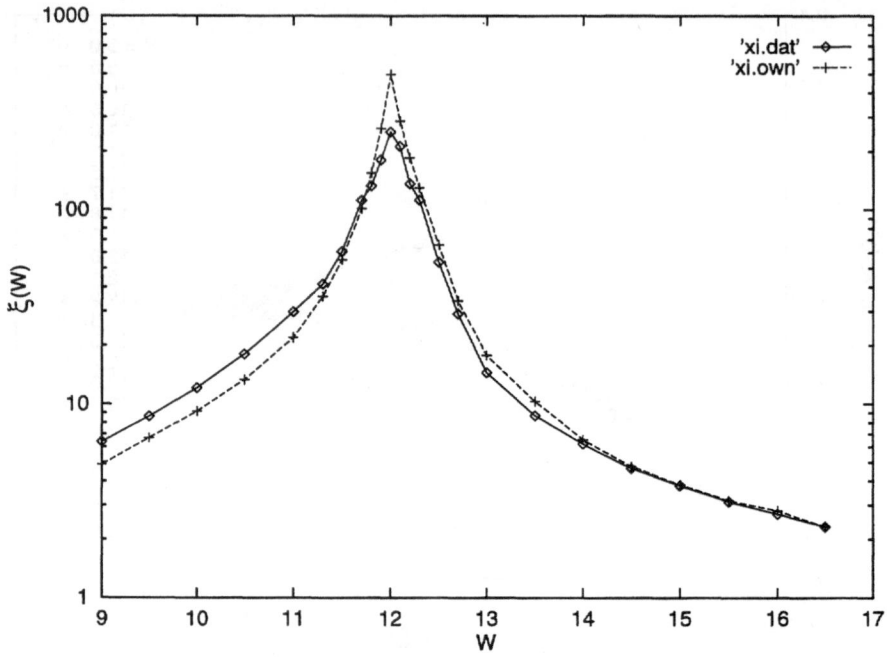

Fig. 4. Scaling parameters ξ which have been used for the construction of the scaling curve in Fig. 3 (*full line*). Data from a graphic construction of the scaling curve "by hand" are also included (*broken line*)

each disorder, i.e., changing the data in xi.own, scaling the raw data accordingly with scaling.for and plotting the resulting scaling curve with scacurve.gnu. Experience shows that this graphical construction of a master curve is often more successful than the mean least squares fit which provided the data in xi.dat. An example is presented in Fig. 4 in which the scaling parameters from the mean least squares procedure are compared with results of such a scaling "by hand". A divergence of the characteristic length $\xi(W)$ at the critical disorder $W^* \approx 12$ can be clearly seen. However, the divergence is significantly rounded due to the numerical procedure. It can also be seen from Fig. 4 that the data which have been obtained "by hand" yield a much smoother curve. It must be pointed out that the graphical construction of the respective master curve corresponding to Fig. 3 has been performed as explained above, i.e., trying to improve the smoothness of the scaling curve but not looking at the smoothness of the $\xi(W)$ curve in Fig. 4.

We note that the deviation of the scaling parameters in the metallic regime in Fig. 4 is not significant. This deviation is due to the fact that in principle both branches of the scaling curve in Fig. 3 can be derived independently. Therefore it is necessary to determine an asymptotic behavior of the scaling curve in the metallic regime for small disorder. This can be performed associating ξ with the resistivity of the 3D system [7] yielding $f(x) = x^{2-d}$ in agreement with (51). We did not invoke this asymptotic relation for the construction of the scaling curve because we have not included raw data for a sufficiently small disorder.

Finally we note that the critical exponent ν can be derived according to (59) from the data in Fig. 4. A nonlinear fit to the data xi.own yielded a value $\nu \approx 1.4$ in agreement with previous investigations [1, 14] of isotropic 3D systems. Of course, the value cannot be very accurate because we have not used as many raw data as in those investigations. Nevertheless it is another corroboration of the universality of one-parameter scaling that the critical exponents of isotropic and anisotropic systems agree.

For completeness we note that in practice another determination of the critical exponent has proven more accurate. It follows from (60) that close to the critical disorder the system-size dependence of the raw data is determined by that contribution to the sum in (60), which is governed by the critical exponent. This allows a straightforward determination of the critical exponent ν from the raw data without any necessity of constructing a scaling curve.

5   Present Status of the Results from Transfer-Matrix Calculations

The status of the numerical work concerning the disorder-induced metal–insulator transition can be briefly summarized.

Numerous and extensive calculations were performed for the (real) Anderson Hamiltonian on a simple cubic lattice with different distributions of the lattice site energies. Results have been comprised in [1]. Basically, for energies inside the unperturbed band, $|E| < 6$, the above one-parameter scaling hypothesis was confirmed. The critical exponents of the conductivity and of the localization length were found to be equal within the achieved accuracy. Their numerical values were determined as $\nu = 1.45 \pm 0.15$. Previously obtained values, which differed for different distributions [15], were too strongly influenced by the above-discussed numerical rounding of the divergence of the scaling parameter at the phase transition. An intricate method of evaluating the raw data based on a quantitative criterion for the exclusion of inaccurate data points close to the transition [14] yielded $\nu \approx 1.35$ for the box, the Gaussian, and the binary distribution, i. e., independent of the distribution of the site energies, again within the accuracy. Thus, universality can be considered to have been explicitly demonstrated for this model, which belongs to the so-called orthogonal universality class of time-reversal invariant systems.

In 2D the numerical data obtained so far for time-reversal invariant systems include square, triangle, and honeycomb lattices [16]. Irrespective of the number of nearest neighbors all the results are fully consistent with the expectation that ν diverges with an essential singularity for vanishing disorder. All of the quantum states are therefore localized in 2D when the disorder is finite.

The divergence of the critical exponent in 2D systems was recently corroborated by applying the transfer-matrix method to bifractal systems [17]. The cross section of the investigated quasi-1D bifractals is a fractal, e. g., the Sierpinski gasket described in this book in the chapter by Bunde et al. or a cluster constructed by diffusion-limited aggregation. These fractals are stacked regularly along the direction in which the recursion (62) is performed. Therefore a straightforward application of the transfer-matrix method is feasible for these bifractals, which are characterized by dimensions $2 < d < 3$. Thus the dimensionality dependence of the critical disorder W^* and the critical exponent ν could be obtained. For $d \to 2$ the vanishing W^* and the diverging ν corroborated that $d = 2$ is the lower critical dimension of the Anderson model. Moreover the behavior of ν was shown to agree with a prediction obtained from the ε expansion within the non-linear σ model [18].

Some numerical work has been performed which concerned the "unitary universality class" of systems that are not time-reversal invariant. They describe a charged particle in a magnetic field in the presence of disorder. Although from general considerations one expects that the critical behavior of the Anderson transition should change upon a change in the symmetry class of the system (from orthogonal to unitary), the numercially obtained data indicate that within the accuracy the critical exponent is the same, namely $\nu = 1.35 \pm 0.15$, independent of the strength of the (homogeneous) magnetic field [19]. This was obtained for different models. It means that universality is seen in the numerical results also in this case. The surprising conclusion is that the change in the symmetry causes only minor changes in the critical behavior at the Anderson transition.

On the other hand, when the potential disorder is removed, and instead the magnetic field is randomized in order to produce a localization–delocalization transition, the critical behavior changes. First of all, due to the absence of the disorder, the position of the transition is situated near the band edge, in an energy region where the density of states changes rapidly [20]. Second, the critical exponent is found to be $\nu = 1.05 \pm 0.15$. This is surprising because the system still belongs to the unitary universality class. When in addition to the random magnetic field the random potential is again introduced, the critical exponent turns out to be the same as the one obtained in the case of the homogeneous magnetic field. It seems that the influence of the scalar random potential on the critical properties is much stronger than the changes in the symmetry. Within the conventional wisdom of phase transitions this is at the least somewhat unexpected. Considerably more work has to be done in the future, in order to clarify this issue.

In 2D a homogeneous magnetic field has dramatic effects on the localization properties. First of all, the electronic spectrum splits into Landau bands. They are well separated as long as the disorder is not too large. In this limit careful numerical work was carried out, showing that for each Landau band the one-parameter scaling hypothesis is valid. The localization length was found to diverge at the centres of the Landau bands with an exponent $\nu = 2.34 \pm 0.04$ [21]. This result was obtained by completely different methods for a variety of different potential models, including white-noise Gaussian as well as very long-range correlated randomness. It constitutes one of the most striking and precise demonstrations of universality of the disorder-induced localization–delocalization transition.

For a random magnetic field in 2D the present conclusions from the numerical and analytical work are controversial. They range from the statement that all of the states are again localized [22, 23] to the suggestion of a Kosterlitz–Thouless transition [24]. Again much more, and particularly more precise, work has to be done for this problem.

Spin-orbit scattering induces another change of symmetry. As a consequence, one expects changes in the critical behavior. Indeed, in 2D, instead of only localized states, a localization–delocalization transition occurs. It has been investigated numerically to some extent [25, 26, 27]. However, due to the enormous requirements of computer power, the reliability of the results is not comparable with that of the above-described cases. Also for this problem considerable numerical efforts are necessary in order to answer satisfactorily questions confirming the "validity of one-parameter scaling" or the "universality".

References

[1] B. Kramer and A. MacKinnon, Rep. Progr. Phys. **56**, 1469 (1993)

[2] A.A. Abrikosov and I.A. Ryzhkin, Adv. Phys. **27**. 147 (1978)

[3] K. Ishii, Suppl. Progr. Theor. Phys. **53**, 7 (1973)

[4] J.B. Pendry, Adv. Phys. **43**, 461 (1994)

[5] V.I. Oseledec, Trans. Moscow Math. Soc. **19**, 197 (1968)

[6] A. MacKinnon, Z. Phys. B **59**, 385 (1985)

[7] A. MacKinnon and B. Kramer, Z. Phys. B **53**, 1 (1983)

[8] G. Czycholl, B. Kramer, and A. MacKinnon, Z. Phys. B **43**, 5 (1981)

[9] A.J. O'Connor, Commun. Math. Phys. **45**, 63 (1975)

[10] M. Kappus and F. Wegner, Z. Phys. B **45**, 15 (1981)

[11] E.N. Economou, *Green's Functions in Quantum Physics*, Springer Series in Solid-State Sciences 7 (Springer, Berlin 1983)

[12] A. MacKinnon, J. Phys. C **13**, L1031 (1980)

[13] P.W. Abrahams, E. Abrahams, D.C. Licciardello, and T.V. Ramakrishnan, Phys. Rev. Lett. **42**, 673 (1979)

[14] E. Hofstetter and M. Schreiber, Europhys. Lett. **21**, 933 (1993)

[15] B. Kramer and M. Schreiber in *Fluctuations and Stochastic Phenomena in Condensed Matter*, ed. L. Garrido, Springer Lecture Notes in Physics 268, (Springer, Berlin 1987), p. 351

[16] M. Schreiber and M. Ottomeier, J. Phys. Condens. Matter **4**, 1959 (1992)

[17] M. Schreiber and H. Grussbach, Phys. Rev. Lett. **76**, 1687 (1996)

[18] F. Wegner, Nucl. Phys. B **316**, 663 (1989)

[19] M. Henneke, B. Kramer, and T. Ohtsuki, Europhys. Lett. **27**, 389 (1994)

[20] T. Ohtsuki, Y. Ono, and B. Kramer, J. Phys. Soc. Japan **63**, 685 (1994)

[21] B. Huckestein and B. Kramer, Phys. Rev. Lett. **64**, 1437 (1990)

[22] A.G. Aronov, A.D. Mirlin, and P. Wölfle, Phys. Rev. B **49**, 16609 (1994)

[23] T. Sugiyama and N. Nagaosa, Phys. Rev. Lett. **70**, 1980 (1993)
[24] S.-C. Zhang and D. P. Arovas, Phys. Rev. Lett. **72**, 1886 (1994)
[25] S.N. Evangelou and T. Ziman, J. Phys. C **20**, L235 (1988)
[26] S.N. Evangelou, in *Quantum Coherence in Mesoscopic Systems,* ed. B. Kramer, NATO ASI Series B 254, (Plenum, New York 1991), p. 435
[27] T. Ando, Phys. Rev. B **40**, 5325 (1989)

Quantum Monte Carlo Investigations for the Hubbard Model[*]

Hans-Georg Matuttis[1] and Ingo Morgenstern[2]

[1] Institut für Computeranwendungen I, Universität Stuttgart, Pfaffenwaldring 27, D–70569 Stuttgart, Germany, e-mail: hg@ical.uni-stuttgart.de
[2] Institut für Physik II, Universität Regensburg, Universitätsstrasse 31, D–93040 Regensburg, Germany

Abstract. Quantum Monte Carlo (QMC) simulations are a powerful means to obtain information about quantum mechanical systems with strong interactions. The chapter shows how the quantum Monte Carlo method is applied for strongly correlated systems at finite temperature, which can be described by the Hubbard model. The partition function is decomposed using the Trotter–Suzuki transformation (TS), the interaction is decoupled using the Hubbard–Stratonovich transformation. The problems that arise due to numerical instabilities and the negative sign problem, which is due to the Monte Carlo sampling technique, are briefly mentioned.

1 Introduction

1.1 The Hubbard Model

It is not possible to describe the complex behavior of Hubbard-like Hamiltonians and their simulation in a single chapter without simplifications. Therefore the diagrams throughout this article show "classical particles" moving and interacting, disregarding the occurrence of collective phenomena which are important for the "real" physics of the problem.

A general Hamiltonian for a system of N electrons which interact in real space via the Coulomb interaction in an external potential V^{ext} can be written as (spin indices are omitted)

$$H^{\text{general}} = -\underbrace{\frac{\hbar}{2m} \sum_{i=1}^{N} \nabla_i^2}_{E_{\text{Kin}}} + \underbrace{\frac{1}{2} \sum_{i\neq j=1}^{N} \frac{e^2}{|r_i - r_j|}}_{E_{\text{Int}}} + \underbrace{\sum_{i=1}^{N} V^{\text{ext}}(r_i)}_{E_{\text{Pot}}} \quad , \qquad (1)$$

where r_i represents the coordinates of the electrons in real space. For tight-binding electrons in a solid, r_i will not be continuous, only certain "lattice points" can be occupied and the Coulomb interaction will be screened. To describe interacting electrons in the tight binding approximation, we

1. use the occupation number formalism to allow only discrete sites as particle positions;

---

[*] Software included on the accompanying diskette.

2. neglect the external potential, in order our system to become translation invariant;
3. retain only the on-site interaction.

We obtain the Hubbard Hamiltonian

$$H^{\text{Hub}} = -t \underbrace{\sum_{i,j,s} c_{i,s}^+ c_{j,s}}_{E_{\text{Kin}}} + U \underbrace{\sum_i n_{i\uparrow} n_{i\downarrow}}_{E_{\text{Int}}} - \mu \underbrace{\sum_i (n_{i\uparrow} + n_{i\downarrow})}_{\text{Chem. Pot}} \quad , \tag{2}$$

where the operator $c_{i,s}^+$ creates a particle with spin s at site i and $c_{i,s}$ annihilates a particle with spin s at site i. $n_{i,s} = c_{i,s}^+ c_{i,s}$ is the number operator, whose eigenvalue is the occupation number of site i with a fermion of spin s. The summation index is over all lattice sites, the hopping term E_{Kin} occurs between nearest-neighbor sites i, j, and the Hubbard on-site interaction E_{Int} is felt by electrons with opposite spins on the same site. The last term in (2) is analogous to $-PV$ in classical physics and indicates the injection of particles into the system if $\mu > 0$, whereas particles are drawn away from the system if $\mu < 0$; the situation $\mu = 0.5$ corresponds to a half-filled system $\langle n_{i\uparrow} \rangle = \langle n_{i\downarrow} \rangle = 0.5$. Fig. 1 shows the two cases with Hamiltonian (1) respectively Hamiltonian (2). Fig. 2 shows the behavior of the system for three different electron densities.

Coulomb interaction, continuous r_i Hopping   Hubbard interaction

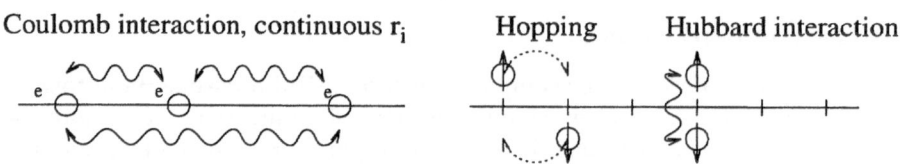

Fig. 1. Simplified behavior of the electrons for the Hamiltonian (1) shown *left*, and the Hamiltonian (2) shown *right*. The diagrams are one-dimensional simplifications of the two-dimensional situation

In the following, the energy and the temperature will be given in units of $t = 1$ ($k_B = 1$). The hopping term is diagonal in momentum space, the interaction term is diagonal in real space. Therefore, perturbation expansions are valid only for certain regimes of doping.

The Hubbard Hamiltonian is the generic model for tight-binding Hamiltonians. In its most simple form it is the workhorse for strongly correlated electrons, in the same way as the Ising model is used for "spin" systems. However it can be generalized via:

- Additional interactions; in the extended Hubbard Model, nearest-neighbor (nn) interactions $V \sum_{i,j,s,s'} n_{is} n_{js'}$, ($i$ is nearest neighbor to j) are included.
- Additional hopping terms, $t - t'$ hopping; hopping between next-nearest (nnn) neighbor sites modifies the kinetic energy and the density of states.

- Additional bands; the 3-band Hubbard model (Emery model) is supposed to describe the electronic structure of the CuO planes in high temperature super conductors (different hopping terms between copper and oxygen sites, different interaction strength in copper and oxygen sites, different chemical potentials on the copper and oxygen sites).

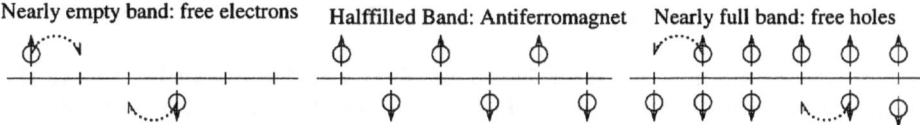

Nearly empty band: free electrons   Halffilled Band: Antiferromagnet   Nearly full band: free holes

Fig. 2. For low electron densities, collisions between the electrons in the Hubbard model are improbable, so they will behave as free particles. For medium densities (near-half filling) and "strong" interaction, the Hamiltonian exhibits antiferromagnetism. For nearly full bands, the "hopping" of electrons leads to double-occupied sites, which is energetically costly and therefore unprobable, the system becomes insulating

1.2   What to Compute

Depending on what physical phenomena one is interested in, different observables can be computed using quantum Monte Carlo (QMC) simulations:

- Correlation functions in real space indicate whether the net interaction is repulsive or attractive, and whether there is any ordering in the system.
- Magnetic structure factors indicate the magnetic ordering of the system.
- The distribution $n(\mathbf{k})$ gives the "Fermi surface" and indicates how important many particle effects are (see Fig. 3).
- The dependence of the filling on the chemical potential and the interaction strength can be computed directly using QMC.
- Densities of states using Greens functions in imaginary time can also be computed, but the details of these computations would lead too far (see Sect. 4).

When statistical Monte Carlo methods are used, only systems in thermal equilibrium can be simulated, so that currents and super-currents (where the resistance of the conductor vanishes) cannot be observed directly. Therefore, the breakdown of the resistivity and the Meissner effect cannot be verified in the QMC simulations.

1.3   Quantum Simulations

Many analytical and numerical methods have been used in connection with the Hubbard model (for didactic introduction see [2] Chapt. 20; for a more technical introduction of the analytic methods see [3]).

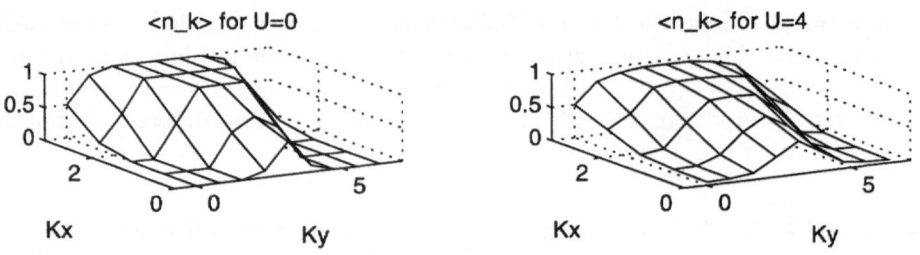

Fig. 3. $N(k)$ for a 8×8 system for $U = 0$ and $U = 4$ at $\beta = 4$ after [1]. Due to the depletion of the distribution for small $|k|$ at $U = 4$, the electrons cannot be treated as a Fermi liquid

A general problem in the simulation of quantum systems is the treatment of the problem in such a way that instead of handling operators one deals with matrix elements, which are real numbers. Exact diagonalization is a "conceptually straightforward" approach (the numerical realization is less straightforward, see [4] and references therein) but works only for system sizes of about 4×4 lattice points. A new concept is stochastic diagonalization, whereby states are sampled by a Monte-Carlo procedure [5].

To set up a quantum Monte Carlo simulation, we need

1. a procedure to deal with the noncommuting parts of the Hamiltonian (Trotter–Suzuki decomposition);
2. a procedure to decouple the interaction, so that we obtain effective single-particle states for a fixed configuration (Hubbard–Stratonovich transformation);
3. a Boltzmann weight, which can be obtained form the partition function;
4. a method to evaluate the matrix elements for the computation of observables (real numbers on the computer). For our algorithm, the most useful of these matrix elements are the single-particle Greens functions.

The Boltzmann weight for the algorithm that is introduced in the following chapter is computed by a determinant of matrices, so that the algorithm is called Determinantal quantum Monte Carlo algorithm (DQMC).

2   Grand Canonical Quantum Monte Carlo

In this section, the grand canonical quantum Monte Carlo (GQMC) algorithm will be explained. This algorithm allows the simulation of the Hubbard model at finite temperatures in the grand canonical ensemble.

2.1   The Trotter–Suzuki Transformation

The Hubbard Hamiltonian consists of a kinetic term and an interaction term (with chemical potential). The computation of the partition function at finite temperature,

$$Z = \mathrm{Tr}\Big(\exp(-\beta H)\Big) \quad ,$$

involves the computation of an operator exponential.

In quantum physics one has frequently to deal with operators of the shape $\exp(-\beta(A + B))$, where A and B are hermitian parts of a Hamiltonian, β is real or complex. Typical examples are the partition function or Greens functions. Usually it is easy to find a basis where A or B are diagonal, but if A and B do not commute there will be no tractable diagonal basis for $A + B$.

The Trotter–Suzuki (TS) transformation, also called the generalized Trotter formula or Lie–Trotter decomposition, is a method for decomposing the above exponential operators. It was introduced into computer simulations by M. Suzuki [6];

$$e^{A+B} = \lim_{n \to \infty} \left(e^{A/n}e^{B/n}\right)^n \tag{3}$$

for arbitrary, bounded operators A, B.

For the numerical implementation the limit $(n \to \infty)$ in (3) must be approximated by a finite discretization. The exponent with the prefactor β is decomposed into a product with small $\mathrm{d}\tau$, the factors are called slices:

$$e^{-\beta(A+B)} \approx \left(e^{-\mathrm{d}\tau A}e^{-\mathrm{d}\tau B}\right)^n, \quad \mathrm{d}\tau = \beta/n \ . \tag{4}$$

The error can be computed by inserting a parameter λ into the exponent. Afterwards the expression is replaced by the integral over the derivative with respect to λ, and the norm is estimated. One can show that [7]

$$e^{\mathrm{d}\tau(A+B)} = e^{\mathrm{d}\tau A}e^{\mathrm{d}\tau B} + \mathcal{O}(\mathrm{d}\tau^2) \ , \tag{5}$$

with Landau's symbol for the order of the magnitude $\mathcal{O}$. This approximation is called "first order" because it is correct up to order $\propto \mathrm{d}\tau$. A decomposition of second order (correct up to order $\propto \mathrm{d}\tau^2$) is

$$e^{\mathrm{d}\tau(A+B)} = e^{\mathrm{d}\tau A/2}e^{\mathrm{d}\tau B}e^{\mathrm{d}\tau A/2} + \mathcal{O}(\mathrm{d}\tau^3) \ . \tag{6}$$

In contrast to the decomposition of first order, this decomposition yields real symmetric exponentials for real symmetric A, B.

It is a general feature of numerical approximations (also for integrations, differentiations, solutions of partial differential equations) that the symmetric formulae are of higher accuracy than the asymmetric ones.

If one wants to increase the accuracy for a decomposition of a certain order, numerical costs increase. Nevertheless, for long products the necessary work for a decomposition of first order is nearly the same as for a decomposition for second order:

first order;

$$e^{n\mathrm{d}\tau(A+B)} \approx \underbrace{e^{\mathrm{d}\tau A}e^{\mathrm{d}\tau B}}_{first\ \text{slice}}\underbrace{e^{\mathrm{d}\tau A}e^{\mathrm{d}\tau B}}_{second\ \text{slice}}\cdots\underbrace{e^{\mathrm{d}\tau A}e^{\mathrm{d}\tau B}}_{nth\ \text{slice}} \tag{7}$$

second order;

$$e^{nd\tau(A+B)} \approx \underbrace{e^{d\tau A/2}e^{d\tau B}e^{d\tau A/2}}_{first\ slice}\underbrace{e^{d\tau A/2}e^{d\tau B}e^{d\tau A/2}}_{second\ slice}\cdots\underbrace{e^{d\tau A/2}e^{d\tau B}e^{d\tau A/2}}_{nth\ slice}$$

$$= e^{d\tau A/2}e^{d\tau B}e^{d\tau A}e^{d\tau B}e^{d\tau A}\cdots e^{d\tau B}e^{d\tau A/2}\ . \tag{8}$$

To evaluate the first-order decomposition in (7) one needs $2n - 1$ products and for the second-order decomposition in (8) one needs $2n$ products because $e^{d\tau A/2}e^{d\tau A/2} = e^{d\tau A}$. For the physically interesting case of large n (low temperatures) this is practically the same for first as for second order.

Several verbalizations are commonly used to describe the occurrence of the product. The number of the slices is considered as the "coordinate" in an additional dimension. Sometimes, the TS formula is called "*equivalence theorem*", "an m-dimensional quantum system mapped onto an $(m + 1)$-dimensional classical system." The system is classical inasmuch as there are only commuting operators $\frac{A}{n}$, $\frac{B}{n}$ (for large n) left. Nevertheless, the term "classical system" does not mean that the system can be mapped onto something like a harmonic oscillator or another classical "workhorse" of theoretical physics in high dimensions. The coordinate for the $(m + 1)$ "slices" is also called Trotter time, which is imaginary time for many quantum simulations.

Another verbalization is [6], "The additional dimension plays the role of path integrals in a discrete space ...", so that formally, all QMC algorithms can be viewed as numerical path integral methods. Nevertheless, the dynamics and the implementation of different algorithms varies considerably.

A decomposition of fourth order can be found in [7]; one decomposition up to fourth order that does not rely on the troublesome computation of a double commutator is the so-called "fractal" decomposition [8]. Very often, the Baker–Haussdorff formula (see [9])

$$e^A e^B = e^{A+B+\frac{1}{2}[A,B]+\cdots}$$

is used in high-energy physics in the same sense as the Trotter–Suzuki decomposition is used in solid-state physics.

2.2   The Hubbard–Stratonovich Transformation

In DQMC, the Hubbard–Stratonovich transformation (HS) is used to decouple the interaction and thereby to transform the many-particle system with interaction into a problem without interaction. The transformation is also called auxiliary field transformation, because auxiliary potentials are built up on the lattice in such a way that the fluctuating potential is able to model the interaction U for the many-particle problem.

Instead of simulating interacting electrons, one simulates electrons with spin up/down in a fluctuating potential. The sign of the potential is determined by the

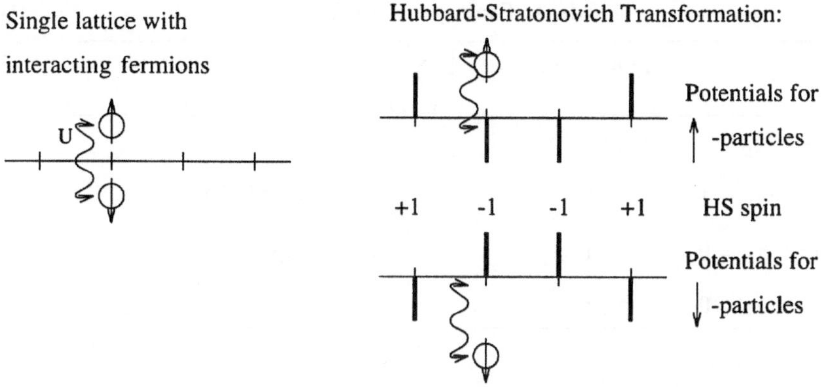

Single lattice with

interacting fermions

Hubbard-Stratonovich Transformation:

Potentials for

$\uparrow$ -particles

+1   -1   -1   +1   HS spin

Potentials for

$\downarrow$ -particles

Fig. 4. Decoupling of the interaction U using the Hubbard–Stratonovich transformation: Instead of simulating one grid with interacting spin-up and spin-down particles, one simulates two grids where the particles with spin-up/spin-down move in a fluctuating potential. The prefactor of the potentials of size λ depends on the spin of the electrons (up, down) and the Hubbard–Stratonovich spins $\sigma_i = \pm 1$ on site i

sign of the Hubbard–Stratonovich spin σ (Fig. 4). One has to sum over several configurations of this potential to obtain the effect of the interaction,

$$\sum_{\sigma} e^{d\tau V_\uparrow(\sigma)n\uparrow} e^{d\tau V_\downarrow(\sigma)n\downarrow} \quad \text{instead of} \quad e^{d\tau U n\uparrow n\downarrow} \quad .$$

The Hubbard–Stratonovich transformation (HS) for a single site can be derived in the following way. We will show, that

$$e^{-d\tau U n_\uparrow n_\downarrow} = \frac{1}{2} \sum_{\sigma=\pm 1} e^{\lambda\sigma(n_\uparrow - n_\downarrow)} e^{-d\tau U/2(n_\uparrow + n_\downarrow)} \quad , \tag{9}$$

where λ is a free parameter. Comparing both expressions in Table 1 for all different values for the electron densities $n_\uparrow, n_\downarrow \in \{1, 0\}$, one obtains [1]

$$\cosh(\lambda) = e^{d\tau |U|/2} \quad . \tag{10}$$

For more sites, a site index i has to be introduced, the exponential of an operator becomes the exponential of a diagonal matrix of operators.
In [1] and in the remaining chapter, the notation Tr_σ is used instead of $\frac{1}{2} \sum_\sigma$.
To avoid the occurrence of a complex λ, different HS transformations are necessary for positive and negative U. For negative U, (10) has no real solution.
  The advantage of this decoupling strategy is:

---

[1] In [1], this is written as $\lambda = 2\mathrm{atanh}\sqrt{\tanh(d\tau U/4)}$, "atanh" is misprinted as "atan".

Table 1. Evaluation of (9)

	$e^{-d\tau U n_\uparrow n_\downarrow}$	$\dfrac{1}{2}\displaystyle\sum_{\sigma=\pm1} e^{\lambda\sigma(n_\uparrow - n_\downarrow)} e^{-d\tau U/2(n_\uparrow + n_\downarrow)}$
$n_\uparrow = 1, n_\downarrow = 1$	$e^{-d\tau U}$	$e^{-d\tau U}$
$n_\uparrow = 0, n_\downarrow = 1$	1	$\cosh(\lambda)e^{-d\tau U/2}$
$n_\uparrow = 1, n_\downarrow = 0$	1	$\cosh(-\lambda)e^{-d\tau U/2}$
$n_\uparrow = 0, n_\downarrow = 0$	1	1

- The interacting system is reduced to a noninteracting system with potentials for a fixed configuration of σ, so that the Wick theorem can be applied. This can be seen by separating the right-hand side expression in (9) for spin-up and spin-down particles:

$$\frac{1}{2}\sum_{\sigma=\pm1} e^{\lambda\sigma(n_{i\uparrow} - n_\downarrow)} e^{-d\tau U/2(n_\uparrow + n_\downarrow)} = \frac{1}{2}\sum_{\sigma=\pm1} e^{(\lambda\sigma - d\tau U/2)n_\uparrow} e^{(\lambda\sigma - d\tau U/2)n_\downarrow} \quad .$$

- It is not necessary to sum over all σ configurations, but it is sufficient to sum over the most important ones using Monte Carlo. In the literature, the summation is often called "integrating out the interaction degrees of freedom".

The Hubbard–Stratonovich spins are very often termed "Ising spins" or "Ising fields". The terminology is slightly misleading, because the dynamics of the DQMC algorithm is in no way related to dynamics of an Ising simulation. Although the configurations of the Ising-spins are ±1, the resulting fluctuating potentials are elements of a diagonal matrix which enter in a matrix product, they do not interact in the same way as a spin model.

In the special case of the interaction on discrete lattice points, the HS transformation can be written as a sum. For continuous fields, there is an integral instead of a sum. In the case of the grand canonical ensemble, the chemical potential is added in the exponent to control the filling. The HS transformation may only be applied *after* the TS transformation, so that every slice has its own spin configuration. The fluctuating potentials correspond to a field which helps to model the interaction, so that the Hubbard–Stratonovich transformation in the QMC algorithms is also called the auxiliary field method.

2.3   The Partition Function

The Boltzmann weight for a fixed configuration of HS spins can be derived from the partition function. In the derivations, the identity matrix will be denoted as 1. The hopping matrix will be denoted as K.

First, the exponential of the Hamiltonian is rewritten as a product of slices
(11), which is just the definition of the exponential. Then the slices are decomposed using the Trotter–Suzuki decomposition (12). In (14) the interaction is
decoupled using σ_{il} as the Hubbard–Stratonovich spin on the lth slice on site i,
and in (14) the decomposition is separated for spin-up and spin-down particles:

$$Z = \mathrm{Tr}\left(e^{-\beta H}\right) = \mathrm{Tr}\left(\prod_l e^{-d\tau H}\right) \tag{11}$$

$$\approx \mathrm{Tr}\left(\prod_l e^{-d\tau K} e^{-d\tau\left(U\sum_i n_{i\uparrow} n_{i\downarrow} - \mu\sum_i (n_{i\uparrow} + n_{i\downarrow})\right)}\right) \tag{12}$$

$$= \mathrm{Tr}_\sigma\left(\mathrm{Tr}\left(\prod_l e^{-d\tau K} e^{\lambda\sigma_{i,l}(n_{i\uparrow} - n_{i\downarrow}) - d\tau(\mu - U/2)(n_{i\uparrow} + n_{i\downarrow})}\right)\right) \tag{13}$$

$$= \mathrm{Tr}_\sigma\left(\mathrm{Tr}\left(\prod_l e^{-d\tau K_\uparrow} e^{n_{i\uparrow}\left(+\lambda\sigma_{i,j} - d\tau(\mu - U/2)\right)}\right.\right.$$

$$\left.\left.\cdot \prod_l e^{-d\tau K_\downarrow} e^{n_{i\downarrow}\left(-\lambda\sigma_{i,j} - d\tau(\mu - U/2)\right)}\right)\right) \quad . \tag{14}$$

With the notation for the "slices" $B_l^\uparrow, B_l^\downarrow$

$$B_l^\uparrow = e^{-d\tau K} e^{V^\uparrow(l)}, \qquad V_{ij}^\uparrow(l) = \delta_{ij}\left(+\lambda\sigma_{i,l} - d\tau(\mu - U/2)\right), \tag{15}$$

$$B_l^\downarrow = e^{-d\tau K} e^{V^\downarrow(l)}, \qquad V_{ij}^\downarrow(l) = \delta_{ij}\left(-\lambda\sigma_{i,l} - d\tau(\mu - U/2)\right), \tag{16}$$

one can define the operators

$$D_l^\uparrow = e^{d\tau c_i^+ K_{i,j} c_j}\, e^{c_i^+ V_{i,i}^\uparrow(l) c_i}\,, \qquad D_l^\downarrow = e^{d\tau c_i^+ K_{i,j} c_j}\, e^{c_i^+ V_{i,i}^\downarrow(l) c_i}\,, \tag{17}$$

so that the partition function can be written as

$$Z = \mathrm{Tr}_\sigma \mathrm{Tr}\left(\prod_l D_l^\uparrow\right)\left(\prod_l D_l^\downarrow\right). \tag{18}$$

The bilinear forms in fermion operators $c_i^+ \ldots c_j$ can be formally "diagonalized"
so that the trace over the fermions can be taken explicitly; this takes up the
whole appendix in [1].

Finally, one can prove that one can compute the partition function by computing the determinant of the matrices B_l instead of computing the trace of
the operator exponentials D_l. In equation (19), $(\det O_{\uparrow,\sigma} \det O_{\downarrow,\sigma})$ is introduced
as a convenient abbreviation for the "statistic weight" of a σ configuration of
Hubbard–Stratonovich spins.

$$Z = \mathrm{Tr}_\sigma \mathrm{Tr}\left(\prod_l D_l^\uparrow\right)\left(\prod_l D_l^\downarrow\right)$$

$$= \text{Tr}_\sigma \left(\prod_l \det(1 + B_L^\uparrow \ldots B_1^\uparrow) \right) \left(\prod_l \det(1 + B_L^\downarrow \ldots B_1^\downarrow) \right)$$

$$= \text{Tr}_\sigma (\det O_{\uparrow,\sigma})(\det O_{\downarrow,\sigma}) \ . \tag{19}$$

2.4 The Monte Carlo Weight

We will not give a justification of the Monte Carlo method (for that purpose, see [10]), but in Table 2 we show the analogies between classical and quantum Monte Carlo methods in the derivation of the statistical weight by analogy. The heat-bath algorithm is usually used for DQMC simulations (P_{HB} in Table 2).

The product $(\det O_\uparrow(\sigma) \det O_\downarrow(\sigma))$ may be negative, therefore the Boltzmann weight for a HS configuration σ is defined using the absolute value as

$$P(\sigma) = |\det O_\uparrow \det O_\downarrow| \ \ .$$

Therefore, in order to compute an observable $\langle A \rangle$ one has to compute separately $\langle A^+ \rangle$ and $\langle A^- \rangle$ for positive and negative statistical weights, and with the percentage of positive and negative statistical weights $\langle \text{Sign}^+ \rangle$, $\langle \text{Sign}^- \rangle$, one can compute an observable as

$$\langle A \rangle = \frac{\langle A^+ \rangle - \langle A^- \rangle}{\langle \text{Sign}^+ \rangle - \langle \text{Sign}^- \rangle} \ \ .$$

Computing observables without the use of this "minus sign" would lead to non-physical observables (see Fig. 5).

Table 2. Analogies between classical and Quantum Monte Carlo methods in the derivation of the statistical weight

	Classical MC	Quantum MC
States to sample	Classical state i	σ (HS spins)
Partition function Z	$\sum_i e^{\beta E_i}$	$\text{Tr}_\sigma (\det O_{\uparrow,\sigma})(\det O_{\downarrow,\sigma})$
Boltzmann weight	$e^{-\beta E_i}$ for i	$(\det O_{\uparrow,\sigma})(\det O_{\downarrow,\sigma})$ for σ
Subsequent states in MC	$i \to i'$	$\sigma \to \sigma'$
Relative probabilities	$P_{(i \to i')} = e^{-\beta(E_{i'} - E_i)}$	$P_{(\sigma \to \sigma')} = \dfrac{(\det O_{\uparrow,\sigma'})(\det O_{\downarrow,\sigma'})}{(\det O_{\uparrow,\sigma})(\det O_{\downarrow,\sigma})}$
Heat bath algorithm	$P_{HB} = \frac{P(i \to i')}{1 + P(i \to i')}$	$P_{HB} = \frac{P(\sigma \to \sigma')}{1 + P(\sigma \to \sigma')}$

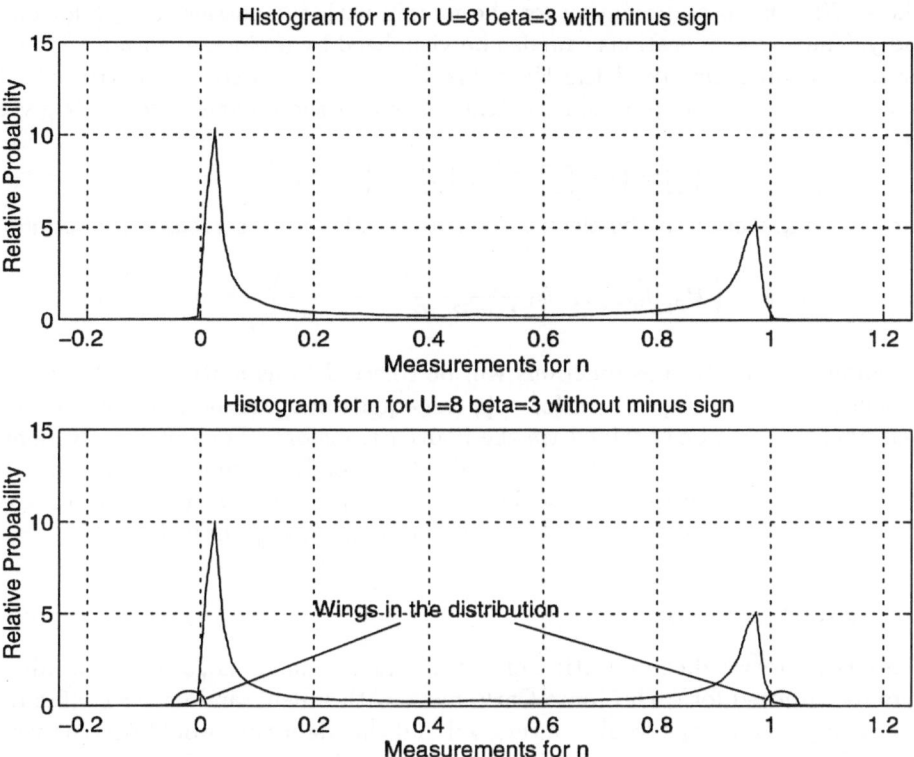

Fig. 5. Distribution of the n_i for $U = 8, \beta = 3$ with correct (first part of figure) and wrong (second part of figure) use of the sign. Nonphysical "tails" in the distribution of the electron density occur outside $[0, 1]$

3   Equal-Time Greens Functions

In the following derivations the index for the electron spin will be omitted if the quantity is evaluated for both spin directions in the same sense. Thermodynamic averages, which are usually denoted by $\langle \ldots \rangle$ in textbooks are physical quantities. Monte Carlo measurements are denoted by $\langle\!\langle \ldots \rangle\!\rangle$, and they are not physical quantities, only their average is.[2]

The equal-time Greens function $\langle\!\langle c_i \, c_j^+ \rangle\!\rangle$ in the grand canonical algorithm can be derived as

$$\langle\!\langle c_i \, c_j^+ \rangle\!\rangle = \left(\frac{1}{1 + B_m B_{m-1} \ldots B_1} \right)_{ij} = G(m)_{ij} \tag{20}$$

[2] In [1], this is expressed using the notation of $\langle \ldots \rangle$ for nonphysical "single-bracket averages" and physical "double-bracket averages" $\langle\!\langle \ldots \rangle\!\rangle$.

where $G(m)$ is an abbreviation for the matrix with the entries $\langle c_i\ c_j^+ \rangle$ for all i and j. The above operations can also be visualized by a block diagram; $\backslash$ stands for the diagonal matrix of the HS fields, $\square$ for the full square matrix of the exponential of the hopping matrix. The matrix of the Greens function is given by

$$\square = (\backslash + \square \cdot \backslash \cdots \square \cdot \backslash \cdot \square \cdot \backslash)^{-1} \quad .$$

For the computation of the electron densities and spatial correlation functions,

$$\langle c_i^+ c_j\ \rangle = \left(B_m B_{m-1} \ldots B_1 \frac{1}{1 + B_m B_{m-1} \ldots B_1} \right)_{ij} = \delta_{ij} - G(m)$$

is computed. The Greens functions will be referred to as matrix $G(m)$ for convenience. From the $\langle c_i\ c_j^+ \rangle$ or $\langle c_i^+ c_j\ \rangle$, the equal-time Greens functions in momentum space can be obtained via the Fourier transform. For isotropic systems, the Greens functions in real space for a fixed HS configuration are averaged over corresponding lattice sites before they are Fourier transformed to save storage. G will be used to denote the the Greens functions on a general slice.

3.1   Single Spin Updates

The straightforward computation of G takes about $2mL^3$ flops. If this product were recomputed for each Monte Carlo move with a new random HS configuration, a high percentage would be rejected and the algorithm would become very costly. Due to the diagonal nature of the interaction term, it is possible to decide whether a single HS spin has to be flipped and how this changes the Greens function without recomputing the whole expression for G. This takes $\propto L^2$ flops [11].

3.2   Numerical Instabilities

There is a product in the denominator of

$$G(m) = (1 + B_m B_{m-1} \ldots B_1)^{-1}$$

which is a product of exponentials of matrices. For $U = 0$ and a system size of 4×4 sites, the inverse of the Greens function for several β can be seen in Fig. 6.

For $\beta = 4$, G is computed as the inverse of a diagonal dominant matrix, whereas for increasing β, the entries of the inverse of G increase exponentially. For increasing U, the entries increase and the structure of the matrices gets lost.

The numerically accurate computation of the inverse of $(1 + B_m \ldots B_1)$ is difficult for a number of reasons. For $U = 0$, the product of the $B_m \ldots B_1$ is just the exponential of the hopping matrix, with singular values distributed between $\exp(4\beta t)$ and $\exp(-4\beta t)$, t is the hopping element. The addition of the "1" constitutes a cutoff for all singular values smaller than 1, each singular value of size ϵ is replaced by a singular value of size $(\epsilon + 1)$. This means that for the

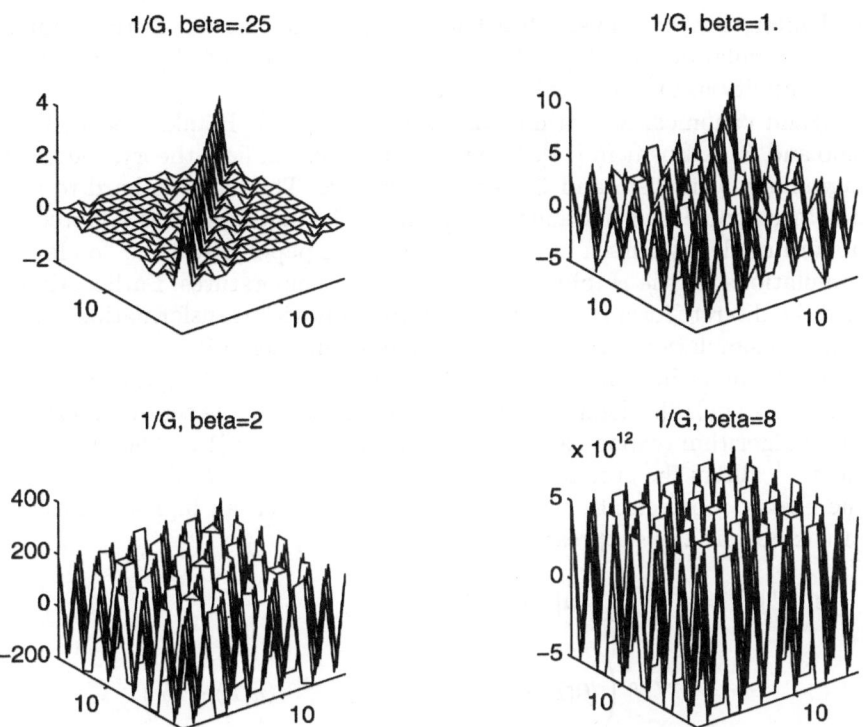

Fig. 6. Inverse of the Greens function for a 4×4 $U = 0$ system without interaction. The x,y coordinates are the rows and columns, the z coordinate is the matrix entry. Due to the viewpoint of the plotting routine, the *upper left* graph corresponds to a diagonal (dominant) matrix. The size of the matrix entries increases exponentially with increasing β

computation of the inverse of $(1 + B_m \ldots B_1)$, the addition of the "1" cannot be considered as a small perturbation to $B_m \ldots B_1$. The inverse of $B_m \ldots B_1$ could be easily computed as $(B_m \ldots B_1)^{-1} = B_1^{-1} \ldots B_m^{-1}$, but its largest singular values are not present in the inverse of $(1 + B_m \ldots B_1)^{-1}$. This means that numerical perturbation theory or related methods cannot be employed for this problem.

4   History and Further Reading

The Monte Carlo method was introduced by Metropolis et al. forty years ago [12]. M. Suzuki introduced the Trotter decomposition and performed the first "real" quantum Monte Carlo simulation on a quantum spin system in 1977 [6].

The worldline algorithm for Fermion simulations was presented by Hirsch et al. [13]. It looks conceptually easier than the determinantal method presented in this chapter, but has severe drawbacks if it is used for "realistic" simulations. Most simulations for the Hubbard model in the field of high-temperature superconductivity use the determinantal algorithms.

The grand canonical algorithm was introduced by R. Blankenbecler, D. J. Scalapino and R. L. Sugar in [14]. Intended for the use in field theory, the paper uses Lagrangian formalism and Grassman variables. The paper is hard to read for solid-state people, but a standard reference. The algorithm was adapted for the Hubbard model by Hirsch in 1985 in [1] and this paper is still the "bible" for QMC simulations for the Hubbard model at finite temperatures. Earlier, Hirsch presented the discrete form for the Hubbard–Stratonovich transformation, which speeds up the simulation and makes the numerics simpler [15].

The treatment of the numerical instabilities is presented in [16], the treatment of the minus sign in [17]. The most complete introduction of the zero-temperature (projector) algorithm (canonical ensemble) can be found in [18]. The projection operator $e^{-\theta H}$ filters the ground state from a test function $|T\rangle$. This can be seen from the energy representation of the problem, in which all higher states $|n\rangle$, $n > 0$ are exponentially suppressed:

$$
e^{-\theta H}|T\rangle = e^{-\theta H} \sum_n \langle n|T\rangle \cdot |n\rangle
$$

$$
= e^{-\theta E_0}\left(\langle 0|T\rangle \cdot |0\rangle + \sum_{n>0} e^{-\theta(E_n-E_0)}\langle n|T\rangle \cdot |n\rangle\right) \quad .
$$

A nice review, which presents the concepts of zero- and finite-temperature algorithms, can be found in [11]. The formulae for the computation of time-dependent Greens functions for finite temperatures are given in [1], while for the ground state-algorithm [19] may be refered to. Methods to extract dynamical information (spectral functions, densities of states) from time-dependent Greens functions can be found in [19] and [20].

Appendix A: Statistical Monte Carlo Methods

The program `Montecarlotest.m` computes the probability distribution for a certain phase space using statistical Monte Carlo methods. It is a didactical attempt to show "why Monte Carlo works". A phase space (coordinate i) is set up: every state (point in the phase space) is assigned a certain "eigenenergy" E_i. At temperature T, the probability of the system occupying state i is given by the Boltzmann factor

$$
P_i = e^{-E_i/T} \quad .
$$

The phase space may be too large (e.g., continuous) to sample all possible states. Instead, one moves randomly from a coordinate i to the coordinate i' with the relative probability

$$
P_{\text{rel}}(i \to i') = e^{-(E_{i'}-E_i)/T} .
$$

If $E_{i'} < E_i$, one moves in any case. In the following program, the "move" in phase space is realized by randomly (using the random number generator rand) choosing a new coordinate itest from the vector x. The corresponding energies are stored in the vector energy.

The "general" area of the walk is stored in the vector xbin. It can be seen that the entries in xbin (the probabilities of visiting a certain area in the phase space) correspond to the Boltzmann probability. There, the statistical Monte Carlo method is able to sample a phase space without using all possible configurations. It can also be seen that through statistical fluctuations the accuracy of the method is limited. To increase the accuracy, more sweeps have to be computed. The probabilities are not normalized. The set of commands which are commented with %, allow the creation of Postscript-files.

Apart from generating all "physical" configurations using an "infinitely" long Monte Carlo run, one has to take several things into account when using "finite" computer time.

1. Thermalization. For a "realistic" simulation, before measurements can be made the thermodynamic system has to be in equilibrium. Imagine a ferromagnetic spin system above the critical temperature: if you set up an initial configuration with ordered spins, you have to allow the computer some time to destroy this ordering to obtain a "physical" configuration.

2. Independent configurations. If you move around in phase space, the measurements have to be taken at points "far enough apart" so that the measurements are not correlated (i.e., they are "statistically independent"). In Fig. 7, failure to achieve this would correspond to just sitting in an energy "valley", for example near coordinate 44 and generating configurations in the neighborhood. In that case, the Monte Carlo probability distribution would just be a spike near 44, the system would not visit the whole phase space.

3. Detailed balance. Apart from the fact that the whole phase space must be sampled, it must be sampled in such a way that the simulation does not create more offers to sample a certain region. In the example program Montecarlotest.m, the coordinate to be sampled is chosen randomly from the whole phase space. Choosing a "normal distribution" around some coordinate would violate detailed balance.

4. Error bars. Monte Carlo simulations generate measurements with statistical error bars which decrease with the number of measurements N_M as $1/\sqrt{N_M}$. This means that for useful results, sufficient Monte Carlo measurements have to be generated.

Appendix B: OCTAVE

Due to the large amount of linear algebra in the DQMC algorithms, the example programs are written in OCTAVE. OCTAVE is an interpreter for a matrix-oriented programming language. It is very fast on matrix operations, but due to

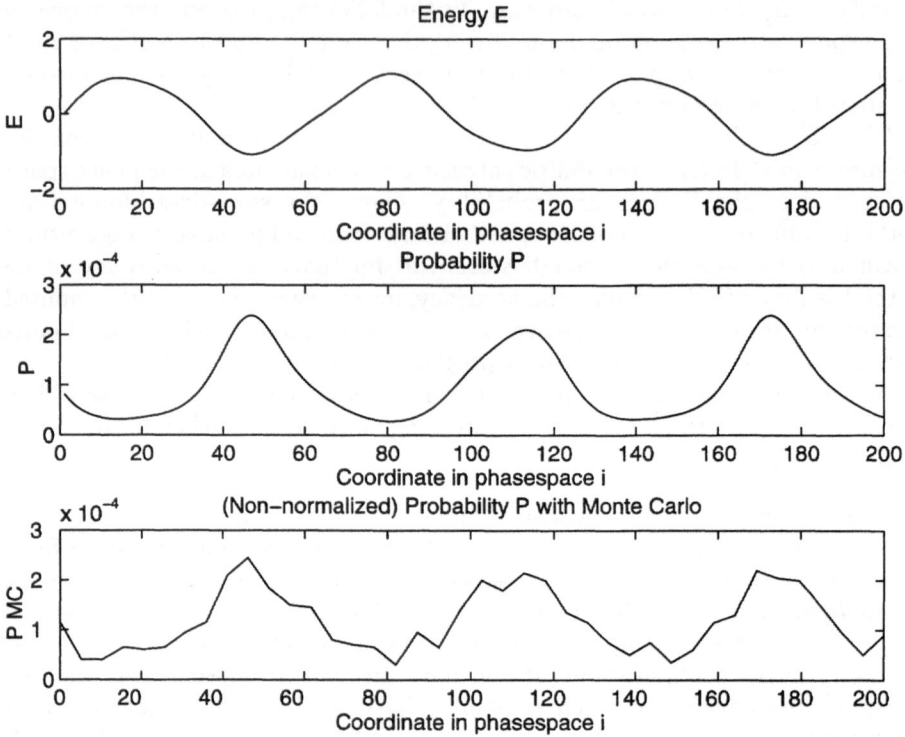

Fig. 7. Energy in phase space E, thermodynamic probability P for an arbitrary phase space and relative probability of hitting this part of phase space using Monte Carlo approach (*below*)

interpreting instead of compiling the code it is comparatively slow on inner DO and IF operations.

In OCTAVE, the visualization of data at any stage of the computation is possible. This allows fast debugging of complicated numerical codes because errors can be traced down using the resident data in the memory even after a program crash.

There is a vast (and increasing) number of "intrinsic functions" which can be applied to two-dimensional arrays, such as matrix decompositions, Fourier transforms etc., which makes the "language" also attractive for the processing of output data of "serious" computer simulations.

The language uses the overloading of operators in such a way that matrix and vector products are automatically recognized from the type of variables in real and complex arithmetic. Using the wrong type of data, such as row-wise instead of column-wise vectors, leads to error messages. The language can handle complex numbers in such a way that $\sqrt{-1}$ is automatically assigned the value $0 + i$.

There are some slight drawbacks, due to the "simple" structure of the language:

1. Only double precision is available, so that algorithms such as the Newton–Raphson algorithm for the iterative improvement of the solution of a set of equations cannot be computed in mixed precision.
2. The language can only handle two-dimensional arrays, so that arrays of higher dimensions must be unrolled.
3. Programming errors which result in the wrong kind of data structures, e.g., matrices instead of scalar values, are very often propagated through the program, because OCTAVE applies all following operations in the matrix sense.

OCTAVE is the public domain version of MATLAB, written by John W. Eaton at the Institute for Chemistry at the University of Texas. The source code is available via ftp at `ftp.che.utexas.edu` in the directory `/pub/octave`. Binaries are also available for most UNIX systems from other public domain file servers.

Appendix C: Exercises

The programming exercises are supposed to give an introduction to determinant quantum Monte Carlo methods from the point of view of matrix computations.

Exercise 1: Matrix exponentials

Ex. 1.1: (Hopping and boundary conditions) Set up the hopping matrix "by hand" for nearest neighbor hopping in
a) 1 dimension for a system of N lattice sites,
b) 2 dimensions for a system of $N_x \times N_y$ lattice sites.
c) Write a program which is able to perform these task.
The results should be visualized using the OCTAVE online graphics.

Ex. 1.2: (Linear Algebra first year) If you don't remember it, look up the theorem: "The eigenvalues of a hermitian/real symmetric matrix are real" in a reasonably good introductory text for linear algebra. Set up hermitian and real symmetric matrices of reasonable size and test this theorem experimentally using the `eig` function of OCTAVE.

Ex. 1.3: Using Ex. 1.2, think of a way to compute the exponential of a matrix and write a program in OCTAVE. Verify your results using the built-in matrix-exponential functions `expm`, `expm1`, `expm2`, `expm3`. and read up the documentation on `expm` in the OCTAVE handbook (see Appendix B). If you have access to MATLAB, study the help file for these functions.

Ex. 1.4: Write a program that compares the matrix exponential obtained by using the diagonalization with the matrix exponential obtained by using the Taylor series.

Exercise 2: The Trotter–Suzuki decomposition

Ex. 2.1: Set up a real symmetric matrix A. Compute $e^{-\beta A}$ for $\beta = 10, 1,$ $0.1, 0.01 \ldots$ What does the matrix look like for decreasing β?

Ex. 2.2: Set up real symmetric matrices A, B. Compute $e^{-\beta(A+B)}$ for different β directly and by using the TS decompositions of first and second order for different discretizations $d\tau = \beta/n$. Try to verify that more accurate decompositions give more accurate approximations for $e^{-\beta(A+B)}$. Compute the difference in the relative norm

$$\frac{\|A_{\text{exact}} - A_{\text{approx}}\|}{\|A_{\text{exact}}\|}$$

using the built-in OCTAVE function norm: `normrel=norm(Aexact-Aapprox)/` `norm(Aexact)`. Why does the TS transformation "work"? What do $e^{-\beta/nA}$, $e^{-\beta/nB}$ look like for large n? Remember that diagonal matrices commute.

Exercise 3: The Grand Canonical Quantum Monte Carlo

Ex. 3.1: Compare the lecture and the example program `ghd2main.m` and make sure that you understand which formulae correspond to which program parts.

Ex. 3.2: (Filling) Set the interaction u to zero and run the program for one sweep. Modify the chemical potential `mu` and see how the filling depends on the chemical potential. Remember that the electron densities can be obtained from the diagonal of the matrices `grup`, `grdn`. The filling is given by the densities on site i, $\langle n_i \rangle = \langle c_i^+ c_i \rangle = \langle 1 - c_i c_i^+ \rangle$; $\langle 1 - c_i c_i^+ \rangle$ can be obtained as diagonal element of `grup,grdn`.
Verify that you can use $\mu=$mu to fill ($\mu \to \infty$) or empty ($\mu \to -\infty$) the system.

Ex. 3.3: Rewrite the code so that you obtain the Greens functions in second order instead of first order. Two additional matrix multiplications are enough. This can be accomplished either in `compute_Greensf.m` or after the computation of the Greens function by wrapping:

$$G(m)^{1.\text{ord}} = (1 + B_m B_{m-1} \ldots B_1)^{-1}$$
$$G(m)^{2.\text{ord}} = e^{-\tau K/2} (1 + B_m B_{m-1} \ldots B_1)^{-1} e^{\tau K/2} \ .$$

Ex. 3.4: (Antiferromagnetism) What is the effect on the diagonal elements in the case of half filling if the interaction U is increased? What is the physical result?

Ex. 3.5: (Numerical Instabilities) Impose a symmetry on the HS Spins and see whether you can find this symmetry in the Greens function. See how far you can increase U and β before the symmetry vanishes.

Solutions to some of the exercises

Ex. 1.1: See example program `Hoppingmatrix.m`

Ex. 1.3: If matrix A is real symmetric, it can be decomposed into $A = U \cdot E \cdot U^T$ with unitary U and real diagonal E. Therefore, $\exp(A) = U \cdot \exp(E) \cdot U^T$

where exp(E) can be evaluated element-wise. (Proof: look at the Taylor series and remember that $UU^T = 1$ for unitary matrices.)

Ex. 1.4: You can also use the intrinsic functions expm2 (Taylor series) and expm3 (Eigenvalue decomposition) from OCTAVE.

Ex. 2.1: Real symmetric matrices A: for random matrices, a real symmetric random matrix can be set up in the following way:

```
beta=              % Insert a proper value
A=rand(20)         % set up a 20x20 random matrix
A=.5*(A+A')        % Matrix A plus its transpose
mesh(exp(-beta*A)) % make a mesh-plot of the matrix
```

For decreasing beta, the exponential of $-\beta * A$ should approach the identity matrix.

References

[1] J. E. Hirsch, Phys. Rev. B **31**, 4403 (1985)

[2] *Magnetismus von Festkörpern und Grenzflächen*, Vorlesungsmanuskript des 24. IFF–Ferienkurses von 1993, KFA Jülich (1993)

[3] K. Elk and W. Gasser, *Die Methode der Greenschen Funktionen in der Festkörperphysik*, (Akademie-Verlag, Berlin 1979)

[4] E. Dagotto, Rev. Mod. Phys. **66**, 763 (1994)

[5] H. de Raedt and W. von der Linden, Phys. Rev. B **45**, 8787 (1991)

[6] M. Suzuki, S. Miyashita, and A. Kuroda, Prog. Theor. Phys. **58**, 5 (1977)

[7] H. de Raedt, *Product Formula Algorithms for Solving the Time Dependent Schrödinger Equation*, Computer Physics Reports 7, (North-Holland, Amsterdam 1987)

[8] M. Suzuki, Phys. Lett. A **146**, 6319 (1992)

[9] J.B. Kogut, Rev. Mod. Phys. **51**, 4 (1979)

[10] K. Binder and D.W. Heermann, *Monte Carlo Simulation in Statistical Physics*, 2nd ed. (Springer, Berlin 1992)

[11] E.Y. Loh and J.E. Gubernatis, in *Electronic Phase Transitions*, eds. W. Hanke and Yu.K. Kopaev, (North-Holland, Amsterdam 1992), p. 177

[12] N. Metropolis, A.W. Rosenbluth, M.N. Rosenbluth, H. Teller, and E. Teller, J. Chem. Phys. **21**, 1087 (1953)

[13] J.E. Hirsch, R.L. Sugar, D.J. Scalapino, and R. Blankenbecler, Phys. Rev. B **26**, 5033 (1982)

[14] R. Blankenbecler, D.J. Scalapino, and R.L. Sugar, Phys. Rev. D **24**, 2278 (1981)

[15] J.E. Hirsch, Phys. Rev. B **28**, 4049 (1983)

[16] E.Y. Loh, J.E. Gubernatis, R.T. Scalettar, R.L. Sugar, and S.R. White, Stable Matrix Multiplication Algorithms for the Low-Temperature Numerical Simulations of Fermions, in *Interacting Electrons in Reduced Dimensions*, eds. D. Baeriswyl and D.K. Kampbell, (Plenum, New York 1989)

[17] E.Y. Loh, J.E. Gubernatis, R.T. Scalettar, S.R. White, D.J. Scalapino, and R.L. Sugar, Phys. Rev. B **41**, 9301 (1990)

[18] M. Imada and Y. Hatsugai, J. Phys. Soc. Japan **58**, 3752 (1989)

[19] W. von der Linden, Physics Reports **220**, 53 (1992)
[20] S.V. Meshkov and D.V. Berkov, Int. J. Mod. Phys. C **5**, 987 (1994)
[21] W. von der Linden, Applied Physics A **60**, 155 (1994)

Quantum Dynamics in Nanoscale Devices*

Hans De Raedt

Institute for Theoretical Physics and Materials Science Centre, University of
Groningen, Nijenborgh 4, NL-9747 AG Groningen, The Netherlands,
e-mail: deraedt@rugth2.th.rug.nl

Abstract. The purpose of this lecture is to introduce the general concepts for building
algorithms to solve the time-dependent Schrödinger equation and to discuss ways of
turning these concepts into unconditionally stable, accurate, and efficient simulation
algorithms. The approach is illustrated using results of a computer-simulation study of
charged-particle interferometry, combining features of both the Aharonov–Bohm and
Hanbury–Brown–Twiss experiment.

1 Introduction

Progress in nanoscale lithography has made it possible to perform "electron-
optics" experiments in solid-state devices [1, 2]. In an ideal device the motion of
the electrons is not affected by interactions with impurities, phonons etc., i.e.,
the electrons travel ballistically, just as they would do in ultra-high vacuum. In
real devices, typical distances for ballistic motion can be as large as $250\lambda_F$, λ_F
being the Fermi wavelength of the electrons [3].

A similar, but otherwise unrelated, breakthrough is the development of atom-
size field-electron-emission sources. Recent experiments using these atom-size
tips [4, 5] have demonstrated that they act as unusual electron beam sources,
emitting electrons at fairly low applied voltages (a few thousand volts or less)
with a small angular spread (of a few degrees). These properties make such electron sources very attractive for applications to electron microscopy, holography,
and interferometry.

From a physical point of view, both these nanoscale structures have at least
one important common feature: the characteristic dimensions of these devices
are comparable to the wavelength (typically the Fermi wavelength λ_F) of the
relevant particles (typically electrons). Under this stringent condition, a classical,
"billiard-ball" description of the particle motion is no longer valid. A calculation
of the device properties requires a full quantum-mechanical treatment.

The dynamic properties of a nonrelativistic quantum system is governed by
the time-dependent Schrödinger equation (TDSE)

$$i\hbar\frac{\partial}{\partial t}|\Phi(t)\rangle = \mathcal{H}|\Phi(t)\rangle \tag{1}$$

---

* Software included on the accompanying diskette.

where $|\Phi(t)\rangle$ represents the state of the system described by the Hamiltonian $\mathcal{H}$ (here and in the following we use $\mathcal{H}$ to denote the differential operator and H for the hermitian matrix representing $\mathcal{H}$). In analogy with ordinary differential equations, the formal solution of the matrix differential equation

$$\frac{\partial}{\partial x} U(x) = HU(x) \quad , \quad U(0) = I \ , \tag{2}$$

where I denotes the $M \times M$ unit matrix and H is a $M \times M$ matrix, is given by

$$U(x) = e^{xH} \tag{3}$$

and is called the exponential of the matrix H. In quantum physics and quantum statistical mechanics, the exponential of the Hamiltonian is a fundamental quantity. All methods for solving these problems compute, one way or another, (matrix elements of) the exponential of the matrix H. In the case of real-time quantum dynamics $x = -it/\hbar$ whereas for quantum statistical problems $x = -\beta = -1/k_BT$.

Formally, the exponential of a matrix H can be defined in terms of the Taylor series

$$e^{xH} = \sum_{n=0}^{\infty} \frac{x^n}{n!} H^n \tag{4}$$

just as if H were a number. For most problems of interest, there will not be enough memory to store the matrix H (typical applications require matrices of dimension $10^5 \times 10^5$ or larger) and hence there also will be no memory to store the full matrix e^{xH}. So let us concentrate on the other extreme, the calculation of an arbitrary matrix element $\langle\psi|e^{xH}|\psi'\rangle$. Although from a mathematical point of view, formal expansion (4) is all that is really needed, when it comes to computation (4) is quite useless. The reason is not so much that it is a Taylor series but rather that it contains powers of the matrix, indicating that simply summing the terms in (4) may be very inefficient (and indeed it is).

There is one particular case in which it is easy to compute the matrix element $\langle\psi|e^{xH}|\psi'\rangle$ namely if all the eigenvalues and eigenvectors are known. Indeed, from (4) it follows that

$$e^{xH}|\Phi_j\rangle = \sum_{n=0}^{\infty} \frac{x^n}{n!} H^n|\Phi_j\rangle = \sum_{n=0}^{\infty} \frac{x^n}{n!} E_j^n|\Phi_j\rangle = e^{xE_j}|\Phi_j\rangle \tag{5}$$

where (here and in the following) E_n denotes the n-th eigenvalue of the matrix H and $|\Phi_n\rangle$ is the corresponding eigenvector. We will label the eigenvalues such that $E_0 \leq E_1 \leq \ldots \leq E_{M-1}$ where M is the dimension of the matrix H. From (5) is follows that

$$\langle\psi|e^{xH}|\psi'\rangle = \sum_{j=0}^{M-1} \langle\psi|\Phi_j\rangle\langle\Phi_j|\psi'\rangle e^{xE_j} \ . \tag{6}$$

Of course, (6) is almost trivial but it is important to keep in mind that, except for some pathological cases, there seems to be no other practical way to compute the matrix element $\langle \psi | e^{xH} | \psi' \rangle$ without making approximations (assuming H is a large matrix). In general we don't know the solution of the eigenvalue problem of the matrix H, otherwise we would already have solved the most difficult part of the whole problem. Therefore (6) is not of practical use.

Solving the time-dependent Schrödinger equation for even a single particle moving in a non-trivial (electromagnetic) potential is not a simple matter. The main reason is that for most problems of interest, the dimension of the matrix representing $\mathcal{H}$ is quite large and although the dimension of the matrices involved is certainly not as large as in the case of typical many-body quantum systems, exact diagonalization techniques are quite useless. Indeed, a calculation of the time-development of the wave function by exact diagonalization techniques requires the knowledge of *all* eigenvectors and *all* eigenvalues (i.e. for a matrix of dimension $10^5 \times 10^5$ one needs $\approx 10^{13}$ Mb or more RAM to store these data). Thus, we need algorithms that do not use more than $\mathcal{O}(M+1)$ storage elements. Diagonalization methods that only require $\mathcal{O}(M + 1)$ memory locations are of no use either because they can only compute a (small) part of the spectrum. Methods based on importance sampling concepts cannot be employed at all because there is no criterion to decide which state is important or which is not: The "weight" of a state $e^{-itE_j/\hbar}$ is a complex number of "size" one.

Although from a numerical point of view the TDSE looks like any other differential equation which one should be able to solve by standard methods (e.g. Runge–Kutta) this similarity is misleading. Standard methods are based on (clever) truncations of the Taylor series expansion. It is easy to convince oneself that for the TDSE this implies that these numerical algorithms do not conserve the norm of the wave function [6]. This, from a physical point of view, is unacceptable because it means that during the numerical solution of the TDSE, the number of particles will change. Moreover, it can be shown [6] that this implies that these methods are not always stable with respect to rounding and other numerical errors. For completeness it should be mentioned that the Cranck–Nicholson algorithm does conserve the norm of the wave function and is unconditionally stable. However, except for one-dimensional problems, in terms of accuracy and efficiency it cannot compete with the algorithms to be discussed below [6].

A key concept in the construction of an algorithm for solving the TDSE is the so-called unconditional stability. An algorithm for solving the TDSE is unconditionally stable if the norm of the wavefunction is conserved *exactly*, at *all* times [6]. From a physical point of view, unconditional stability is obviously an essential requirement. If an algorithm is unconditionally stable the errors due to rounding, discretization etc. never run out of hand, irrespective of the choice of the grid, the time step, or the number of propagation steps. Recall that the formal solution of the TDSE is given by

$$|\Phi(m\tau)\rangle = e^{-im\tau H}|\Phi(t = 0)\rangle \qquad (7)$$

where $m = 0, 1, \ldots$ counts the number of time-steps τ. Here and in the following we absorb $\hbar$ in τ.

A simple, general recipe for constructing an unconditionally stable algorithm is to use unitary approximations to the (unitary) time-step operator $U(\tau) = e^{-i\tau H}$ [6]. The Trotter–Suzuki product formula approach, to be discussed in the next section, provides the necessary mathematical framework for constructing unconditionally stable, accurate, and efficient algorithms to solve the TDSE [6].

2 Theory

In all cases that we know of, the Hamiltonian is a sum of several contributions and each contribution itself is usually simple enough so that we can diagonalize it ourselves by some (simple) transformation. The Hamiltonian for a particle in a potential provides the most obvious example: we can write the Hamiltonian as a sum of the free-particle Hamiltonian and a potential energy. It is trivial to diagonalize both parts independently, but it is usually impossible to diagonalize the sum.

The question we can now put to ourselves is the following. Suppose we can diagonalize each of the terms in H by hand. Then, it is very reasonable to assume that we can also compute the exponential of each of the contributions separately (see the discussion in the previous section). Is there then a relation between the exponentials of each of the contributions to H and the exponential of H and if so, can we use it to compute the latter?

The answer to this question is affirmative and can be found in the mathematical literature of the previous century. The following fundamental result due to Lie [7] is the basis for the Trotter–Suzuki method for solving quantum problems. It expresses the exponential of a sum of two matrices as an infinite ordered product of the exponentials of the two individual matrices:

$$e^{x(A+B)} = \lim_{m\to\infty} \left(e^{xA/m} e^{xB/m} \right)^m \tag{8}$$

where, for our purposes, A and B are $M \times M$ matrices. The result (8) is called the Trotter formula. A first hint for understanding why (8) holds comes from comparing the two Taylor series

$$\begin{aligned} e^{x(A+B)/m} &= I + \frac{x}{m}(A + B) + \frac{1}{2}\frac{x^2}{m^2}(A + B)^2 + \mathcal{O}(x^3/m^3) \\ &= I + \frac{x}{m}(A + B) \\ &\quad + \frac{1}{2}\frac{x^2}{m^2}(A^2 + AB + BA + B^2) + \mathcal{O}(x^3/m^3) , \end{aligned} \tag{9}$$

and

$$e^{xA/m} e^{xB/m} = I + \frac{x}{m}(A + B) + \frac{1}{2}\frac{x^2}{m^2}(A^2 + 2AB + B^2) + \mathcal{O}(x^3/m^3) . \tag{10}$$

It is clear that for sufficiently large m, both expansions will agree up to terms of $\mathcal{O}(x^2\||[A,B]\||/m^2)$.[1] Thus, for sufficiently large m (how large depends on x and $\||[A,B]\||$),

$$e^{x(A+B)/m} \approx e^{xA/m}e^{xB/m} \; . \tag{11}$$

A mathematically rigorous treatment shows that [9]

$$\|e^{x(A+B)/m} - e^{xA/m}e^{xB/m}\| \leq \frac{x^2}{2m^2}\||[A,B]\|| e^{|x|(\|A\|+\|B\|)/m} \tag{12}$$

demonstrating that for finite m, the difference between the exponential of a sum of two matrices and the ordered product of the individual exponentials vanishes as x^2/m. As expected, (12) also reveals that this difference is zero if A and B commute: if $[A,B] = 0$ then $e^{x(A+B)} = e^{xA}e^{xB}$. For the case at hand $x = -im\tau$ the upper bound in (12) can be improved considerably to read [6]

$$\|e^{-i\tau(A+B)} - e^{-i\tau A}e^{-i\tau B}\| \leq \frac{\tau^2}{2}\||[A,B]\|| \; . \tag{13}$$

Except for the fact that we assumed that $H = A + B$, the above discussion has been extremely general. This suggests that one can apply the Trotter–Suzuki approach to a wide variety of problems and indeed one can. We have only discussed the most simple form of the Trotter formula. There now exist a vast number of extensions and generalizations of which we will consider only three.

The Trotter formula is readily generalized to the case of more than two contributions to H. Writing $H = \sum_{i=1}^{p} A_i$ it can be shown that [6, 9]

$$\|e^{-i\tau(A_1+...+A_p)} - e^{-i\tau A_1}...e^{-i\tau A_p}\| \leq \frac{\tau^2}{2} \sum_{1 \leq i < j \leq p} \||[A_i, A_j]\|| \; , \tag{14}$$

showing that any decomposition of the Hamiltonian qualifies as a candidate for applying the Trotter–Suzuki approach. This is an important conclusion because the flexibility of choosing the decomposition of H can be exploited to construct efficient algorithms. From the above discussion it is also clear that at no point an assumption was made about the "importance" of a particular contribution to H. This is the reason why the Trotter–Suzuki approach can be used where perturbation methods break down.

The product formula (11) is the simplest one can think of. We use it to define an approximate time-step operator

$$U_1(\tau) = e^{-i\tau A_1}...e^{-i\tau A_p} \; . \tag{15}$$

The hermitian conjugate of this operator is given by

$$U_1^\dagger(\tau) = e^{i\tau A_p}...e^{i\tau A_1} \tag{16}$$

from which it follows that

$$U_1(\tau)U_1^\dagger(\tau) = I \; . \tag{17}$$

---

[1] The norm of a matrix X is defined by $\|X\| = M^{-1/2}(\mathrm{Tr}X^\dagger X)^{1/2}$.

For simplicity we have assumed that H has been written as a sum of hermitian contributions, i.e., $A_i = A_i^\dagger$ for $i = 1, \ldots, p$. Result (17) implies that $(U_1(\tau))^{-1} = U_1^\dagger(\tau)$; hence $U_1(\tau)$ is a unitary approximation to the time-step operator $e^{-i\tau H}$. Thus, if we succeed in implementing $U_1(\tau)$, the resulting algorithm will be unconditionally stable by construction. The upperbound in (14) shows that the error made by replacing $e^{-i\tau H}$ by $U_1(\tau)$ will, in the worst case, never exceed a constant multiplied by τ^2. Therefore $U_1(\tau)$ is said to be a first-order approximant to the time-step operator.

For many applications it is necessary to employ an algorithm that is correct up to fourth order in the time step. Approximations correct up to second order are obtained by symmetrization [6, 8, 9]

$$U_2(\tau) = U_1^T(\tau/2)U_1(\tau/2) \tag{18}$$

where the U_1^T is the transpose of U_1. Trotter–Suzuki formula-based procedures to construct algorithms that are correct up to fourth order in the time step are given in [6]. From a practical point of view, a disadvantage of the fourth-order methods introduced in [6] is that they involve commutators of various contributions to the Hamiltonian. Recently Suzuki proposed a symmetrized fractal decomposition of the time-evolution operator [9]. Using this formula, a fourth-order algorithm is easily built from a second-order algorithm by applying [9]

$$U_4(\tau) = U_2(p\tau)U_2(p\tau)U_2((1 - 4p)\tau)U_2(p\tau)U_2(p\tau) \ , \tag{19}$$

where $p = 1/(4 - 4^{1/3})$ and $U_n(\tau)$ is the nth order approximation to $U(\tau)$, i.e., $U(\tau) = U_n(\tau) + \mathcal{O}(\tau^{n+1})$. It is trivial to show that all of the above approximations are unitary operators, hence the corresponding algorithms will be unconditionally stable. Note that once we have programmed a first-order algorithm, writing the code to implement the second- and fourth-order algorithms will normally only take a few seconds.

3   Data Analysis

The amount of data generated by a TDSE solver can be tremendous: the wave function is known at each time step so that in principle the TDSE solver can generate $\mathcal{O}(16mM)$ bytes of data in a single run. In typical applications, $M \approx 10^6$ and $m > 1000$. Clearly it may be difficult to store all these data. Therefore it is more appropriate to process the data as it is generated and compress it as much as possible.

A very appealing method for looking at the data is to make say 100 snapshots of the (coarse-grained) probability distribution and to use visualization techniques to produce digital videos [10, 11, 12]. Simply looking at these videos can bring a lot of insight; however, to be on the save side this insight should be confronted with the results of more advanced, numerical processing of the data.

The numerical processing of the raw data generated by the TDSE solver depends to a considerable extent on the details of the actual application. Therefore

I will not dwell on this subject in full generality but confine myself to a discussion of a simple, widely applicable method for extracting information about the spectrum of the model Hamiltonian from the raw data.

The idea is straightforward. Consider the matrix element $\langle \Phi(t=0)|\Phi(t)\rangle$ and write $|\Phi(t)\rangle$ in terms of the (unknown) eigenvalues and eigenvectors of H to obtain

$$f(t) \equiv \langle \Phi(t=0)|\Phi(t)\rangle = \sum_{j=0}^{M-1} |\langle \Phi(t=0)|\Phi_j\rangle|^2 e^{-itE_j}. \tag{20}$$

From (20) it is clear that the Fourier transform of $f(t)$ with respect to t will give direct information on all the E_js for which the overlap $|\langle \Phi(t=0)|\Phi_j\rangle|^2$ is not negligible. In other words, if we keep all the values of $f(t=m\tau)$ and compute its Fourier transform, we obtain the local (with respect to the initial state $\Phi(t=0)$) density of states.

4   Implementation

In general there will be many possibilities for writing down different decompositions of a given Hamiltonian. From a theoretical point of view, the choice of the decomposition is arbitrary. In practice, however, this flexibility can be exploited to a considerable extent to tailor the algorithm to the computer architecture on which the algorithm will execute. Of particular interest are decompositions that vectorize well and have a large intrinsic degree of parallelism.

We now illustrate the application of the theory presented above to the case of a charged (spinless) nonrelativistic particle in an external, static magnetic field **B**. The Hamiltonian reads

$$\mathcal{H} = \frac{1}{2m^*}\left(\mathbf{p} - e\mathbf{A}\right)^2 + V \tag{21}$$

where m^* is the effective mass of the particle with charge e, $\mathbf{p} = -i\hbar\nabla$ is the momentum operator, **A** represents the vector potential and V denotes the potential. For many applications it is sufficient to consider the choice $\mathbf{B} = (0,0,B(x,y))$ and $V = V(x,y)$. Then the problem is essentially two-dimensional and the motion of the particle may be confined to the x–y plane. For numerical work, there is no compelling reason to adopt the Coulomb gauge ($\mathrm{div}\mathbf{A} = 0$). A convenient choice for the vector potential is $\mathbf{A} = (A_x(x,y),0,0)$ where

$$A_x(x,y) = -\int_0^y B(x,y)\,\mathrm{d}y \ . \tag{22}$$

We will solve the TDSE for the Hamiltonian (21) with the boundary condition that the wave function is zero outside the simulation box, i.e., we assume perfectly reflecting boundaries.

For computational purposes it is expedient to express all quantities in dimensionless units. Fixing the unit of length by λ, wavevectors are measured in units of $k = 2\pi/\lambda$, energies in $E = \hbar^2 k^2 / 2m^*$, time in $\hbar/E$ and the vector potential in $e\lambda/\hbar$. Expressed in these dimensionless variables Hamiltonian (21) reads

$$\mathcal{H} = -\frac{1}{4\pi^2}\left\{ \left[\frac{\partial}{\partial x} - iA_x(x,y)\right]^2 + \frac{\partial^2}{\partial y^2}\right\} + V(x,y) \ . \tag{23}$$

An essential step in the construction of a numerical algorithm is to discretize the derivatives with respect to the x and y coordinates (of course, if the problem is defined on a lattice instead of in continuum space this step can be omitted). For many purposes, it is necessary to use a difference formula for the first and second derivatives in (23) that is accurate up to fourth order in the spatial mesh size δ. Using the standard four and five point difference formula [13] the discretized r.h.s. of (23) reads

$$
\begin{aligned}
H\Phi_{l,k}(t) = \frac{1}{48\pi^2\delta^2}\Big\{ &\left[1 - i\delta\left(A_{l,k} + A_{l+2,k}\right)\right]\Phi_{l+2,k}(t) \\
&+ \left[1 + i\delta\left(A_{l-2,k} + A_{l,k}\right)\right]\Phi_{l-2,k}(t) \\
&- 16\left[1 - \frac{i\delta}{2}\left(A_{l,k} + A_{l+1,k}\right)\right]\Phi_{l+1,k}(t) \\
&- 16\left[1 + \frac{i\delta}{2}\left(A_{l-1,k} + A_{l,k}\right)\right]\Phi_{l-1,k}(t) \\
&+ \Phi_{l,k+2} + \Phi_{l,k-2} - 16\Phi_{l,k+1} - 16\Phi_{l,k-1}(t) \\
&+ \left[60 + \delta^2 A_{l,k}^2 + 48\pi^2\delta^2 V_{l,k}\right]\Phi_{l,k}(t)\Big\} + \mathcal{O}(\delta^5) \ , \tag{24}
\end{aligned}
$$

where $\Phi_{l,k}(t) = \Phi(l\delta, k\delta, t)$ and $A_{l,k} = A_x(l\delta, k\delta)$. The discretized form (24) will provide a good approximation to the continuum problem if δ is substantially smaller than the smallest physical length scale. For the case at hand there are two such scales. One is the de Broglie wavelength of the particle (which by definition is equal to λ) and the other is the (smallest) magnetic length defined by $l_B^2 = \min_{(x,y)} |\hbar/eB(x,y)|$. From numerical calculations (not shown) it follows that $\delta = 0.1 \min(1, l_B)$ yields a good compromise between accuracy and the CPU time required to solve the TDSE.

Straightforward application of the product-formula recipe to (24) requires a cumbersome matrix notation. This can be avoided in the following way [6]. Defining

$$|\Phi(t)\rangle = \sum_{l=1}^{L_x}\sum_{k=1}^{L_y} \Phi_{l,k}(t)c_{l,k}^+|0\rangle \ , \tag{25}$$

where L_x and L_y are the number of grid points in the x and y direction respectively and $c_{l,k}^+$ creates a particle at lattice site (l, k), (25) can be written as

$$|\Phi(m\tau)\rangle = e^{-im\tau H}|\Phi(t=0)\rangle \tag{26}$$

where

$$H = \frac{1}{48\pi^2\delta^2} \sum_{l=1}^{L_x-2} \sum_{k=1}^{L_y} \Big\{ \Big[1 - i\delta(A_{l,k} + A_{l+2,k})\Big] c_{l,k}^+ c_{l+2,k}$$

$$+ \Big[1 + i\delta(A_{l,k} + A_{l+2,k})\Big] c_{l+2,k}^+ c_{l,k} \Big\}$$

$$- \frac{1}{3\pi^2\delta^2} \sum_{l=1}^{L_x-1} \sum_{k=1}^{L_y} \Big\{ \Big[1 - \frac{i\delta}{2}(A_{l,k} + A_{l+1,k})\Big] c_{l,k}^+ c_{l+1,k}$$

$$+ \Big[1 + \frac{i\delta}{2}(A_{l,k} + A_{l+1,k})\Big] c_{l+1,k}^+ c_{l,k} \Big\}$$

$$+ \frac{1}{48\pi^2\delta^2} \sum_{l=1}^{L_x} \sum_{k=1}^{L_y-2} (c_{l,k}^+ c_{l,k+2} + c_{l,k+2}^+ c_{l,k})$$

$$- \frac{1}{3\pi^2\delta^2} \sum_{l=1}^{L_x} \sum_{k=1}^{L_y-1} (c_{l,k}^+ c_{l,k+1} + c_{l,k+1}^+ c_{l,k})$$

$$+ \frac{1}{48\pi^2\delta^2} \sum_{l=1}^{L_x} \sum_{k=1}^{L_y} (60 + \delta^2 A_{l,k}^2 + 48\pi^2\delta^2 V_{l,k}) + \mathcal{O}(\delta^5) , \qquad (27)$$

and where $c_{l,k}$ annihilates a particle at lattice site (l,k).

Hamiltonian (27) describes a particle that moves on a two-dimensional lattice by making nearest- and next-nearest-neighbor jumps. This interpretation suggests that H should be written as a sum of terms that represent groups of independent jumps [6]. A convenient choice is

$$A_1 = \frac{1}{48\pi^2\delta^2} \sum_{l\in X_1} \sum_{k=1}^{L_y} \Big\{ \Big[1 - i\delta(A_{l,k} + A_{l+2,k})\Big] c_{l,k}^+ c_{l+2,k}$$

$$+ \Big[1 + i\delta(A_{l,k} + A_{l+2,k})\Big] c_{l+2,k}^+ c_{l,k} \Big\} ;$$

$$X_1 = \{1, 2, 5, 6, 9, 10, \ldots\} ,$$

$$A_2 = \frac{1}{48\pi^2\delta^2} \sum_{k=1}^{L_y} \sum_{l\in X_2} \Big\{ \Big[1 - i\delta(A_{l,k} + A_{l+2,k})\Big] c_{l,k}^+ c_{l+2,k}$$

$$+ \Big[1 + i\delta(A_{l,k} + A_{l+2,k})\Big] c_{l+2,k}^+ c_{l,k} \Big\} ;$$

$$X_2 = \{3, 4, 7, 8, 11, 12, \ldots\} ,$$

$$A_3 = \frac{-1}{3\pi^2\delta^2} \sum_{k=1}^{L_y} \sum_{l\in X_3} \Big\{ \Big[1 - \frac{i\delta}{2}(A_{l,k} + A_{l+1,k})\Big] c_{l,k}^+ c_{l+1,k}$$

$$+ \Big[1 + \frac{i\delta}{2}(A_{l,k} + A_{l+1,k})\Big] c_{l+1,k}^+ c_{l,k} \Big\} ;$$

$$X_3 = \{1, 3, 5, 7, 9, 11, \ldots\} \quad,$$

$$A_4 = \frac{-1}{3\pi^2\delta^2} \sum_{k=1}^{L_y} \sum_{l\in X_4} \left\{ \left[1 - \frac{i\delta}{2}(A_{l,k} + A_{l+1,k})\right] c_{l,k}^+ c_{l+1,k} \right.$$

$$\left. + \left[1 + \frac{i\delta}{2}(A_{l,k} + A_{l+1,k})\right] c_{l+1,k}^+ c_{l,k} \right\} \quad;$$

$$X_4 = \{2, 4, 6, 8, 10, 12, \ldots\} \quad,$$

$$A_5 = \frac{1}{48\pi^2\delta^2} \sum_{k\in X_5} \sum_{l=1}^{L_x} (c_{l,k}^+ c_{l,k+2} + c_{l,k+2}^+ c_{l,k}) \quad;$$

$$X_5 = \{1, 2, 5, 6, 9, 10, \ldots\} \quad,$$

$$A_6 = \frac{1}{48\pi^2\delta^2} \sum_{k\in X_6} \sum_{l=1}^{L_x} (c_{l,k}^+ c_{l,k+2} + c_{l,k+2}^+ c_{l,k}) \quad;$$

$$X_6 = \{3, 4, 7, 8, 11, 12, \ldots\} \quad,$$

$$A_7 = \frac{-1}{3\pi^2\delta^2} \sum_{k\in X_7} \sum_{l=1}^{L_x} (c_{l,k}^+ c_{l,k+1} + c_{l,k+1}^+ c_{l,k}) \quad;$$

$$X_7 = \{1, 2, 5, 6, 9, 10, \ldots\} \quad,$$

$$A_8 = \frac{-1}{3\pi^2\delta^2} \sum_{k\in X_8} \sum_{l=1}^{L_x} (c_{l,k}^+ c_{l,k+1} + c_{l,k+1}^+ c_{l,k}) \quad;$$

$$X_8 = \{3, 4, 7, 8, 11, 12, \ldots\} \quad,$$

$$A_9 = \frac{1}{48\pi^2\delta^2} \sum_{k=1}^{L_y} \sum_{l=1}^{L_x} (60 + \delta^2 A_{l,k}^2 + 48\pi^2\delta^2 V_{l,k}) \quad, \tag{28}$$

and

$$U_1(\tau) = \prod_{n=1}^{9} e^{-i\tau A_n} \tag{29}$$

is the first-order approximant from which the algorithm, correct up to fourth-order in the spatial (δ) and temporal (τ) mesh size, can be built.

Inspection of A_n for $n = 1, \ldots, 9$ shows that each of the terms commutes with all the other terms in the sum over k and l. This is because each of these terms corresponds to a jump of the particle between a pair of two isolated sites. For the purpose of implementation, this feature is of extreme importance [6]. To illustrate this point it is sufficient to consider the first of the exponents in (29) and use the fact that all terms commute to rewrite it as

$$e^{-i\tau A_1} = \prod_{k=1}^{L_y} \prod_{l\in X_1} \exp\left(\frac{-i\tau}{48\pi^2\delta^2}\left\{ \left[1 - i\delta(A_{l,k} + A_{l+2,k})\right] c_{l,k}^+ c_{l+2,k} \right.\right.$$

$$\left.\left. + \left[1 + i\delta(A_{l,k} + A_{l+2,k})\right] c_{l+2,k}^+ c_{l,k} \right\}\right) . \tag{30}$$

Furthermore, each of the exponents in the product (30) describes a two-site system, and the exponent of the corresponding 2×2 matrix can be worked out analytically [6]. In general

$$
\exp\left(\tau \alpha c_{l,k}^{+} c_{l',k'} + \tau \alpha^{*} c_{l',k'}^{+} c_{l,k}\right) = \left(c_{l,k}^{+} c_{l,k} + c_{l',k'}^{+} c_{l',k'}\right) \cos \tau |\alpha|
$$
$$
- i\left(\alpha^{*-1} c_{l,k}^{+} c_{l',k'} + \alpha^{-1} c_{l',k'}^{+} c_{l,k}\right) \sin \tau |\alpha| . \quad (31)
$$

The rather formal language used above easily translates into a computer program. All that (28)–(31) imply is that for each factor in product formula (29), one has to pick successive pairs of lattice points, get the values of the wave function for each pair of points and perform a plane rotation using matrices of the form

$$
M = \begin{pmatrix} \cos \tau |\alpha| & -i\alpha^{-1} \sin \tau |\alpha| \\ -i\alpha^{*-1} \sin \tau |\alpha| & \cos \tau |\alpha| \end{pmatrix} . \quad (32)
$$

For each of the nine exponentials[2], the order in which the pairs of points are processed is irrelevant. Therefore, the computation of each of the nine factors can be done entirely parallel, fully vectorized, or mixed parallel and vectorized depending on the computer architecture on which the code will execute. Further technical details on the implementation of this algorithm can be found elsewhere [14].

5   Application: Quantum Interference of Two Identical Particles

Trotter–Suzuki-based TDSE solvers have been employed in the study a variety of problems including wave localization in disordered and fractals [6, 15], electron emission from nanotips [16, 17, 10], Andreev reflection in mesoscopic systems [18, 11], the Aharonov–Bohm effect [14, 12], quantum interference of charged identical particles [19, 12], etc. Appealing features of the TDSE approach are that is extremely flexible in the sense that it can handle arbitrary geometries and (vector) potentials and that its numerical stability and accuracy are such that for all practical purposes the solution is exact.

Trotter–Suzuki-formula-based algorithms can and also have been used to solve the TDSE for few-body quantum systems, including a 26-site S=1/2 Heisenberg model [20]. The application of the TDSE approach is mainly limited by the storage needed for the (complex-valued) wave function.

In this section we will use the TDSE approach to study some aspects of quantum interference of charged identical particles. Recently Silverman [21, 22] proposed and analyzed a thought experiment that combines both the features of the Aharonov–Bohm (AB) and Hanbury–Brown and Twiss (HBT) experiments.

---

[2] The case $n = 9$ is even simpler than the other eight cases but for the sake of brevity, a discussion of this detail is omitted.

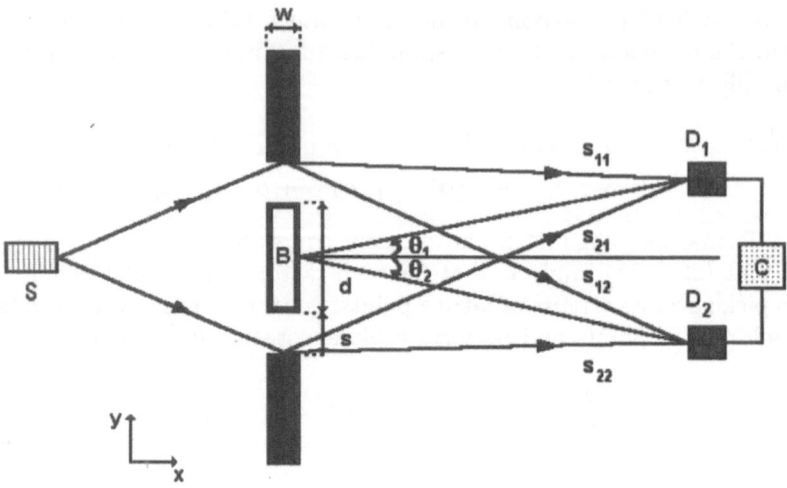

Fig. 1. Schematic view of the combined Aharonov–Bohm and Hanbury–Brown–Twiss apparatus. Charged fermions or bosons leave the source S, pass through the double slit and arrive at detectors D_1 and D_2. The signals of these detectors are multiplied in correlator C. The particles do not experience the magnetic field B enclosed in the double-slit apparatus

The former provides information on the effect of the magnetic field on correlations of two *amplitudes*. The latter on the other hand yields direct information on the correlations of two *intensities*, i.e., of correlations of *four* amplitudes.

A schematic view of the AB-HBT apparatus is shown in Fig. 1. Charged fermions or bosons leave the source S, pass through the double slit and arrive at detectors D_1 and D_2. In order for the particle statistics to be relevant at all, it is necessary that in the detection area the wave functions of two individual particles overlap. For simplicity, it is assumed that the particles do not interact. The particle statistics may affect the single-particle as well as two-particle interference. The former can be studied by considering the signal of only one of the two detectors. Information on the latter is contained in the cross-correlation of the signals of both detectors. Below we report some of our results [19] for the AB-HBT thought experiment, as obtained from the numerically exact solution of the time-dependent Schrödinger equation (TDSE) using the algorithm described above.

In practice we solve the two-particle TDSE subject to the boundary condition that the wave function is zero outside the simulation box (a grid of 1024×513 points), i.e., we assume perfectly reflecting boundaries. The algorithm that we use is accurate to fourth order in both the spatial and temporal mesh size [14]. Additional technical details can be found elsewhere [14]. Physical properties are calculated from the two-particle amplitude $\Phi(\mathbf{r}, \mathbf{r}', t) =$

$\phi_1(\mathbf{r}, t)\phi_2(\mathbf{r}', t)\pm\phi_2(\mathbf{r}, t)\phi_1(\mathbf{r}', t)$ where $\phi_1(\mathbf{r}, t)$ and $\phi_2(\mathbf{r}, t)$ are the single-particle amplitudes and the plus and minus signs correspond to the case of bosons and fermions respectively.

Let us first reproduce Silverman's analysis [21, 22]. Assume that the double-slit apparatus can be designed such that the probability for two identical particles (fermions or bosons) passing through the same slit can be made negligibly small. The two slits then act as the two sources in the HBT experiment with one modification: due to the presence of the vector potential the waves can pick up an extra phase shift. According to Silverman [21, 22], it immediately follows that the signal generated by the cross-correlator will *not* show any dependence on the confined magnetic field. The AB shifts for the direct process and the one in which the identical particles have been interchanged mutually cancel. This cancelation is independent of the fact that the particles are fermions or bosons [23].

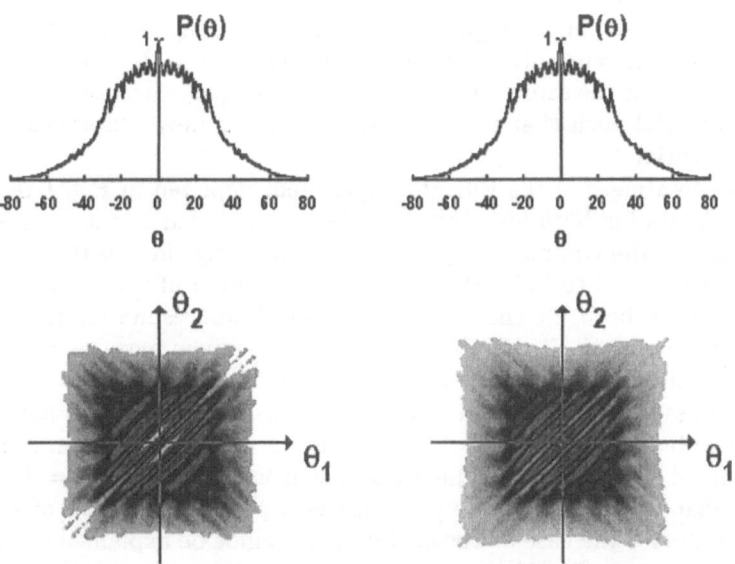

Fig. 2. Simulation results for single (*top*) and correlated (*bottom*) detector signal for $B = 0$, obtained from the solution of the TDSE for the initial state as described in the text. *Left* Signals generated by fermions. *Right* Signals generated by bosons. The corresponding pictures for $B = B_0$ are identical and not shown [3]

The basic assumption of Silverman's analysis is easily incorporated into a computer experiment. The initial two-particle wave function is a properly sym-

---

[3] B_0 is the magnetic field for which the Aharonov–Bohm shift of the interference pattern is equal to π.

metrized product of single-particle wave functions which, for simplicity, are taken to be Gaussians. Each Gaussian is positioned such that during propagation it effectively "hits" only one slit. The single (top) and correlated (bottom) signals, received by detectors placed far to the right of the slits for $B = 0$ for fermions (l.h.s) as well as for bosons (r.h.s.) are shown in Fig. 2.

For fermions the correlated signal for $\theta_1 = \theta_2$ vanishes, as required by the Pauli principle. This feature is hardly visible, due to the resolution we used to generate the pictures but it is present in the raw data. Within four digit accuracy, the corresponding data for $B = B_0$ (or, as a matter of fact, for any B) are identical to those for $B = 0$ [19]. Comparison of the cross-correlated intensities (bottom part) clearly lends support to Silverman's conclusion [21], [22]. However, it is also clear that the single-detector signals (upper part) do *not* exhibit the features characteristic of the AB effect. Under the conditions envisaged by Silverman, not only is there no AB effect in the cross-correlated signal, there is no AB effect at all.

The absence of the AB effect can be traced back to Silvermans's assumption that the slits can be regarded as sources, thereby eliminating the second, topologically different, alternative for a particle to reach the detector. A different route to arrive at the same conclusion is to invoke gauge invariance to choose the vector potential such that the two particles would never experience a nonzero vector potential.

A full treatment of the thought experiment depicted in Fig. 1 requires that *all* possibilities for *both* identical particles are included in the analysis. This is easily done in the computer experiment by changing the position and width of the Gaussians used to build the initial wave function of the fermions or bosons such that they both hit the two slits. Some of our results for the case of two bosons are shown in Fig. 3. Comparison of the upper parts of Fig. 3 provides direct evidence of the presence of the AB effect.

The cross-correlated boson intensities (r.h.s. of the bottom part of Fig. 3) clearly exhibit an AB-like effect. The positions of the maxima and minima are interchanged if the magnetic field changes from $B = 0$ to $B = B_0$. We have verified that the shift of these positions is a periodic function of the field B. These results for the case of boson statistics cannot be explained on the basis of Silverman's theory [21, 22].

In general we find that there is only a small quantitative difference between the fermion and boson single-detector signals: the interference fringes of the fermions are less pronounced than in the case of bosons, another manifestation of the Pauli principle. The differences in the cross-correlated fermion intensities, due to B, are not as clear as in the boson case. Substracting the $B = 0$ from the $B = B_0$ signal and plotting the absolute value of this difference (not shown) clearly shows that also the cross-correlated fermion intensity exhibits features that are characteristic of the AB effect [19]. The high symmetry in all the correlated signals shown is due to our choice $B = 0$ or B_0. The fact that we recover this symmetry in our simulation data provides an extra check on our method. If B is not a multiple of B_0, this high symmetry is lost but the salient features

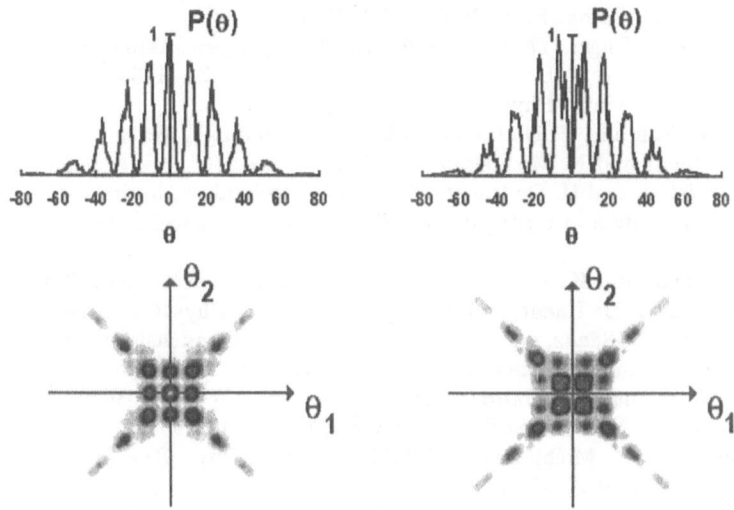

Fig. 3. Simulation results for single- (*top*) and correlated (*bottom*) detector signal generated by two bosons, as obtained from the solution of the TDSE for the initial state described in the text. *Left* $B = 0$. *Right* $B = B_0$

of the signals remain the same. From our numerical experiments, we conclude that in an AB-HBT experiment, an AB shift of the interference pattern will be observed in both the single- and two-detector experiments. The AB effect (in both experiments) is more pronounced for bosons than for fermions.

Acknowledgements

Most of the material used for this paper is taken from work done in collaboration with K. Michielsen. I would like to thank K. Michielsen for a critical reading of the manuscript. Financial support by the "Stichting voor Fundamenteel Onderzoek der Materie (FOM)", which is financially supported by the "Nederlandse Organisatie voor Wetenschappelijk Onderzoek (NWO)", the "Stichting Nationale Computer Faciliteiten (NCF)", and the EEC is gratefully acknowledged.

References

[1] S. Washburn, Nature **343**, 415 (1990)
[2] H. van Houten, C.W.J. Beenakkern, and B.J. van Wees, *Semiconductors and Semimetals* (M.A. Reed, New York 1990)
[3] H. van Houten et al., Phys. Rev. B **39**, 3556 (1989)
[4] Vu Thien Binh and J. Marien, Surface Science **102**, L539 (1988)
[5] Vu Thien Binh, J. Microscopy **152**, 355 (1988)

[6] H. De Raedt, Comp. Phys. Rep. **7**, 1 (1987)

[7] S. Lie and F. Engel, *Theorie der Informationsgruppen* (Teubner, Leipzig 1988)

[8] H. De Raedt and B. De Raedt, Phys. Rev. A **28**, 3575 (1983)

[9] M. Suzuki, J. Math. Phys. **32**, 400 (1991)

[10] K. Michielsen and H. De Raedt, "Electron Focussing", Multimedia Presentation

[11] H. De Raedt and K. Michielsen, "Andreev Reflection", Multimedia Presentation

[12] K. Michielsen and H. De Raedt, "Quantum Mechanics", Multimedia Presentation

[13] M. Abramowitz and I. Stegun, *Handbook of Mathematical Functions* (Dover, New York 1964)

[14] H. De Raedt and K. Michielsen, Computers in Physics **8**, 600 (1994)

[15] P. de Vries, H. De Raedt, and A. Lagendijk, Comp. Phys. Commun. **75**, 298 (1993)

[16] N. García, J.J. Sáenz, and H. De Raedt, J. Phys.: Condens. Matter **1**, 9931 (1989)

[17] H. De Raedt and K. Michielsen, in *Nanosources and Manipulation of Atoms Under High Fields and Temperatures: Applications*, eds. Vu Thien Bienh, N. García, and K. Dransfeld, NATO-ASI Series, (Kluwer, Dordrecht 1993)

[18] H. De Raedt, K. Michielsen, and T.M. Klapwijk, Phys. Rev. B **50**, 631 (1994)

[19] H. De Raedt and K. Michielsen, *Annalen der Physik* (in press)

[20] P. de Vries and H. De Raedt, Phys. Rev. B **47**, 7929 (1993)

[21] M.P. Silverman, Am. J. Phys. **61**, 514 (1993)

[22] M.P. Silverman, *And Yet It Moves: Strange Systems and Subtle Questions in Physics* (Cambridge, New York 1993)

[23] M.P. Silverman, private communication

Quantum Chaos

Hans Jürgen Korsch and Henning Wiescher

Fachbereich Physik, Universität Kaiserslautern, D-67653 Kaiserslautern, Germany,
e-mail: korsch@physik.uni-kl.de

Abstract. The study of dynamical quantum systems, which are classically chaotic, and the search for quantum manifestations of classical chaos, require large-scale numerical computations. Special numerical techniques developed and applied in such studies are discussed; the numerical solution of the time-dependent Schrödinger equation, the construction of quantum phase-space densities, quantum dynamics in phase space, the use of phase-space entropies for characterizing localization phenomena, etc. As an illustration, the dynamics of a driven one-dimensional anharmonic oscillator is studied, both classically and quantum mechanically. In addition, spectral properties and chaotic tunneling are addressed.

1 Classical and Quantum Chaos

During the last three decades it has become evident that the dynamics of simple Hamiltonian systems can be remarkably complex. Typical examples of such 'simple' systems are a point mass in a two-dimensional time-independent potential, or – even simpler – an explicitly time-dependent system with a single degree of freedom. In many important cases, such a system can be considered as time-periodic. The best-studied case is certainly the celebrated forced or parametrically excited harmonic oscillator. Numerous papers have been published that analyze the classical or quantum dynamics of such a harmonic oscillator in much detail. One should be aware of the fact, however, that the harmonic oscillator is a very special case: the classical equations of motion are linear. For all other systems this is not the case. Their behavior is studied in 'nonlinear dynamics'.

It is well-known that deterministic classical systems show erratic, irregular behavior. Moreover, this chaotic dynamics is a generic property. Typical systems show an intricate mixture of regular and irregular motion, whose structural organization can be most conveniently displayed by means of Poincaré sections of phase space. Rather than analyzing a full trajectory in the higher-dimensional phase space, one considers only its intersections with a reasonably chosen surface. In this way, the dynamics can be treated as a discrete mapping.

Such a discretization is of particular simplicity for time-periodic systems with one degree of freedom. Here one can look at the system stroboscopically, i.e., at times $t_n = nT$, $n = 0, 1, \ldots$, where T is the period. Figure 1 shows such a phase-space plot for a forced nonharmonic oscillator (see Sect. 4 for details). A synoptic plot of several trajectories with different initial conditions is shown. One observes regular regions in which the phase-space points generated by a trajectory trace out lines, the so-called invariant curves. Here the motion is regular. In addition, we find points that cover an area in phase space. In fact, all these points in Fig. 1 are computed from a *single* trajectory, a chaotic one. Classically, a chaotic trajectory – or more generally chaotic dynamics – is defined by an exponential separation of initially nearby trajectories in the long time limit (more precisely a positive Lyapunov exponent). Numerical studies (computer 'experiments') of this type are very helpful for studying chaotic dynamics and PC programs are available for many systems of interest in physics [28]. More details on the theory of the dynamics of Hamiltonian systems in the context of quantum dynamics can be found in textbooks (e.g., [44, 35, 21]).

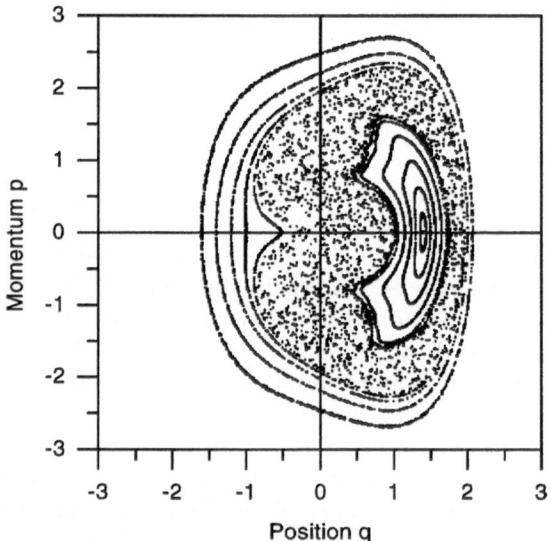

Fig. 1. Stroboscopic Poincaré section for a classically driven anharmonic oscillator

Quantum dynamics, however, is governed by the Schrödinger equation

$$i\hbar\,\dot{\psi} = H\,\psi\;,\tag{1}$$

which is a *linear* equation, and it is therefore questionable if such a time evolution can be chaotic. For example, it is straightforward to show that – for a finite (N) dimensional Hilbert space – the time-dependent coefficients in a basis set expansion $\psi(t) = \sum_n c_n(t)\phi_n$ satisfy a finite system of coupled linear equations. Moreover, by separating real and imaginary parts, e.g., $c_n = q_n + ip_n$, these differential equations can be written as the canonical equations of motion of a classical N-dimensional harmonic oscillator, which is certainly *not* chaotic.

Nevertheless, classical mechanics is the limit of quantum mechanics for $\hbar \to 0$, and therefore it is of fundamental importance to understand this highly nontrivial limit, and considerable work has been done. This fascinating field of contemporary research is denoted as *quantum chaos* [22] or *postmodern quantum mechanics* [24] and various excellent books [21, 40, 23, 14, 34] as well as recent conference proceedings [11, 15, 8, 26] summarize the results.

In order to find the quantum manifestations (if any) of classical chaos, much of the recent research is supported by large-scale computations (for small $\hbar$ the dimension of the Hilbert space is large, the wave functions are highly oscillatory, a long time propagation is of interest, etc) and special techniques for analyzing the system's behavior have been developed. Here, we will discuss some of these methods and illustrate their application to the seemingly simple case of time-periodic systems with one degree of freedom (see [9] for an overview of the properties of such systems).

2   Quantum Time Evolution

The time evolution of a quantum state ψ is determined by the Schrödinger equation (1), which can be solved numerically by numerous methods. Among the most popular and efficient ones is an expansion in a discrete basis set, which converts the Schrödinger equation into a set of coupled linear differential equations, and – with an increasing number of applications – the direct solution as a partial differential equation in the coordinate representation, e.g.,

$$i\hbar \frac{\partial \psi(\mathbf{x}, t)}{\partial t} = H\psi(\mathbf{x}, t) = \left(-\frac{\hbar^2}{2m} \nabla^2 + V(\mathbf{x}, t)\right) \psi(\mathbf{x}, t) \ . \tag{2}$$

In addition, a mixed treatment is also possible (and sometimes also the most efficient strategy), whereby some of the degrees of freedom are treated by a basis set expansion and the remaining ones are dealt with by solving a set of coupled differential equations.

One of the most powerful techniques for solving equation (2) is the so-called split-operator method [13], whereby the time propagator U for the Hamiltonian – split into a kinetic and potential energy part –

$$H\psi = \left(\frac{\mathbf{p}^2}{2m} + V(\mathbf{x})\right) \psi = (K + V)\,\psi \tag{3}$$

can be approximated for a (small) time step δ by

$$U(\delta) = \mathrm{e}^{-\frac{i}{\hbar} H \delta} \approx \mathrm{e}^{-\frac{i}{2\hbar} K \delta} \mathrm{e}^{-\frac{i}{\hbar} V \delta} \mathrm{e}^{-\frac{i}{2\hbar} K \delta} \ . \tag{4}$$

Observing that the operator V is diagonal in the coordinate representation (i.e., a simple multiplication by the number $V(\mathbf{x})$), whereas the operator K is diagonal in the momentum representation (i.e., a multiplication by the number $\mathbf{p}^2/2m$), the time propagation can be easily carried out by switching between the two representations by means of a fast Fourier transformation [39].

Different propagation schemes have been developed; for a critical comparison see [29]. Some more recent techniques are the staggered-time algorithm [49], the (unitary) fourth order method [12], a multigrid method [2], and the (t, t') method [37] based on an extended phase-space description for explicitly time-dependent systems.

Here we will restrict ourselves to one degree of freedom, i.e., the numerical solution of

$$i\hbar \frac{\partial \psi(x,t)}{\partial t} = -\frac{\hbar^2}{2m} \frac{\partial^2 \psi(x,t)}{\partial x^2} + V(x,t)\psi(x,t) \tag{5}$$

with boundary conditions $\psi(x_{\min}, t) = \psi(x_{\max}, t) = 0$. We describe in some detail a numerical method, the so-called Goldberg algorithm [16, 27], which works very well in this case.

First we discretize the coordinate x and construct the solution ψ_j only at points $x_j = x_{\min} + j\epsilon$, for $j = 1, \ldots, j_{\max}$ and $x_{j\max} = x_{\max}$.

Using a discrete expression for the second derivative

$$\left(\frac{\partial^2 \psi}{\partial x^2}\right)_j = \frac{1}{\varepsilon^2}[\psi_{j+1} - 2\psi_j + \psi_{j-1}] + O(\varepsilon^2) , \tag{6}$$

the action of the Hamiltonian is given by

$$H\psi_j = -\frac{\hbar^2}{2m\varepsilon^2}[\psi_{j+1} - 2\psi_j + \psi_{j-1}] + V_j\psi_j . \tag{7}$$

Discretizing the time in equidistant steps δ, i.e., $t_n = t_0 + n\delta$, the wavefunction ψ_{n+1} at time t_{n+1} is obtained from ψ_n by

$$\psi_j^{n+1} = e^{-\frac{i}{\hbar}H\delta} \psi_j^n . \tag{8}$$

Here it is *not* possible to approximate the exponential operator by $1 - iH\delta/\hbar$, because such a nonunitary approximation leads to instabilities of the temporal evolution. A uniform approximation is given by the Cayley form

$$e^{-\frac{i}{\hbar}H\delta} = \frac{1 - \frac{i}{2\hbar}H\delta}{1 + \frac{i}{2\hbar}H\delta} + O(\delta^3) , \tag{9}$$

which is correct up to second order in δ. Inserting now (9) into (8) and moving the denominator of (9) to the left-hand side in (8), we obtain the iteration scheme

$$\psi_{j+1}^{n+1} + (i\lambda - v_j^n - 2)\psi_j^{n+1} + \psi_{j-1}^{n+1} = -\psi_{j+1}^n + (i\lambda + v_j^n + 2)\psi_j^n - \psi_{j-1}^n \tag{10}$$

with $\lambda = 4m\epsilon^2/\hbar\delta$ and $v_j^n = 2m\varepsilon^2 V(x_j, t_n)/\hbar^2$. These difference equations are stable and unitary, however implicit. The solution of the tridiagonal matrix equation (10) is a standard problem of numerical mathematics, which is solved by recursion [16, 27] using the boundary conditions $\psi_0^n = \psi_{j_{max}}^n = 0$. This method can be easily implemented numerically and allows a fast real-time solution of the Schrödinger equation on a PC. For the time-independent case, PC programs with graphical representation are available [3] and can be used for illustrating phenomena of elementary quantum mechanics, e.g., motion of wavepackets in various potentials; dynamics of coherent and squeezed harmonic oscillator states; tunneling through potential barriers, etc.

A few remarks on a reasonable choice of the parameters will be helpful. First, the mesh width ϵ determines the smallest wavelength that can be accurately described on a discrete grid. A typical choice is $\epsilon = \hbar/5p_{max}$, where p_{max} is the largest classical momentum. A reasonable choice of the time step is $\delta = m\epsilon^2/\hbar$, leading to a balance of the errors induced by time and space discretization (see [16, 27, 39] for more details).

It should be noted that a direct extension of the Goldberg algorithm to systems with more degrees of freedom requires a matrix inversion at each time step (the solution by recursion is no longer possible) and is therefore inefficient. Special techniques have been developed, however, which reduce the problem to intertwined one-dimensional ones; for details see [42, 43].

3   Quantum State Tomography

3.1   Phase-Space Distributions

In classical dynamics, the equations of motion (or experimental measurements) yield the trajectory $q(t)$ of the system for given initial conditions. This trajectory contains all information, but important dynamical features are only visible if they are carefully extracted, e.g., by plotting the trajectory in adequate variables. Typically, one analyzes the dynamics in phase space (p, q), where the momentum $p(t)$ can be obtained from $q(t)$ by differentiation. In addition, one can restrict oneself to a section of phase space, the Poincaré section, as, for example, in the stroboscopic plot shown in Fig. 1 for a driven anharmonic oscillator. Such a plot reveals the dynamical properties of the system for all initial conditions and it provides a *global* description of the dynamical features of the system.

In quantum mechanics the situation is similar. As described above, one can numerically generate the wavefunction $\psi(x, t)$ in coordinate space as a function of time, but the essential dynamical features are still to be determined. The absolute square $|\psi(x, t)|^2$ yields the probability of finding the particle at a given position, and by means of a Fourier transform of $\psi(x, t)$ to momentum space one obtains the overall probability for the momentum. In order to obtain information about the momentum distribution at a given position, one can use the Gabor (or Fourier window) transformation

$$\langle p, q | \psi \rangle = \sqrt[4]{\frac{s}{\pi\hbar}} \int \exp\left[\frac{-s(x-q)^2}{2\hbar} - i\frac{p}{\hbar}\left(x - \frac{q}{2}\right) \right] \psi(x)\,dx \ . \qquad (11)$$

This is a Fourier transform to momentum (p) space, which is weighted by a Gaussian window centered at position q. The so-called squeezing parameter s controls the width of the window. Up to a multiplicative factor, $\langle p, q|\psi\rangle$ is equal to the momentum distribution for $s = 0$, and the coordinate representation is reproduced for large s. Equation (11) can also be expressed as a projection of the wavefunction onto a so-called minimum uncertainty wavepacket (also called a coherent state):

$$\phi_{p,q}(x) = \sqrt[4]{\frac{s}{\pi\hbar}} \exp\left[\frac{-s(x-q)^2}{2\hbar} + i\frac{p}{\hbar}\left(x - \frac{q}{2}\right)\right] , \tag{12}$$

a state with mean values q and p for position and momentum, respectively, and the uncertainties $\Delta p = \sqrt{\hbar s/2}$, $\Delta q = \sqrt{\hbar/2s}$, $\Delta p\,\Delta q = \hbar/2$. The squeezing parameter $s = \Delta p/\Delta q$ determines the ratio of the uncertainties and can be adapted to the problem under investigation. The absolute square

$$\rho_H(p,q) = |\langle p, q|\psi\rangle|^2 = \left|\int \phi_{p,q}^*(x)\psi(x)\,dx\right|^2 \tag{13}$$

with normalization

$$\frac{1}{2\pi\hbar}\int \rho_H(p,q)\,dp\,dq = 1 \tag{14}$$

is called the Husimi density [25]. It provides a quantum mechanical (quasi-) probability distribution in phase space for a given wavefunction ψ and is very useful in an analysis of the classical-quantum correspondence in dynamical systems [45].

3.2 Phase-Space Entropy

The overall degree of localization in phase space can be obtained from the average information content measured by the phase-space entropy,

$$S = -\frac{1}{2\pi\hbar}\int \rho_H(p,q)\,\ln\rho_H(p,q)\,dp\,dq . \tag{15}$$

This entropy satisfies the inequality $S \geq 1$ [50], which corresponds to the uncertainty relation $\Delta p\,\Delta q \geq \hbar/2$. The quantity e^S measures the number of minimum uncertainty states populated by the wavepacket and $A = 2\pi\hbar\,e^S$ is the space area covered by the phase-space distribution. It is instructive to calculate the entropy of a minimum uncertainty wavepacket (12) with squeezing parameter s_0 analyzed by minimum uncertainty states with squeezing s_1. The result is simply

$$S = 1 + \ln\frac{s_0 + s_1}{2\sqrt{s_0 s_1}} . \tag{16}$$

Note that for $s_0 = s_1$ one obtains the smallest possible entropy $S = 1$, as expected for a minimum uncertainty state [50].

As an example, Figs. 2 and 3 show the time evolution of the Husimi distribution of a minimum uncertainty wavepacket with $s_0 = s_1 = 1$ and $s_0 = s_1 = 2$ moving in a harmonic potential with unit mass and frequency. In Fig. 2, the wavefunction is a coherent state of the harmonic oscillator and moves without changing its form. In Fig. 3, we have a squeezed oscillator state, which changes its form and uncertainty product as monitored by the entropy S.

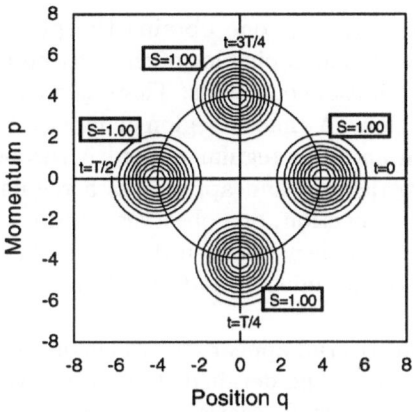

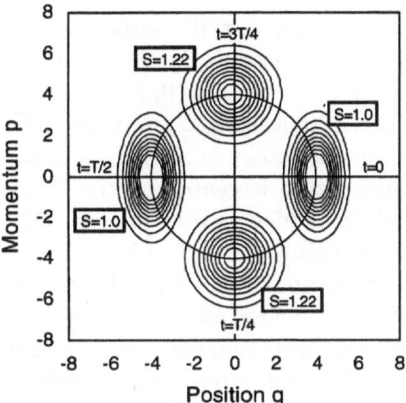

Fig. 2. Contour plot of the Husimi distribution (squeezing parameter $s_1 = 1$) for an initial minimum uncertainty wavepacket ($s_0 = 1$) moving in a harmonic oscillator with unit mass and frequency

Fig. 3. Contour plot of the Husimi distribution (squeezing parameter $s_1 = 2$) for an initial minimum uncertainty wavepacket ($s_0 = 2$) moving in a harmonic oscillator with unit mass and frequency

4   Case Study: A Driven Anharmonic Quantum Oscillator

Various paradigmatic systems are investigated repeatedly in the literature to explore the properties of quantum systems, which are classically chaotic. Here we study the time evolution of a wavepacket for a forced quartic oscillator

$$H(p, q, t) = \frac{p^2}{2m} + bq^4 - fq \cos(\omega t) ,$$ (17)

which is time-periodic with period $T = 2\pi/\omega$. We choose parameter values $b = 0.25$, $f = 0.5$ and $\omega = 1$, a case for which the classical-quantum correspondence [5, 6] and the semiclassical EBK quantization of regular quasienergy states [4, 46, 47] has been investigated recently (see also [31], where different parameters are chosen).

4.1   Classical Phase-Space Dynamics

The classical dynamics for this system is typical and shows a mixed regular and chaotic behavior depending on the initial conditions. Solving the classical equations of motion

$$\frac{dp}{dt} = -\frac{\partial H}{\partial q} = -4bq^3 + f\cos(\omega t) \quad , \quad \frac{dq}{dt} = \frac{\partial H}{\partial p} = \frac{p}{m} \tag{18}$$

for a specified initial condition $(p(0), q(0)) = (p_0, q_0)$ one obtains the phase-space trajectory $(p(t), q(t))$. Figure 1 shows a stroboscopic plot of the trajectory at times $t_n = nT$, $n = 0, 1, 2, \ldots$, for selected initial conditions. There is a clear division of phase space into three different regions. A chaotic region generated by a single trajectory is sharply separated from an outer regular region. A second regular region is centered on a T-periodic trajectory and appears as a regular island embedded in a chaotic sea. By closer inspection, one observes additional smaller chains of stability islands close to the boundary between the inner island and the chaotic sea. The phase-space area of the inner island is 2.25 and the chaotic sea covers an area of 7.85.

For different choices of the parameters, the overall appearance (a chaotic sea bounded by an outer regular region) is the same. The detailed structure of the inner stability islands changes, however, and shows characteristic bifurcations. The case studied here is selected because of its structural simplicity.

4.2   Quantum Phase-Space Dynamics

A minimum uncertainty wavepacket $\psi_{p_0,q_0}(x, 0)$, see (12), localized initially at a position (p_0, q_0) in phase space is propagated in time using a value $\hbar = 0.05$ (note that in the dimensionless units used here – by, for example, setting the field frequency equal to unity – the Planck constant of the system depends on the parameters and can therefore be adjusted). At times $t_n = nT$, the Husimi density

$$\rho_H(p, q; p_0, q_0; t_n) = \left| \int \phi_{p,q}^*(x)\psi_{p_0,q_0}(x, t_n)dx \right|^2 \tag{19}$$

is computed on a grid of 50×50 points in (p, q) space. The grid covers the same region as the classical phase space shown in Fig. 1.

Let us first study the quantum phase-space dynamics for a wavepacket localized at $(p_0, q_0) = (0, 0.6)$ initially, which is inside the classically chaotic region (the squeezing parameter of $s = 5$ is adapted to a harmonic approximation to the center of the inner regular island). In the computation, the Goldberg algorithm (see Sect. 2) is used with mesh size $\varepsilon = 0.005$ and time step $\delta = 0.001$. Figures 4 and 5 show the Husimi densities for the first 30 periods as contour plots. Dark regions mark large probabilities and correspond to strong localization of the wavefunction.

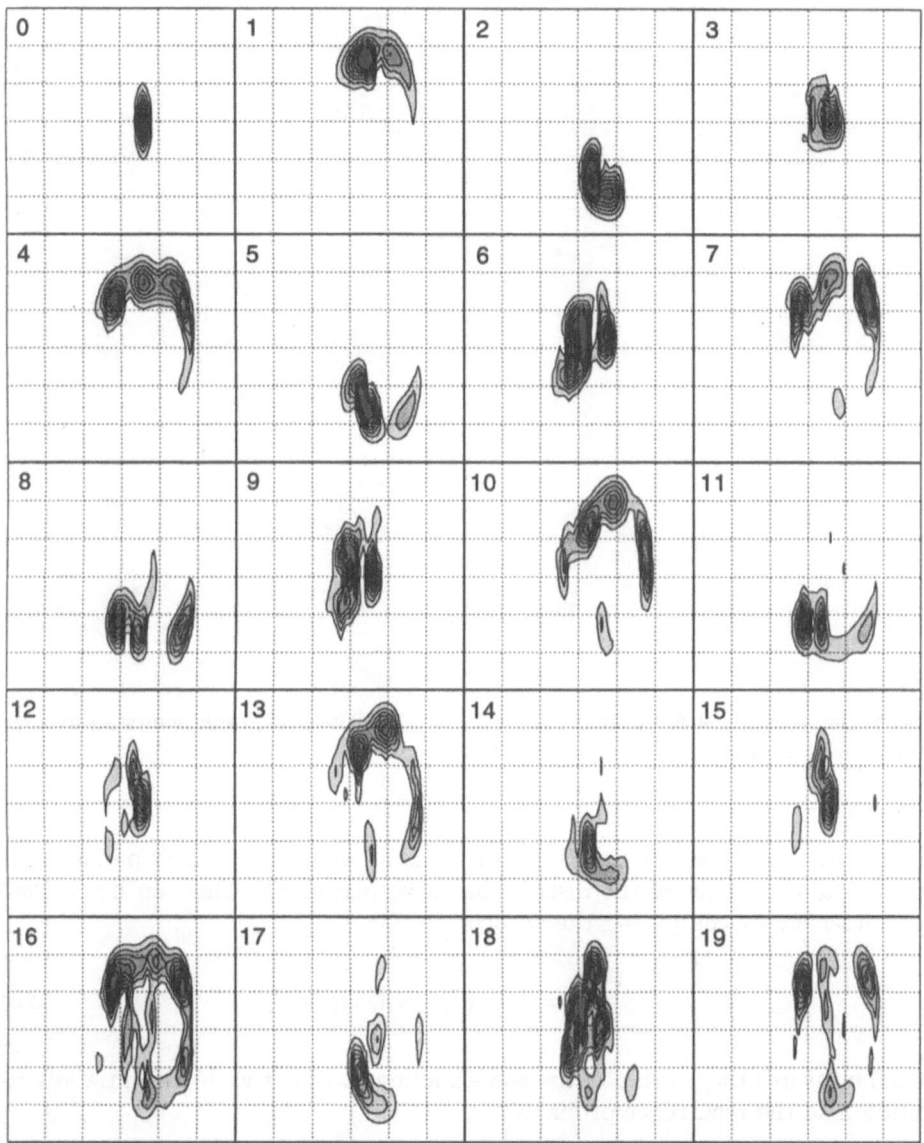

Fig. 4. Contour plots of the first 20 Husimi distributions at times $t_n = nT$. Shown is the region $[-3 : 3]$ with q on the horizontal and p on the vertical axis. Picture number 0 shows the initial minimum uncertainty state at $(p_0, q_0) = (0, 0.6)$ with squeezing parameter $s = 5$

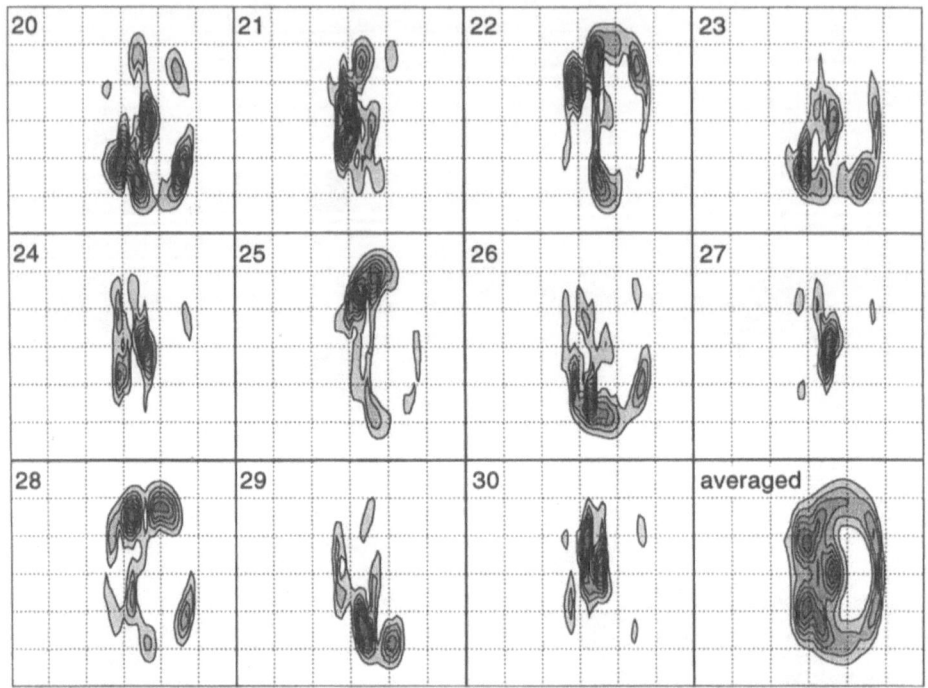

Fig. 5. Same as Fig. 4 for times t_n, $n = 20, \ldots, 30$. The last graph shows an average over periods 20 to 120

The first impression from Figs. 4 and 5 is of an approximately periodic circulation of the center of the distribution with period $3T$. This can be checked quantitatively by computing the autocorrelation

$$C(t, t_0) = \int \psi^*(x, t_0)\psi(x, t)\mathrm{d}x \; , \tag{20}$$

which measures the overlap of the wavefunction at time t with the initial distribution, and the recurrence probability

$$P_\mathrm{R}(t, t_0) = |C(t, t_0)|^2 \; . \tag{21}$$

Figure 6 shows the recurrence probability for the first 50 periods. Up to $t = 20\,T$, one observes clear maxima at multiples of three. This periodicity is reflected in the corresponding Fourier spectrum shown in Fig. 7. The strongest peak of the frequency spectrum appears at $\nu = 0.33/T$, which corresponds to period three. The same period-three circulation can be found if an initially Gaussian ensemble is propagated classically.

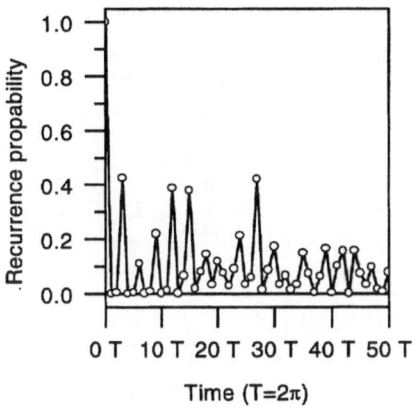

Fig. 6. Probability $C(t, t_0)$ for recurrence of the wavepacket at its initial position

Fig. 7. Fourier spectrum of the recurrence probability shown in Fig. 6

The increasing delocalization of the wavepacket can be measured quantitatively by means of the phase-space entropy (15). Figure 8 shows the entropy

$$S_{p_0, q_0}(t_n) = -\frac{1}{2\pi\hbar} \int \rho_{\mathrm{H}}(p, q; p_0, q_0; t_n) \ln \rho_{\mathrm{H}}(p, q; p_0, q_0; t_n) \, dp \, dq \qquad (22)$$

for the first 120 periods. Starting from the value of 1 for the initial minimum uncertainty state at time zero, the entropy increases within the first 20 periods, then it flattens into a plateau. For long times, the entropy fluctuates almost

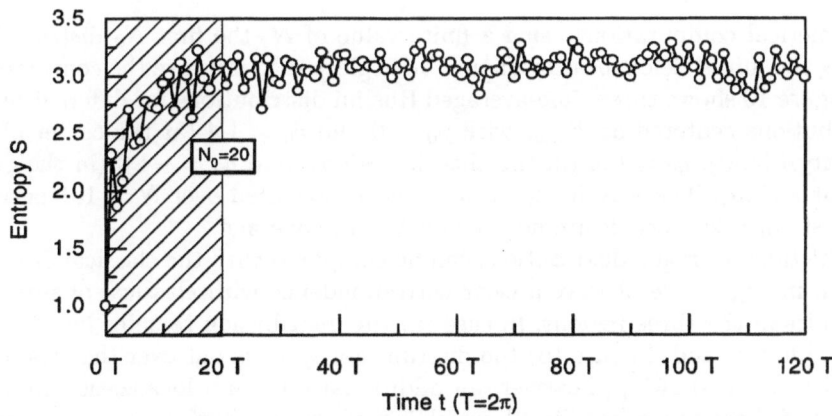

Fig. 8. Time dependence of the phase-space entropy for the distributions shown in Figs. 6 and 7

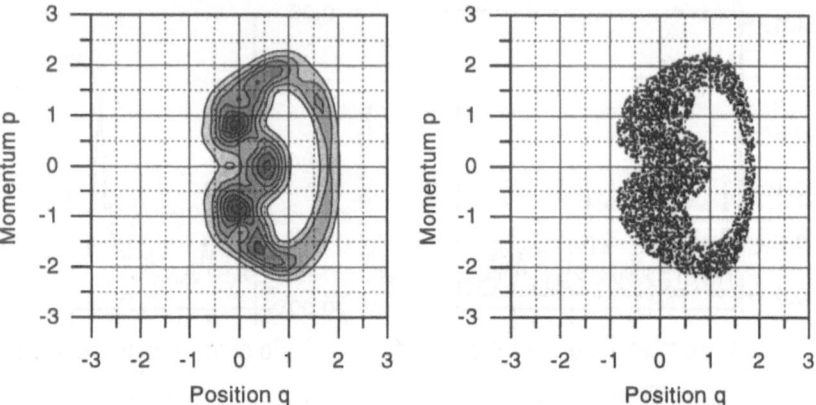

Fig. 9. Contour plot of a time-averaged quantum Husimi distribution for wavepacket (c) and classical Poincaré section for a chaotic trajectory

erratically with an average value of about $\overline{S} = 3.1$, which is somewhat below the value of $S_{\text{cl}} = \ln(A/2\pi\hbar) \approx 3.2$ obtained from the classical chaotic phase-space area $A = 7.85$. The entropy difference is due to the fluctuation of the quantum dynamics in contrast to the classical dynamics, which approaches a uniform limiting distribution (see also the more detailed analysis for a driven-rotor system [33] based on the random vector model).

One can also study the long time average of the Husimi distributions

$$\bar{\rho}_{\text{H}}(p, q; p_0, q_0) = \lim_{N \to \infty} \frac{1}{N - N_0 + 1} \sum_{n=N_0}^{N} \rho_{\text{H}}(p, q; p_0, q_0; t_n) ; \qquad (23)$$

in numerical computations using a finite value of N, the first N_0 distributions during the initial delocalization should be neglected to improve the convergence.

Figure 10 shows three time-averaged Husimi distributions for different initial distributions centered at ψ_{p_0, q_0} with $p_0 = 0$ and $q_0 = 1.4$ (at the center of the stability island), $q_0 = 0.6$ (in the chaotic region), and $q_0 = -1.0$ (in the outer regular region). The wavefunction has been propagated over $N = 101$ periods; the first $N_0 = 20$ periods are not included in the average.

The time-averaged distributions can be compared with the classical Poincaré section in Fig. 1. We observe a clear correspondence with classical phase-space dynamics in the three regions. In case (a) the distribution remains localized on the stability island. In case (b) the distribution spreads out over the classically chaotic region, showing, however, an additional quantum localization in three regions of phase space (see Fig. 9), which is a quantum interference phenomenon. On the other hand, the strong concentration on the region close to $(0, -1)$ in case (c) is a classical effect [4], which is also reproduced by propagation of classical phase-space densities.

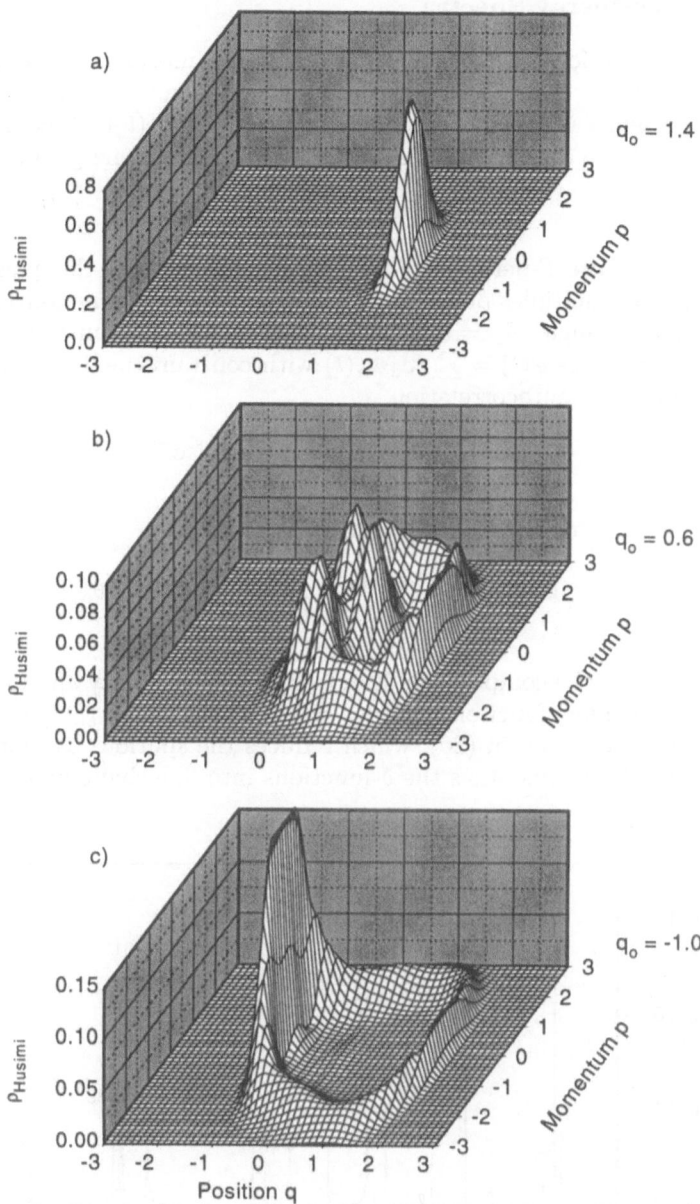

Fig. 10. Time-averaged Husimi distributions of a minimum uncertainty wavepacket located initially at ψ_{p_0,q_0} with $p_0 = 0$ and $q_0 = 1.4$ (at the center of the stability island), $q_0 = 0.6$ (in the chaotic region), and $q_0 = -1.0$ (in the outer regular region)

4.3   Quasienergy Spectra

The time evolution of a wavepacket also provides information about the spectrum of the system, which can be computed from the autocorrelation function (20). In the present case of time-periodic Hamiltonians $H(t + T) = H(t)$, this is the spectrum of the quasienergies ϵ_α defined by the quasienergy states (or Floquet states)

$$\psi_\alpha(t) = e^{-i\epsilon_\alpha t/\hbar} u_\alpha(t) \ , \tag{24}$$

where the u_α are T-periodic. This definition determines the quasienergies only up to integer multiples of $\hbar\omega = h/T$. Often it is therefore convenient to introduce the quasiangles $\theta_\alpha = \epsilon_\alpha T/\hbar$. Expanding the wavefunction in terms of the quasienergy states $\psi(t) = \sum_\alpha c_\alpha \psi_\alpha(t)$ with constant coefficients c_α, the Fourier transform of the autocorrelation

$$C_n = C(t_n, t_0) = \sum_\alpha |c_\alpha|^2 e^{-i\theta_\alpha n} \tag{25}$$

after n periods (compare (20)) yields

$$C(\theta) = \sum_n C_n e^{i\theta n} = 2\pi \sum_\alpha |c_\alpha|^2 \delta(\theta - \theta_\alpha) \ . \tag{26}$$

In practise, the computation is not extended to infinity and for a finite cutoff at n_{max} a window function, e.g., $w_n = (1 - \cos(2\pi n/n_{max}))/2$, must be introduced into the n-sum in (26), which reduces the spurious oscillations produced by the cutoff and smoothes the δ-functions into line shape functions $\mathcal{L}(\theta - \theta_\alpha)$,

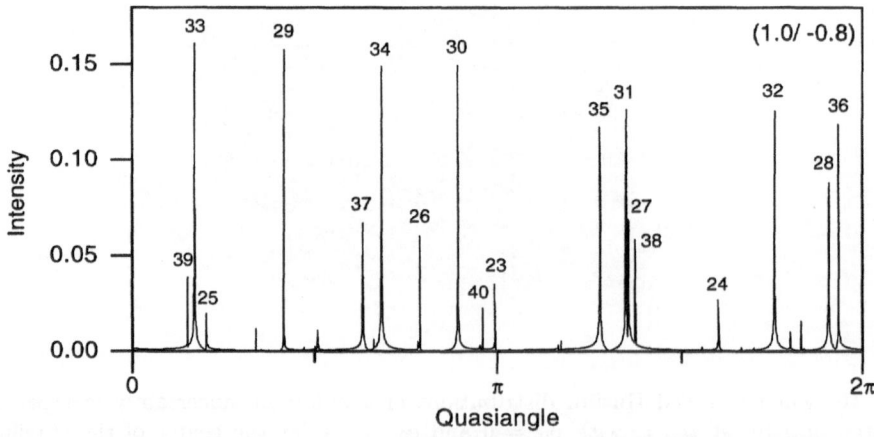

Fig. 11. Fourier transform of the autocorrelation function for a wavepacket placed initially at $(1, -0.8)$, which is inside the chaotic sea and close to the boundary to the outer regular region (see Fig. 1). The peaks appear at the quasiangles θ_α

which are determined by the window function. In addition, the quasienergy function at time $t_0 = 0$ is given by

$$\psi_\alpha(0) \propto \sum_n^N \psi(t_n) w_n e^{i\theta_\alpha n} \ , \tag{27}$$

where the proportionality constant is obtained by normalization. As an example, Fig. 11 shows the Fourier transform of the autocorrelation function for a wavepacket initially placed at $(1, -0.8)$, i.e., inside the chaotic region of Fig. 1. The peaks appear at the quasiangles θ_α.

A subsequent analysis of the Husimi distributions of the corresponding quasienergy states ψ_α provides information about the localization properties on different regions in phase space. Furthermore, by computation of a sufficiently large number of quasienergies or quasiangles, the statistical properties of the quasienergy spectra can be tested. The prediction is, for example, that the nearest-neighbor distance follows a Poisson distribution for those states localizing on a regular region, whereas those corresponding to a chaotic regime are Wigner distributed (see, for example, [23] and references therein).

4.4   Chaotic Tunneling

As an interesting application, one can investigate the influence of a time-periodic field on the dynamics of a wavepacket moving in a bistable potential, e.g. a double minimum potential. One system studied recently by various authors is the potential

$$V(q, t) = bq^4 - dq^2 + fq \cos(\omega t) \ , \tag{28}$$

a frictionless Duffing oscillator, with values $b = 0.5$, $d = 10$, $\omega = 6.07$, and $\hbar = 1$ for the parameters [30].

For a vanishing external driving field, we have a simple time-independent double-well potential and the energy E is conserved. For energies below the barrier, a quantum wavepacket localized in the left potential well tunnels through the barrier – the classically inaccessible region in phase space – and appears on the right-hand side. This process continues and we observe a tunneling oscillation between the two minima. This phenomenon is well understood and can be described semiclassically (e.g., [10]) in terms of the action integral

$$\kappa = \frac{1}{\hbar} \int_{q_-}^{q_+} \sqrt{2m|E - V(q)|} \, dq \tag{29}$$

over the barrier, where $q_\pm$ are the turning points. Well below the barrier, i.e., for large κ, the tunneling probability is $e^{-2\kappa}$ and the tunneling splitting of the almost degenerate energy eigenvalues is given by

$$\Delta E \approx \frac{\hbar\bar{\omega}}{\pi} e^{-\kappa} \ , \tag{30}$$

where $\bar{\omega}$ is the classical frequency in a single well. A superposition of these states oscillates with period $T_{osc} = 2\pi\hbar/\Delta E$ between the two wells.

When the field is switched on, the situation is much more complicated due to the fact that the classical motion is chaotic and there is no conserved quantity in the chaotic region between the wells, and hence no equivalent of the tunneling integral (29). Typically, the two potential minima turn into stability islands and the curve separating the single-well motion from the double well-oscillation at higher energies (the 'separatrix') is destroyed by the interaction with the field. Instead, a chaotic separatrix layer develops, which grows with increasing field strength.

Tunneling through such a chaotic layer is still far from being understood and the theory of such tunneling transitions is an active field of contemporary research (see, for example, [36, 38, 7], or [20, 18, 19, 48] for studies of a driven double-well oscillator and [17, 7, 1]) for related studies of a kicked or driven rotor).

Figure 12 shows a stroboscopic Poincaré section of the system (28) for the parameters given above. Results from three classical trajectories are shown: one trajectory generates the chaotic sea, the other two are regular and move around the left or right island, respectively. Transitions between the left and the right island are classically forbidden. Quantum mechanically, such a transition is allowed, however.

On the right-hand side of Fig. 12, a wavepacket started on the left island (more precisely a minimum uncertainty wavepacket (12) centered at $(p_0, q_0) = (0, -1.5)$ with $s = 1$) is shown after one period T of the driving field. It is obvious, that the distribution is beginning to populate the right stability island. After 58 periods, a considerable part of the distribution is found there, and after

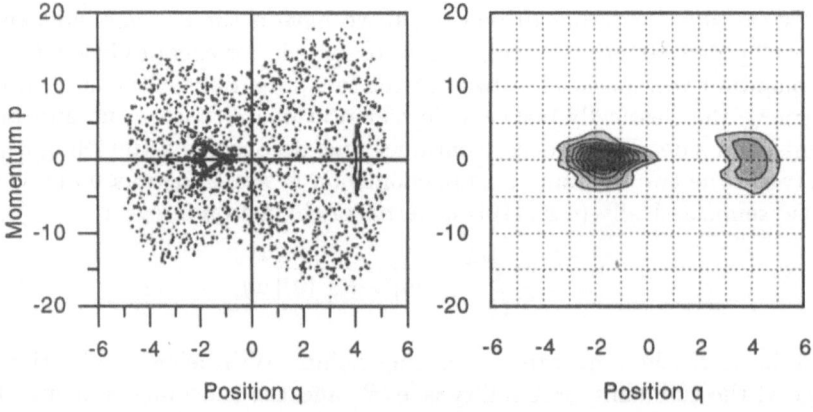

Fig. 12. Poincaré section of classical phase-space (*left*) and quantum Husimi distribution (*right*) of a wavepacket started on the left island after one period

115 periods almost the whole wavepacket has tunneled to the opposite island, as demonstrated in Fig. 13. It is remarkable that the distribution localizes on the regular island again, in spite of the fact, that it has tunneled through the region where the classical dynamics is chaotic.

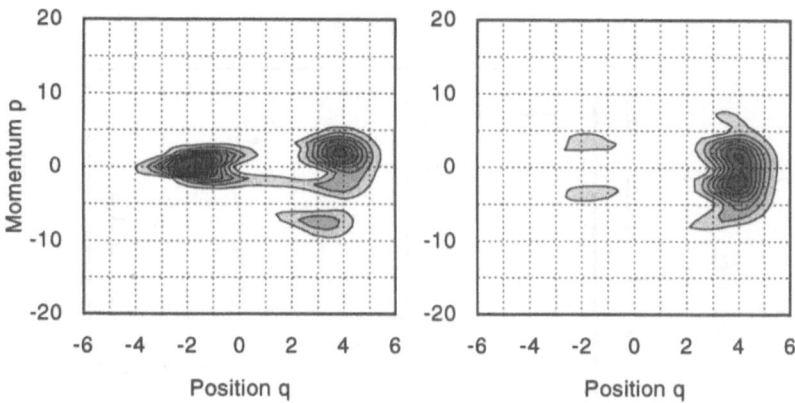

Fig. 13. Quantum Husimi distribution of a wavepacket started on the left island after 58 (*left*) and 115 (*right*) periods

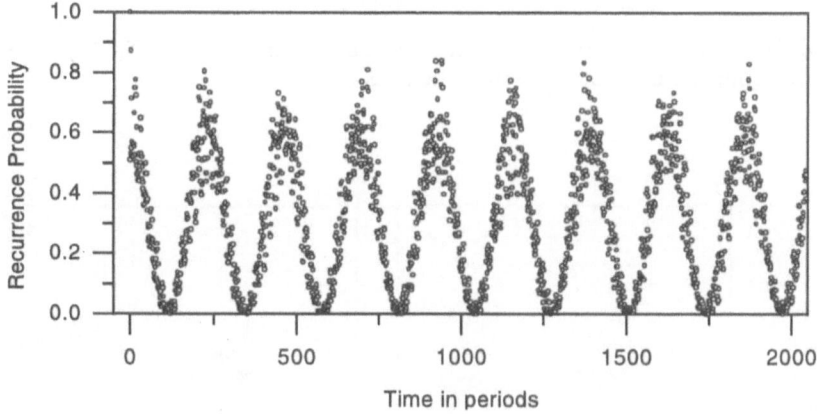

Fig. 14. Recurrence probability as a function of time

The recurrence probability (21) in Fig. 14 shows the continuation of the tunneling process with increasing time. We observe an overall oscillation with period $114\,T$ of a certain fraction of the distribution between the two islands. This 'coherent tunneling' [30] can be explained in terms of quasienergy (Floquet) states of different symmetry localizing on both islands [17, 36, 38, 1]. Their superposition leads to states oscillating between the islands with a period proportional to the inverse of the difference of the two quasienergies. This can be checked by computing the quasienergies and the Husimi distribution from the autocorrelation function as discussed in Sect. 4.4. The results are shown in Figs. 15 and 16.

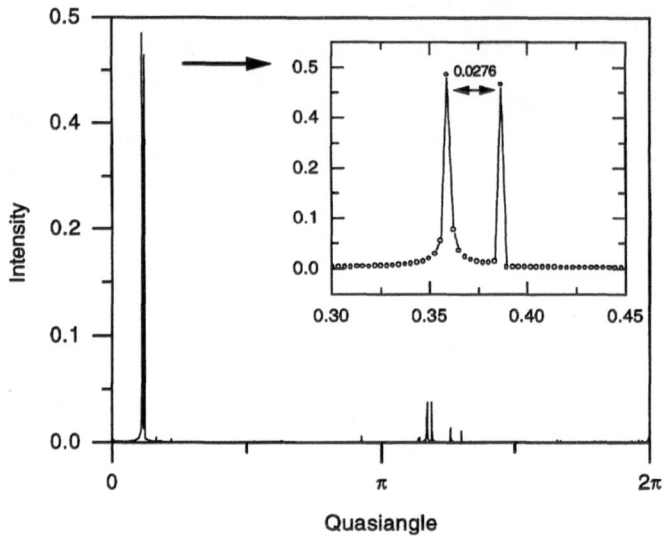

Fig. 15. Quasienergy spectrum of the autocorrelation function (see Fig. 14). The inset shows a magnification of the two strongest peaks

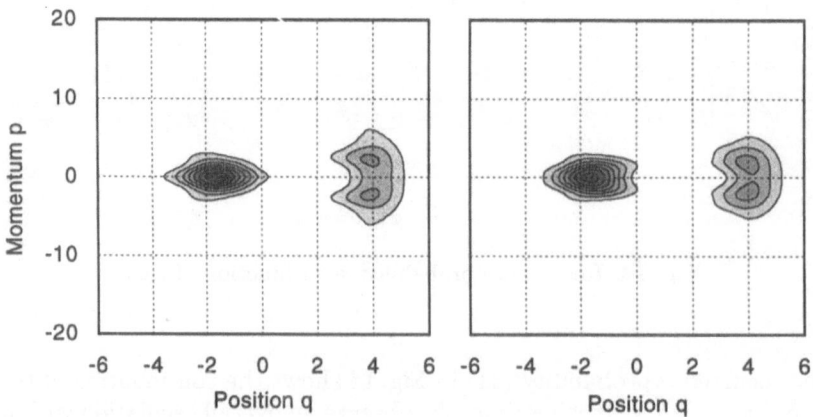

Fig. 16. Husimi distributions of the quasienergy doublet shown in Fig. 15

Recent observations from numerical computations suggest, that the quasienergy splittings for chaotic tunneling do *not* follow the simple semiclassical law (30) when $\hbar$ is varied [41]. Instead, they show a seemingly irregular behavior.

5 Concluding Remarks

In this article we have tried to demonstrate some of the numerical techniques used to explore the manifestation of classical chaos in the corresponding quantum system. We have confined ourselves to the case of a driven anharmonic oscillator and presented some of the techniques using quantum dynamics in phase space. Let us finally give a brief description of a method suggested recently for developing a global picture of the phase-space structure of a quantum system [32]. The basic idea is to compute the long-time average $\overline{S}(p_0, q_0)$ of quantum phase-space entropy (22) for *all* initial positions (p_0, q_0) of the initial wavepacket. This function provides a quantitative measure of the phase-space localization properties of the quantum system in analogy to the classical Poincaré section. Numerical studies for a driven-rotor system [33] have been reported recently [32] and the (semiclassical?) explanation of this phenomenon is not yet clear.

References

[1] V. Averbuckh , N. Moiseyev, B. Mirbach, and H.J. Korsch, Z. Phys. D, in press (1995)

[2] J. Becker, A. Ernesti, and H.J. Korsch, in preparation, (1996)

[3] J. Becker, F. Speckert, H.J. Korsch, and H.-J. Jodl, Phys. u. Didaktik **16**, 127 (1988)

[4] F. Bensch, H.J. Korsch , B. Mirbach, and N. Ben-Tal, J. Phys. A **25**, 6761 (1992)

[5] N. Ben-Tal, N. Moiseyev, and H.J. Korsch, Phys. Rev. A **46**, 1669 (1992)

[6] N. Ben-Tal, N. Moiseyev, S. Fishman, F. Bensch, and H.J. Korsch, Phys. Rev. E **47**, 1646 (1993)

[7] G. Casati, R. Graham, I. Guarneri, and F.M. Izrailev, Phys. Lett. A **190**, 159 (1994)

[8] G. Casati, I. Guarneri, and U. Smilansky, eds., *Quantum Chaos*, Intern. School of Physics Enrico Fermi, Como 1991 (North-Holland, Amsterdam 1993)

[9] G. Casati and L. Molinari, Prog. Theor. Phys. Suppl. **98**, 287 (1989)

[10] M.S. Child, *Semiclassical Mechanics with Molecular Applications* (University Press, Oxford 1991)

[11] P. Cvitanović, I. Percival, and A. Wirzba, eds., *Quantum Chaos – Quantum Measurement*, NATO ASI Copenhagen 1991 (Kluver, Dordrecht 1992)

[12] H. De Raedt and K. Michielsen, Computers in Physics **8**, 600 (1994)

[13] M.D. Feit, J.A. Fleck, Jr., and A. Steiger, J. Comput. Phys. **47**, 412 (1982)

[14] D.H. Feng and Y.-M. Yuan, *Quantum Non-Integrability* (World Scientific, Singapore 1992)

[15] J.-C. Gay, *Irregular Atomic Systems and Quantum Chaos* (Gordon and Breach, Montreux 1992)

[16] A. Goldberg, H.M. Schey, and J.L. Schwarz, Am. J. Phys. **35**, 177 (1967)

[17] R. Grobe and F. Haake, Z. Phys. B **68**, 503 (1987)

[18] F. Grossmann, T. Dittrich, and P. Hänggi, Physica B **175**, 293 (1991)

[19] F. Grossmann, T. Dittrich, P. Jung, and P. Hänggi, J. Stat. Phys. **70**, 229 (1993)

[20] F. Grossmann, P. Jung, T. Dittrich, and P.Hänggi, Phys. Rev. Lett. **67**, 516 (1991); Z. Phys. B **84**, 315 (1991)

[21] M.C. Gutzwiller, *Chaos in Classical and Quantum Mechanics* (Springer, New York 1990)

[22] M.C. Gutzwiller, Scientific American, Jan. (1992), p. 26

[23] F. Haake, *Quantum Signatures of Chaos* (Springer, Berlin 1992)

[24] E.J. Heller and S. Tomsovic, Physics Today, **38** (1993)

[25] K. Husimi, Proc. Phys. Math. Soc. Japan **22**, 264 (1940)

[26] K. Ikeda, ed., *Quantum Chaos* Progr. Theoret. Phys. Suppl. **116** (special issue) (1994)

[27] S. Koonin, *Computational Physics* (Benjamin–Cummings, Menlo Park 1986)

[28] H.J. Korsch and H.-J. Jodl, *Chaos – A Program Collection for the PC* (Springer, Heidelberg 1994)

[29] C. Leforestier, R.H. Bisseling, C. Cerjan, M.D. Feit, R. Friesner, A. Guldberg, A. Hammerich, G. Jolicard, W. Karrlein, H.-D. Meyer, N. Lipkin, O. Roncero, and R. Kosloff, J. Comput. Phys. **94**, 59 (1991)

[30] W.A. Lin and L.E. Ballentine, Phys. Rev. Lett. **65**, 2927 (1990); Phys. Rev. A **45**, 3637 (1992)

[31] B. Mirbach, and H.J. Korsch, J. Phys. A **27**, 6579 (1994)

[32] B. Mirbach, and H.J. Korsch, Phys. Rev. Lett. **75**, 362 (1995)

[33] N. Moiseyev, H.J. Korsch, and B. Mirbach, Z. Phys. D **29**, 125 (1994)

[34] K. Nakamura, *Quantum Chaos, A New Paradigm of Nonlinear Dynamics* (Cambridge University Press, Cambridge 1993)

[35] M. Ozorio de Almeida, *Hamiltonian Systems – Chaos and Quantization* (University Press, Cambridge 1988)

[36] A. Peres, Phys. Rev. Lett. **67**, 158 (1991)

[37] U. Peskin and N. Moiseyev, J. Chem. Phys. **99**, 4590 (1993)

[38] J. Plata and J.M. Gomez Llorente, J. Phys. A **25**, L303 (1992)

[39] W.H. Press, S.A. Teukolsky, W.T. Vetterling, and B.P. Flannery, *Numerical Recipes* (Cambridge University Press, Cambridge 1986)

[40] L.E. Reichl, *The Transition to Chaos* (Springer, New York 1992)

[41] R. Roncaglia, L. Bonci, F.M. Izrailev, B.J. West, and P. Grigolini, Phys. Rev. Lett. **73**, 802 (1994)

[42] J. Schneider, *Subbarrierenfusion als Tunneln von mehrdimensionalen Wellenpaketen.* Thesis, Univ. München (1987)

[43] J. Schneider and H.H. Wolter, in *Heavy Ion Interactions Around the Coulomb Barrier*, eds. C. Signorini, S. Skorka, P. Spolaore, and A. Vitturi, Springer Lecture Notes in Physics 317, (Springer, Berlin 1988); Z. Phys. A **339**, 177 (1991)

[44] M. Tabor, *Chaos and Integrability in Nonlinear Dynamics* (Wiley, New York 1989)

[45] K. Takahashi, Prog. Theor. Phys. Suppl. **98**, 109 (1989)

[46] K.-E. Thylwe and F. Bensch, J. Phys. B **27**, 5673 (1994)

[47] K.-E. Thylwe and F. Bensch, J. Phys. B **27**, 7475 (1994)

[48] R. Utermann, T. Dittrich, and P. Hänggi, Phys. Rev. E **49**, 273 (1994)

[49] P.B. Visscher, Computers in Physics **5**, 596 (1991)

[50] A. Wehrl, Rep. Math. Phys. **16**, 353 (1979)

Numerical Simulation in Quantum Field Theory[*]

Ulli Wolff

Institut für Physik der HUB, Computerphysik, Humboldt Universität,
D-10099 Berlin, Germany, e-mail: uwolff@linde.physik.hu-berlin.de

Abstract. An introductory discussion of quantum field theory as the theoretical framework for particle physics and the standard model is given. For the simplified scalar field sector, the lattice formulation and nonperturbative evaluation by Monte Carlo simulation are described. A FORTRAN package is included that allows the simulation of this model for arbitrary dimension.

1 Quantum Field Theory and Particle Physics

In this introductory section we briefly outline the role of quantum field theory as the generally accepted framework for a theoretical description of elementary particle physics.

1.1 Particles, Fields, Standard Model

A simple example of particle or quanta aspects of fields is given by photons and the Maxwell theory. Let us imagine classical electric and magnetic fields in an (ideal) metallic cavity. The walls represent certain boundary conditions which lead to a denumerably infinite sequence of possible standing waves. Schematically, omitting polarization, we simply label these modes by a discrete set of wavenumber vectors $\{\mathbf{k}\}$ and associated frequencies $\omega(\mathbf{k}) = c|\mathbf{k}|$. The Hamiltonian, that leads to the linear Maxwell equations, is bilinear in the fields. It corresponds to an infinite number of harmonic oscillators. They are decoupled by Fourier normal mode decomposition into independent oscillators with the above mentioned frequencies $\omega(\mathbf{k})$. Now the most naive generalization of canonical quantization of a finite number of harmonic oscillators leads to the quantum mechanical energies E of the Maxwell Hamiltonian

$$E - E_0 = \sum_{\mathbf{k}} n(\mathbf{k})\hbar\omega(\mathbf{k}) \ . \tag{1}$$

To characterize such an eigenstate, occupation numbers $n(\mathbf{k}) \geq 0$ must be specified for all modes. The linearly spaced oscillator spectrum lends itself immediately to a particle interpretation of states with these energies: each mode $\mathbf{k}$ corresponds to a possible photon moving with momentum $\hbar\mathbf{k}$ and possessing energy $\hbar\omega(\mathbf{k})$, and there is a number of $n(\mathbf{k})$ such noninteracting particles. The

---

[*] Software included on the accompanying diskette.

energy E_0 is dismissed as the unobservable infinite zero-point energy of the empty cavity with only energy differences being physical.

This almost trivial correspondence between oscillator levels and the additive energies of noninteracting "objects" has been fruitful in several branches of physics leading to phonons, pseudo particles etc. For particle physics one usually removes the boundary conditions by approaching an infinite volume and thus deals with all of space time as the physical system under study. It is rather easy to set up fields together with a bilinear Hamiltonian, which describe other particles observed in nature with spin, Fermi statistics and certain multiplet structures like electrons, neutrinos, quarks etc.

The field description acquires a much deeper role, if symmetries and interactions are incorporated. In perturbation theory, the entities described by the free (bilinear) field oscillator modes remain intact, and additional small cubic or higher monomials in the fields just add weak interactions between them. There is an intriguing interplay between symmetries and interactions in the sense that certain symmetries (gauge invariance) necessitate the presence of nonlinear terms of a specific structure and thus largely determine the interaction between particles.

In a historical process of learning and applying these principles and injecting experimental information on existing particle varieties and properties, today's standard model of particle physics has emerged. In our present extremely schematic overview it corresponds to a structure of fields $\{\varphi_\alpha(t, \mathbf{r})\}$ and their dynamics in the form of a Hamiltonian or, more commonly, an action $S[\varphi]$. The "internal" index α labels a number of field components per space-time point encoding the observed degrees of freedom. Transformations act on this index by mixing field components. The action S is invariant under the experimentally determined group of such transformations, and it also contains the free parameters of the standard model (of order 20 numbers) that have to be fixed empirically by matching experiment.

1.2   Beyond Perturbation Theory

In the previous subsection we alluded to the perturbative view of quantum field theory. There one has a field of appropriate spin and charge for each species of particle to be described. Interactions are treated in the form of nonlinear but small perturbations. The perturbative terms are usually organized in correspondence with Feynman diagrams giving an intuitive feeling for their effects. The success story including QED, electroweak unification and perturbative quantum chromodynamics (QCD), nowadays integrated into the standard model, is almost entirely based on the application of this method.

It has been known since long from simple model field theories that there may also or in addition be predictions that are not accessible in perturbation theory. Stronger nonlinearities may qualitatively rearrange the spectrum of the field theoretic Hamiltonian, such that a free theory is not a reasonable starting point to perturb around. Often physical theories are benign enough to indicate this already within perturbation theory, for instance by "corrections" becoming of order 100%. A more subtle failure of perturbation theory arises, if some quantity vanishes to all orders which truly has nonzero contributions of the type $\exp(-1/g^2)$, where g is the expansion parameter. Both phenomena are expected to occur in the QCD sector of the standard model. Some of the elementary fields correspond to quarks which phenomenologically do not exist as free particles under normal conditions. This means that, unlike electrons, they never come out of a scattering experiment as isolated asymptotic particles. In QCD the term confinement has been coined for this nonperturbative rearrangement of the spectrum. It means that nonlinear couplings to the gluonic fields, demanded and determined by local $SU(3)$ gauge invariance, produce forces which are strong and not decaying with separation, such that quarks cannot be isolated. As energy is pumped into the system, new particles are created with the consequence that one never sees anything but three quark states (baryons) or quark antiquark states (mesons). For these compounds the strongest component of the gluon force is neutralized.

There clearly arises a demand to extract also the nonperturbative predictions from the standard model and other theories. In this way one hopes to achieve a detailed description of the properties of hadrons like their mass ratios and scattering effects. At present, only numerical methods can offer such results in the sense of a first principles evaluation of the field theory. We hence introduce this method in some detail in the following sections.

2   Lattice Formulation of Field Theory

In this section we introduce the formulation of euclidean quantum field theory on a space time lattice. Connections with statistical mechanics and critical phenomena are pointed out. A comprehensive discussion of this subject can be found in recent textbooks [1, 2, 3].

2.1   Path Integral

Most lattice methods in field theory start from the path integral formulation. We briefly derive it here for a simple quantum mechanical system. We start from its Hamiltonian (in suitable units)

$$\hat{H} = \frac{1}{2}(\hat{p}^2 + \hat{q}^2) + g\hat{q}^4 \ . \tag{2}$$

Here $\hat{p}$ and $\hat{q}$ are the usual canonical operators with commutation relation

$$[\hat{q}, \hat{p}] = i\hbar \ . \tag{3}$$

As a starting point we consider the partition function

$$Z = \mathrm{tr}\left[\mathrm{e}^{-\beta\hat{H}}\right] = \mathrm{tr}\left[\mathrm{e}^{-\epsilon\hat{H}}\mathrm{e}^{-\epsilon\hat{H}}\ldots\mathrm{e}^{-\epsilon\hat{H}}\right]$$

$$= \int dq_1 \dots dq_N \langle q_1 | e^{-\epsilon \hat{H}} | q_2 \rangle \langle q_2 | e^{-\epsilon \hat{H}} | q_3 \rangle \dots \langle q_N | e^{-\epsilon \hat{H}} | q_1 \rangle$$

$$\simeq \int dq_1 \dots dq_N \exp \left\{ -\frac{\epsilon}{\hbar} \sum_i \left[\frac{1}{2} \left(\frac{q_{i+1} - q_i}{\epsilon} \right)^2 + \frac{1}{2} q_i^2 + g q_i^4 \right] \right\}$$

$$\simeq \int Dq(t) \exp \left\{ -\frac{1}{\hbar} \int_0^\beta dt \left[\frac{1}{2} \dot{q}^2 + \frac{1}{2} q^2 + g q^4 \right] \right\} . \tag{4}$$

In this formula we have subdivided the "imaginary time" $\beta = N\epsilon$ into small segments ϵ. The matrix elements $\langle q_i | e^{-\epsilon \hat{H}} | q_{i+1} \rangle$ are only evaluated to leading order in ϵ, hence the $\simeq$ signs. The last line is a symbolic notation meaning the limit $\epsilon \to 0, N \to \infty$ of the previous one. In this limit one imagines the N-fold integration over $q_1 \dots q_N$ to go over into an "integration over all paths". These paths $q(t)$ run from $t = 0$ to $t = \beta$ and are closed, $q_{N+1} = q_1$, $q(0) = q(\beta)$, due to the trace taken.

In the path integral formulation one extracts information on the system in the form of expectation values of functionals of the path $q(t)$,

$$\langle f[q] \rangle = \frac{\int Dq \, e^{-S} f[q]}{\int Dq \, e^{-S}} \tag{5}$$

where the action S is the exponent in (4). A simple example is the two point function that gives information on the spectrum of the original Hamiltonian $\hat{H}$. It is not difficult to show that

$$\langle q(t_1) q(t_2) \rangle \overset{\beta \to \infty}{\simeq} \sum_n e^{-(E_n - E_0)|t_2 - t_1|/\hbar} \, |\langle 0 | \hat{q} | n \rangle|^2 \tag{6}$$

holds, where E_n are energy levels, E_0 refers to the groundstate.

To generalize to field theory one would start from a Hamiltonian written in terms of operators $\hat{\varphi}(\mathbf{r})$ for each point in space[1] instead of $\hat{q}$ together with appropriate momenta conjugate to $\hat{\varphi}$. The quantum mechanical system discussed above may actually be viewed as a zero dimensional field theory. It turns out that for the transition to the path integral the multitude of degrees of freedom just means another index $\mathbf{r}$ to be carried along resulting in an "integral over all fields" $\varphi(t, \mathbf{r})$,

$$Z = \int D\varphi(t, \mathbf{r}) \, e^{-S[\varphi]} . \tag{7}$$

In field theory, again all dynamical information can be obtained from observables

$$\langle F[\varphi] \rangle = \frac{1}{Z} \int D\varphi(t, \mathbf{r}) \, e^{-S[\varphi]} F[\varphi] . \tag{8}$$

---

[1] Suppressing the internal index α.

Instead of spelling out field theoretic Hamiltonians in detail it is more convenient to immediately start from the euclidean action S appearing in the path integral. For the prototype model of scalar φ^4 theory it is given by

$$S = \int \mathrm{d}^4 x \left\{ \frac{1}{2} \sum_\mu (\partial_\mu \varphi)^2 + \frac{m^2}{2} \varphi^2 + \frac{g}{4!} \varphi^4 \right\} . \tag{9}$$

Here $(t, \mathbf{r})$ has been combined into a four-component vector $x_\mu, \mu = 0, 1, 2, 3$. Units with $\hbar = c = 1$ are used from here on. Apart from serving as a simplest example, the φ^4 theory also has a role within the standard model. The famous Higgs sector has scalar fields with this kind of selfinteraction. Although these fields are coupled with fermions and gauge fields, for certain properties of the theory it was considered reasonable to neglect these couplings and study the scalar sector in isolation. Ref.[4] is a well known example for such research.

2.2 Lattice Regularization

At a formal level we have introduced quantum field theory as an integral over all fields over a four dimensional continuum with a weight given by the action S. To be concrete, from now on we consider (7), (8) with action (9). The perturbative treatment would start by neglecting the interaction term proportional to g. Then the integral is gaussian and can be carried out by mild generalization of finite dimensional gaussian integrals. One thus reproduces all results on the harmonic oscillators that make up free fields respectively noninteracting particles. Interactions are then incorporated as power corrections in g. In such calculations one encounters the well known divergences of quantum field theory. They already appeared in a trivial way in the zero point energy E_0 in (1), which receives a finite amount of each degree of freedom, i.e. from each point in space. Note that this is not a problem deriving from infinite volume: the problem is ultraviolet, an infinite number of degrees of freedom *per volume* for a continuum.

The standard procedure in quantum field theory is now to modify the theory in the small while maintaining the large scale features such that all divergences are regularized. With this modification a smallest length scale a is introduced, and only as $a \to 0$ the divergences reappear (cutoff or continuum limit). In the regularized theory one identifies relations between physical quantities, which remain finite in the cutoff limit. These become the predictions of the renormalized theory.

There are various possibilities of regularization. For perturbative calculations dimensional regularization is most popular since it is computationally efficient with a relatively small number of terms at intermediate stages. It can be defined, however, only at the level of Feynman diagrams and hence is strictly perturbative. For both rigorous analyses on the existence of field theories and for numerical computations independent of perturbative expansions, the lattice cutoff is most useful. It allows to regularize to an explicitly finite number of degrees of freedom, an obvious prerequisite to go to the computer.

The lattice cutoff is imposed by restricting the four dimensional space-time continuum to an embedded lattice. The type of lattice is expected to be irrelevant for renormalized predictions in the continuum limit. For technical simplicity one usually takes a simple cubic lattice, where the cartesian components x_μ are restricted to integer multiples of the lattice spacing a. Introducing a finite volume at the same time, usually of linear dimensions large compared to all scales in the problem, one may restrict x to $x_\mu = 0, a, 2a, \ldots, (L-1)a$. Boundary conditions are usually taken periodic, and one thus deals with a volume $(La)^4$ and L^4 real field variables.

The path integral naturally becomes an L^4-fold ordinary integral

$$Z = \int \prod_x \mathrm{d}\varphi(x)\, \mathrm{e}^{-S} \ , \tag{10}$$

where it remains to write S as an ordinary function of the field values on the lattice sites. To this end some discretization is introduced of the same kind as for the approximation of differential equations on a grid. In the simplest case we take replace (9) by

$$S = a^4 \sum_x \left\{ \frac{1}{2} \sum_\mu (\Delta_\mu \varphi)^2 + \frac{m^2}{2} \varphi^2 + \frac{g}{4!} \varphi^4 \right\} \ . \tag{11}$$

with

$$\Delta_\mu \varphi(x) = \frac{1}{a}(\varphi(x + a\hat{\mu}) - \varphi(x)) \ , \tag{12}$$

where $\hat{\mu}$ is a unit vector along the μ'th coordinate axis. For a smooth field the above sum goes over into the integral at a rate proportional to a^2. For the lattice, this is the sense in which the theory, symbolically given by the continuum path integral (7), is modified in the small only. The continuum limit $a \to 0$ of the integral (10), with its growing number of integrations, is a much more nontrivial question.

2.3   Field Theory and Critical Phenomena

As we whish to talk about a continuum limit of a lattice quantum field theory, the question arises as to which physical scale to compare a with. One could think of the only other dimensionful parameter m. This, however, is just a parameter in the action and does not have immediate physical relevance. A physical scale has to be derived from observables. It can be shown [2] that the exponential decay length of the two point correlation function can be related to the (renormalized) mass of particles m_R by

$$\sum_{\mathbf{xy}} \langle \varphi(x)\varphi(y) \rangle \propto \exp(-|x_0 - y_0| m_R) \ . \tag{13}$$

The length $1/m_R = \xi a$ is the distance, over which typical fluctuations of the field are correlated. The continuum limit is approached where the correlation length in lattice units ξ diverges.

At this point the lattice formulation of quantum field theory makes contact with the field of critical phenomena. The partition function (10) can be regarded as a classical statistical system, which undergoes a second order phase transition at those parameter values m, g where $\xi \to \infty$. It is known that critical points possess universal properties like the critical exponents. These quantities depend only on the dimensionality of the lattice and on the number of field components and their symmetry structure. A closer investigation of the correspondence reveals that the renormalized relations that remain finite and meaningful in the continuum limit are such universal properties of the associated phase transition. In other words, they are independent of the detailed structure of the lattice and of the precise choice of the discretization of the action. Obviously, this is a necessary requirement, since these features were arbitrary from the point of view of particle physics.

The lattice formulation has shown, that quantum field theories of elementary particle physics can be mapped on certain four dimensional classical statistical systems. Their variables and symmetries are dictated by experimental observations in particle physics. Universal properties of critical points in these systems correspond to physical predictions in field theory.

2.4   Effective Field Theory

After the discovery of renormalization in QED one first thought that physics is defined as a cutoff limit of regularized theories and that it is hence necessary to really take $a \to 0$ mathematically. But experience with other physical theories actually suggests to be more modest. Presumably, present day field theories are a correct description on today's experimental energy scale and somewhat beyond, but not for arbitrarily high energies and short distances. As the Fermi theory of weak interactions was a low energy approximation to the electroweak sector of the standard model we expect the standard model to approximate something still unknown at some higher but finite energy. Then it is sufficient to take the cutoff energy $1/a$ to about that energy and not necessarily to infinity.

Since the work of Wilson a picture has emerged that is visualized in Fig. 1. We plot the admittedly vague quantity 'physics' versus a logarithmic scale. At a cutoff scale a, small compared to present day physics, we have a number of different cutoff field theories. They can be given by different lattices, for instance. As one derives long distance physics predictions from them with appropriately adjusted parameters, they are able to give very similar predictions, which are hence independent of details of the cutoff as discussed before. Nature, coming from an unknown short distance theory, falls into the same narrow tube of almost fixed long distance behavior. Our ability to make predictive theories at our present scale is based on the fact, that the possible long distance behavior seems to be severely restricted, once the right degrees of freedom and symmetries are

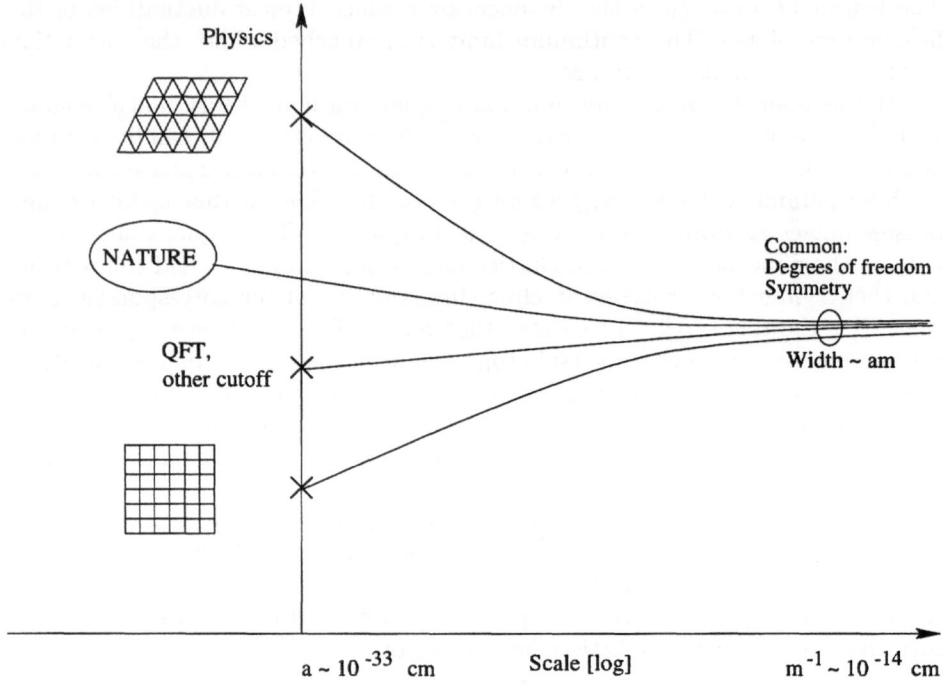

Fig. 1. Description of nature by effective field theories

imposed. Our cutoff-mutilated theories are just some representatives in the same class as nature as far as energies far below the cutoff are concerned [5, 6].

Wilson has argued [7] that this situation is by no means unprecedented. The Navier-Stokes equations of hydrodynamics, for example, could be guessed on the basis of symmetry and some simplicity arguments. Only later was the underlying microscopic (quantum) theory found, which had this limiting behavior. Now some of the purely empirical parameters in the Navier-Stokes theory could be understood to arise from the underlying theory. We may well be in a similar situation with the standard model of particle physics.

3   Stochastic Evaluation of Path Integrals

After the previous motivation, in this section we discuss how to perform nonperturbative numerical calculations in lattice field theory. This amounts to computing numerically the integral in (10) whose dimension is large, for instance 10^4. Standard methods of numerical integration with regular grids in each dimension are obviously not suitable. The equivalence with statistical physics comes to rescue. The real positive weight containing the action $\exp(-S)$ is interpreted as the Boltzmann factor of a classical statistical system with lattice field space as its phase space. The analogy suggests that only a small part of the huge phase space, determined by the Boltzmann weight, is relevant for expectation values of observables of interest. The Monte Carlo method with importance sampling is a tool to sample this subspace. As it is a statistical method, the convergence in the form of the reduction of statistical errors at fixed parameters, will only be proportional to the square root of the invested computer time.

3.1   Monte Carlo Method

For the Monte Carlo [2, 8, 9] evaluation of path integrals like (10) one generates a sequence of lattice field configurations in the memory of a computer starting from an arbitrary $\varphi^{(0)}$

$$\varphi^{(0)} \to \varphi^{(1)} \to \varphi^{(2)} \to \dots. \tag{14}$$

Here each $\to$ corresponds to the modification (update) of the stored configuration by a certain algorithm. By methods to be discussed later it can be achieved that after a certain number of configurations, say n, affected by transient behavior from the arbitrary start, the frequency of appearance of a any given φ in the sequence is proportional to the Boltzmann weight $\exp(-S(\varphi))$. Thus the dominant part of the integrand is incorporated into the sampling of the phasespace which therefore is called importance sampling. The expectation value of any observable is then estimated by an average over a sufficiently long section of the sequence

$$E(F) = \frac{1}{N} \sum_{i=n+1}^{i=n+N} F(\varphi^{(i)}) \overset{N \to \infty}{\Rightarrow} \langle F \rangle \tag{15}$$

The transition from $\varphi = \varphi^{(i)}$ to $\varphi' = \varphi^{(i+1)}$ is stochastic and hence completely characterized by the transition probabilities $T(\varphi, \varphi')$. For the correct sampling with respect to the weight $P(\varphi) = \exp(-S(\varphi))/Z$ the conditions of ergodicity and stability have to be fulfilled. Ergodicity amounts to

$$T^k(\varphi, \varphi') > 0 \text{ for any } \varphi, \varphi', k > k_0 \tag{16}$$

with some integer k_0. This ensures that all phase space can be reached from any start $\varphi^{(0)}$ with finite probability in a finite number of steps. Stability amounts to the desired distribution $P(\varphi)$ being a fixed point under updating,

$$\int \prod_x d\varphi(x) \, P(\varphi) T(\varphi, \varphi') = P(\varphi') . \tag{17}$$

It can be proven [9] that importance sampling follows from these properties.

The efficiency of a valid Monte Carlo algorithm is an independent question. In general, sampled configurations are correlated with each other. These correlations decay exponentially with the separation in the sequence and possess a characteristic decay time τ in units update steps of the algorithm in use. Statistical errors are reduced most by independent estimates.[2] More precisely, there is an error formula for estimates (15) of the following type,

$$\langle [E(F) - \langle F \rangle]^2 \rangle = \frac{\langle [F - \langle F \rangle]^2 \rangle}{N/(2\tau_F)} \ . \tag{18}$$

The numerator is the variance of F, and τ_F in the denominator is an effective autocorrelation time that incorporates the coupling of the observable F to the various (and not only the slowest) modes in the spectrum of T. We see that τ_F is the characteristic scale for the length of the simulation N and this is usually also true for the initial equilibration or thermalization part of the sequence that is discarded. Clearly, $n, N \gg \tau_F$ is necessary. Apart from a few exceptionally favorable cases in certain spin models, all τ_F for known algorithms diverge in the continuum limit with characteristic dynamical exponents, $\tau_F \propto \xi^{z_F}$. A lot of research in the field of Monte Carlo is devoted to the search for algorithms with reduced τ and z.

3.2   Metropolis Algorithm for φ^4

The local Metropolis algorithm is probably the most universal algorithm in lattice field theory. It is almost always available, but for many models it is inferior to other known schemes. Locality means that an update pass is composed of steps T_x for all lattice sites, where only the field at site x is modified and all other values are maintained. In the simplest case of φ^4 one proposes a new value

$$\varphi'(x) \in [\varphi(x) - \delta, \varphi(x) + \delta] \ , \tag{19}$$

where δ is a width parameter to be determined by efficiency and the proposal is drawn from the interval with a flat distribution. Now one computes the action difference

$$\Delta S = S(\varphi') - S(\varphi) \ , \tag{20}$$

where φ' is the configuration with the new proposal at x. Of course, ΔS is only a local sum of terms around x where the configurations differ, with the remaining terms canceling. Now the proposal is accepted as the new configuration in the sequence with probability

$$p = \min(1, \exp(-\Delta S)) \ . \tag{21}$$

A sequence of such steps for all sites (in some order) is called one sweep through the lattice. It fulfills the conditions of ergodicity and stability and is hence a legal Monte Carlo algorithm. The proposal width δ has to be tuned such that the acceptance probability that is monitored is not pathological, that is neither close to 0 nor to 1. It is usually easy to achieve values between 0.3 and 0.7 which are close to optimal empirically.

---

[2] Barring the possibility of anticorrelations that usually cannot be achieved.

Dynamical exponents of $z \approx 2$ are found with this technique. For φ^4 and many other cases including lattice gauge theory, one can substantially improve to $z \approx 1$ by mixing Metropolis with certain microcanonical steps. The latter fail to be ergodic by themselves, but the mixture is. This combination is called overrelaxation technique, and it represents the state of the art for pure lattice gauge theory. For more details, we refer to [10]. In Appendix A a subroutine package for a simulation of scalar theory is described.

4 Summary

In these lecture notes we tried to outline the role of field theory for particle physics, the lattice formulation of field theory and the numerical technique of Monte Carlo simulation for its solution beyond perturbation theory. Since this is a wide area, emphasis was first put on a few principles and the more concrete discussion was restricted to the simplest example of φ^4 scalar field theory. Of course, more interesting with respect to the standard model is the inclusion of gauge fields. For this and other topics we refer to [2] for more introduction and to the proceedings of the annual lattice conferences, for instance [11], for information on the present day status of research. At the Heraeus school a discussion of the ongoing project of the computation of the strong coupling constant α_s was given, for which we refer to the review [12].

To conclude, lattice field theory and numerical simulation continues to be a very active field. Somewhat later than anticipated when it all started about 15 years ago, we are only now more and more coming to simulations that are relevant to experimental data. The computational problem is enormous, and it requires both, any amount of (parallel) computing power available together with a continued research on improving the efficiency of our methods.

Appendix: FORTRAN Monte Carlo Package for φ^4

Included with this volume is a collection of small FORTRAN 77 programs that allow the simulation of φ^4 theory on a torus with the Metropolis algorithm as described in this article.

In numerical context, a parameterization of the theory somewhat different from (11) is customary and advantageous. The action is written as

$$S = \sum_x \left\{ -2\kappa \sum_\mu \phi(x)\phi(x + \hat{\mu}) + \phi^2 + \lambda(\phi^2 - 1)^2 \right\} . \tag{22}$$

By comparing (up to irrelevant constants) with (11) one can see that both forms are equivalent if the following identifications are made:

$$a\varphi = \sqrt{2\kappa}\phi \tag{23}$$

$$g = 2\lambda/\kappa^2 \tag{24}$$

$$(am)^2 = (1 - 2\lambda)/\kappa - 2(D + 1) . \tag{25}$$

The parameter D is the *space* dimension to be set to 3 in the physical case. In the package we allow for a general value to be set as a parameter. An unphysical value $D = 1$ is advisable for short experiments on workstations. With $D = 0$ one can even simulate the anharmonic oscillator.

The new parameterization is in terms of dimensionless parameters with all lengths and energies expressed as multiples of powers of the lattice spacing a (lattice units). For $\lambda = g = 0$ there is no interaction, and criticality is reached for $m = 0$ or $\kappa = \frac{1}{2(D+1)}$. In this case, the theory exists only for κ approaching the critical value from below. The renormalized mass m_R is known exactly and is given in the main program for comparison with simulations. For $\lambda \to \infty$ the integrations become concentrated at $\phi = \pm 1$, and the model becomes identical with the Ising model (2κ is the usual β). Here the critical κ is known with high precision. For $0 < \lambda < \infty$ no exact solution is available, but the critical curve in (κ, λ) is expected to smoothly interpolate. The region above the critical line is the so-called broken phase, where the symmetry $\phi \leftrightarrow -\phi$ is broken. It is more difficult to understand and simulate, and should be considered only later [2].

A very useful test on the correctness of the program is furnished by a nontrivial observable whose expectation value is always known exactly. By scaling all integration variables in the lattice path integral the following equipartition type identity is easy to derive

$$1 = -4\kappa \langle \sum_\mu \phi(x)\phi(x + \hat{\mu}) \rangle - 2(2\lambda - 1)\langle \phi(x)^2 \rangle + 4\lambda \langle \phi(x)^4 \rangle \ . \qquad (26)$$

The site x here is arbitrary due to translation invariance, and one may also sum over it to improve the statistics.

In numerical experiments it is recommended to first study the free case $\lambda = 0$ and to increase κ starting from small values to see the mass m_R in lattice units become small. Then one may slowly raise λ. Note that the local Metropolis algorithm for ϕ^4 will not be efficient in the Ising limit.

A package of routines necessary for some test simulations of the scalar theory is included with the proceedings of the Heraeus school. They are also available by anonymous ftp from linde.physik.hu-berlin.de in directory pub/heraeus. A README file and ample comments supply a further introduction to the programs.

References

[1] M. Creutz, *Quarks, Gluons, Lattices*, (Cambridge University Press, Cambridge 1983)
[2] I. Montvay, G. Münster, *Lattice Field Theory*, (Cambridge University Press, Cambridge 1994)
[3] H.J. Rothe, *Lattice Gauge Theories. An Introduction*, (World Scientific, Singapore 1991)
[4] M. Lüscher and P. Weisz, Phys. Lett. B **212**, 472 (1988)
[5] K.G. Wilson and J.G. Kogut, Phys. Reports **12**, 75 (1974)
[6] J. Polchinski, Nucl. Phys. B **231**, 269 (1984)

[7] K. Wilson, Rev. Mod. Phys. **55**, 583 (1983)

[8] K. Binder and D. Stauffer, in *Monte Carlo Method in Statistical Physics*, 2nd ed., ed. K. Binder (Springer, Berlin 1987) p. 1

[9] A.D. Sokal, *Monte Carlo Methods in Statistical Physics: Foundations and New Algorithms*, Lecture Notes of the Cours de Troisième Cycle de la Physique en Suisse Romande, Lausanne (1989)

[10] U. Wolff, in *Computational Methods in Field Theory*, eds. H. Gausterer, C.B. Lang (Springer, Berlin 1992) p. 127

[11] Nucl. Phys. B (Proc. Suppl.) **42**, Proceedings of Lattice 94 (Bielefeld), (1995)

[12] M.Lüscher, R.Narayanan, R.Sommer, P.Weisz, and U.Wolff, Nucl. Phys. B (Proc. Suppl.) **42**, 139 (1995)

Modeling and a Simulation Method for Molecular Systems

Dieter W. Heermann

Institut für Theoretische Physik, Universität Heidelberg, Philosophenweg 19,
D-69120 Heidelberg, Germany and Interdisziplinäres Zentrum für Wissenschaftliches
Rechnen der Universität Heidelberg
e-mail: heermann@surface.tphys.uni-heidelberg.de

Abstract. In this chapter I introduce a Monte Carlo method for the simulation of systems in continuum that uses global moves for the propagation in phase space. This is then applied to a new general model for the simulation of dense macromolecular systems. It consists basically of ellipsoidal-shaped units strung together to form chains, including branched and side chains. The ellipsoidal-shaped unit can vary in its principal axes and degenerate to a sphere, allowing for flexible modeling of the monomer units. I also touch on the parallelization of such systems.

1 Introduction

Conventional methods for the simulation of dense polymer systems have various shortcomings. Among the most severe is that not all methods generate configurations from a canonical ensemble. A further shortcoming of the methods employing the integration of the equations of motion, in one form or another, is the dependence of the computed observables on the time-step size. In principle an extrapolation to vanishing step size is necessary to obtain an unbiased result.

While Monte Carlo methods do not have the shortcomings listed above they lack the global updating entailed by the integration of motion. To obtain reasonable acceptance rates, the attempted changes in the configuration (moves) are in general local. A segment of a chain is displaced and the new position either rejected or accepted.

The hybrid Monte Carlo method (HMC) is a step towards the development of simulation methods that are exact, in the sense that observables do not depend on step size, that the methods yield the correct statistical-mechanical ensemble and they change a configuration globaly.

2 Brief Review of the Simulation Method

In conventional Monte Carlo (MC) calculations of condensed matter systems such as an N-particle system with a Hamiltonian $\mathcal{H} = \mathcal{U}$ where $\mathcal{U}$ denotes the potential energy, only local moves (displacement of a single particle) are made [1, 2, 3]. Updating more than one particle typically results in a prohibitively low average acceptance probability $\langle P_A \rangle$. This implies large relaxation times and high autocorrelations especially for macromolecular systems. In a Molecular Dynamics (MD) simulation, with $\mathcal{H} = \mathcal{T} + \mathcal{U}$, on the other hand, global moves are made. The MD scheme, however, is prone to errors and instabilities due to the finite time-step size. In order to introduce temperature into the microcanonical context, isokinetic MD schemes are often used [3]. However, they do not yield the canonical probability distribution, unlike Monte Carlo calculations.

The hybrid Monte Carlo (HMC) method [4, 5, 6] combines the advantages of molecular dynamics and Monte Carlo methods: it allows for global moves (which essentially consist in integrating the system through *phase* space); HMC is an exact method, i.e., the ensemble averages do not depend on the step size chosen, algorithms derived from the method do not suffer from numerical instabilities due to finite step size as MD algorithms do, and temperature is incorporated in the correct statistical mechanical sense.

The application of the hybrid Monte Carlo method has been proposed [5] for condensed-matter systems and investigated for atomic fluids. In this chapter the method will be described briefly and applied to macromolecular systems.

In an HMC scheme global moves can be made while keeping the average acceptance probability $\langle P_A \rangle$ for a move high. This can be achieved as follows. One global move in *configuration* space consists in integrating the system through *phase* space for a fixed time t using some discretization scheme $g^{\delta t}$ (δt denotes the step size)

$$g^{\delta t} : \mathbb{R}^{6N} \longrightarrow \mathbb{R}^{6N}$$
$$(x, p) \longrightarrow g^{\delta t}(x, p) =: (x', p')$$

of Hamilton's equations

$$\frac{\mathrm{d}x}{\mathrm{d}t} = \partial_p \mathcal{H} \ ,$$
$$\frac{\mathrm{d}p}{\mathrm{d}t} = -\partial_x \mathcal{H} \ . \tag{1}$$

At the beginning of each global Monte Carlo step the initial momenta are drawn from a Gaussian distribution at inverse temperature β,

$$p_{\mathrm{Gaussian}}(p) \propto \mathrm{e}^{-\beta \mathcal{T}} , \tag{2}$$

and it can be shown [5] that the acceptance probability is then

$$P_A[(x, p) \to g^{\delta t}(x, p)] = \min\{1, \mathrm{e}^{-\beta \delta \mathcal{H}}\} \ , \tag{3}$$

where

$$\delta \mathcal{H} = \mathcal{H}[g^{\delta t}(x, p)] - \mathcal{H}(x, p)$$

is the discretization error associated with the discretization scheme $g^{\delta t}$. Provided the discretization scheme is *time reversible* and *area preserving* detailed balance is satisfied [5]. Thus the HMC algorithm generates a Markov chain with a Boltzmann distribution as the stationary probability distribution. The probability distribution is entirely determined by the detailed balance condition, therefore neither the distribution nor any ensemble averages depend on the step size δt chosen.

However, the average acceptance probability $\langle P_A \rangle$, because of (3), depends on the average discretization error $\langle \delta \mathcal{H} \rangle$ and hence does depend on δt. Increasing the step size will result in a lower average acceptance probability $\langle P_A \rangle$. Varying δt, the average acceptance probability $\langle P_A \rangle$ can thus be adjusted to minimize autocorrelations of observables.

The Metropolis transition probability is really composed of two probabilities [4]:

$$p_M[(x,p) \to (x',p')] = P_C P_A \ ,$$

i.e., a propositional probability for a configuration and one for the acceptance. As long as we accept the configuration with the Hamiltonian $\mathcal{H}$ we can use any other Hamiltonian $\mathcal{H}'$ to derive equations of motion for the proposition. Specifically, we can use $\mathcal{H}' = \alpha \mathcal{H}$. This choice is motivated by the observation [8] that the effect of the discretization of the original Hamiltonian on the equations of motion, is to renormalize the original Hamiltonian. Thus, instead of the original Hamiltonian a new scaled one is solved *exactly*. This observation can be used to scale out to a certain degree the effect of the discretization error on the acceptance rate and accelerate the algorithm.

3   Modeling of Polymer Systems

The simulation of dense macromolecular systems is virtually impossible if one takes into account all degrees of freedom and interactions of a chemically realistic chain [10, 11, 12]. For a single chain one may very well use a chemically realistic description. For long and many chains the computational complexity is overwhelming. Not even the fastest supercomputers on the horizon or beyond will be able to deliver enough computational power to deal with a dense system with all chemical detail present. It is therefore imperative to reduce the complexity in order to make the simulation a tractable approach. This reduction of the complexity of the model is called the coarse-graining of the model. In the coarse-grained approach, the detailed chemistry enters only in the derivation of the potential between new interacting units. These are substitutes for the original detailed chemistry. The system is considered on mesoscopic scales.

A step into this direction is the parametrization of both the intra- and intermolecular interactions by pair potentials. This is an approximation starting from an ab initio quantum mechanical calculation with full chemical detail. The ansatz of united atoms to reduce the number of atoms and thus reduce the complexity of the calculation is also a step in this direction [24]. Formally, the unification of, for example the two H atoms with a C atom for a polyethylene represents a coarse-graining. This has two facets. On the one hand, the reduction in the number of elements decreases the computational complexity. This reduction leads to a possible longer observation time. The reduction in the complexity also allows an increase in the linear dimension of the system.

On the other hand, coarse-graining eliminates out those degrees of freedom that enter into macroscopic properties only through their cooperative effect and substitutes them for effective degrees of freedom. Both facets make feasible the simulation of complex macromolecular systems to predict macroscopic properties.

To discretize the molecular system altogether is another possibility [13]. The continuum chains are mapped to chains on a lattice. This step alone reduces the computational complexity. To further reduce the degrees of freedom one can perform a coarse-graining of a chain. This approach has been successful in treating poly-carbonates [14, 15, 16, 17, 18].

The coarse-graining ansatz [14, 15], in our view, is the only route to simulation of macroscopic properties of polymer materials. It combines the requirement of a simplified model that one can handle computationally with the requirement of representing the degrees of freedom that are necessary for the macroscopic properties one seeks to predict.

Playing the same theme, I describe here a model in which those degrees of freedom that do not enter, or only through their cooperative effect, into macroscopic properties have been eliminated. The new component in the model is the topology, which results in a special geometry. Whereas the usual approach in the united-atom or coarse-grained models is to take the building blocks as spherical symmetric atoms, here ellipsoidal units give rise to nonspherical interaction [25]. Using this building block, chains can be constructed of various connectivities. The model allows the use of spherical and ellipsoidal "atoms" or building blocks to form a chain. Thus the model is able to accommodate and model even complex and asymmetric monomer units in a rather simple way. An example for such a model is given in [23].

4   Coarse-Graining

The starting point of the model is the coarse-graining ansatz as developed by Paul et al, [14, 15, 22]. To present the model along with an application to a specific material, I describe the approach for Bisphenol-A-Polycarbonate (BPA-PC) and a variation of the polycarbonate.

In the coarse-graining approach we want to map larger units of the realistic chain onto one or more units of a new chain such that the interactions between the new units reflect and mimic those of the chemically realistic chain. The units we map can be the repeating unit of the chain, or parts of the repeating unit. The decision on the size or which atoms take part in one unit can be based on the coherence and persistence length of the chain.

Consider three atoms bonded to form a simple chain. We want to keep, for example, the average end-to-end distance or the radius of gyration invariant. The possible variation in the monomer lengths and the monomer angles between the three atoms can be well represented by just two atoms with rescaled monomer length and monomer angle potentials if we also rescale the zero-temperature monomer length and monomer angle.

Once the length scale is fixed, atoms of the unit are taken as base points. From the possible conformations of the chain we find the distributions for the *monomer lengths* and *monomer angles* between the base points along the chemically realistic chains. The distribution contains information on the local structure on scales smaller then the fixed length scale.

To develop a coarse-grained model monomer (in our case for BPA-PC) we proceed in three steps. As input we use the results of ab initio calculations of the geometry and the torsional potentials. Furthermore, we use monomer length and monomer angle distributions that have been determined via Monte Carlo simulations from the ab initio results [23].

1. We compute discrete potentials around one and two atomistic monomer units.
2. From the distributions mentioned above we determine the *bonded* coarse-grained interactions.
3. From the results of item 1. we determine the *non bonded* coarse-grained interaction constants.

One caveat to this approach must be made here. The chains generated by the Monte-Carlo procedure are in vacuum. The intermolecular interaction and its effect on the chain conformations is not taken into account. Only intramolecular interactions are used for the generation of the conformations so far. In principle an intermolecular interaction can be worked into the generation of the conformations.

The ab initio calculations mentioned above showed us that in the case of Bisphenol-A-Polycarbonate, which is considered here, the torsional potential is negligibly small. Thus we introduce no torsional interaction. In general, torsional interaction must and can be included.

5   The Monomer Unit

Our approach starts from the general ellipsoid model for the description of polymer chains [25]. In this model the basic building block is an ellipsoid. Up to now our monomers show rotational symmetry along the backbone. Symmetry-breaking *side groups* can be modeled by sticking degenerate ellipsoids (spheres) to the main ellipsoids. Of course, the ellipsoid itself need not be rotationally symmetric. We can also use oblate ellipsoids and in general ellipsoids with all three principal axes different. This, however, makes the calculation of the non bonded interaction more difficult.

An ellipsoid can easily be adapted to different monomer structures. We only have to change the half axes and radii according to the size of the chemically realistic monomers. In the simplest form we have no side groups at all and the chain consists of rotational ellipsoids whose longer half axes can assume values according a given *monomer-length* distribution. The position of the monomers along the same chain is furthermore determined by the *monomer angles*. The short half axes are chosen so that the volume of the ellipsoids corresponds, in the case of BPA-PC, to the volume of the chemical monomer unit.

6   Bonded Interactions for BPA-PC

The bonded potentials for the lengths of the monomer units and the angle between two of them are obtained in our approach from a coarse-graining procedure [22]. No torsion potential is needed as the distribution of torsion angles is almost uniform. The result of the coarse-graining is a distribution of lengths and angles between the new building blocks. The input for the determination of the coupling is the distribution of lengths (angles) of chemically detailed monomer units [23]. This distribution is obtained from a generation of chains with the detailed chemistry and interaction using Monte–Carlo simulations and shows two clear peaks. It must be kept in mind, however, that the distribution does not contain effects from the packing and intermolecular interaction in a dense system. Consistent with this we may view these interactions as given by a sum of gaussians. The correlations between monomer length and angle and intermolecular interaction are neglected and the coupling constants simply determined from the second moments of the distributions.

In our case of BPA-Polycarbonate we used an average distribution for the backbone atoms of the carbonate-group, i.e. the center of mass of the atoms labeled O_1, C_1 and O_2 of one monomeric unit and the corresponding center of mass of the succeeding monomeric unit (see Fig. 1). The carbonate groups may be regarded as joints along the polymer chain.

Performing a simultaneous fit of two gaussians to the monomer-length distribution as described in [23], we obtain the four parameters

$$\langle l_{01}\rangle \text{ and } \langle l_{01}^2\rangle \tag{4}$$

$$\langle l_{02}\rangle \text{ and } \langle l_{02}^2\rangle \tag{5}$$

representing the two average monomer lengths ($\langle l_{01}\rangle$ and $\langle l_{02}\rangle$) and the two variances ($\langle l_{01}^2\rangle$ and $\langle l_{02}^2\rangle$) characterizing the width of the distributions. For simplification, no crossing of monomers between the two distributions is permitted. The distributions are temperature dependent and the fitting must be carried out for each simulation temperature independently.

Along the same line the parameters for the monomer-angle distribution

$$\langle \theta_0\rangle \text{ and } \langle \theta_0^2\rangle \tag{6}$$

are obtained from the fit of a single gaussian [23].

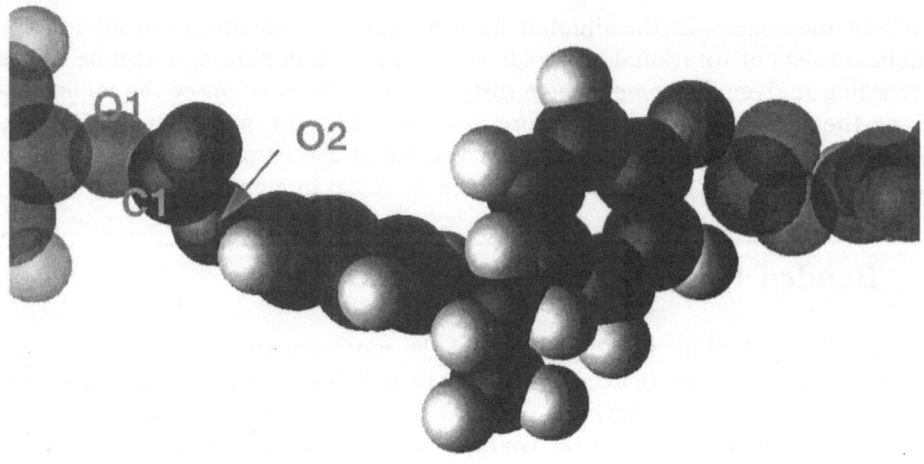

Fig. 1. Position of atoms in the BPA-PC monomer. Two adjacent monomers are shown

These parameters must be fitted to the parameters of the model Hamiltonian

$$
\begin{aligned}
\mathcal{H}_{\text{bond}} = \frac{1}{2} k_1 \sum_i s_i (l_i - l_{01})^2 + \\
\frac{1}{2} k_2 \sum_i (1 - s_i)(l_i - l_{02})^2 + \\
\frac{1}{2} k_\theta \sum_i (\cos \theta_i - \cos \theta_0)^2
\end{aligned}
\tag{7}
$$

where s_i is 1 for monomers belonging to the first distribution (l_{01}) and 0 for monomers of the second distribution (l_{02}). Analysis of cross-correlations between monomer-length and monomer-angle distributions reveals that they can be neglected. Therefore as a first approximation we identify $\langle l_{0i} \rangle$ with l_{0i} and $k_B T / \langle l_{0i}^2 \rangle$ with k_i.

This form of interaction has already been used in other models for Polyethylene [19] and BPA–Polycarbonate [14, 15, 16, 17, 18]. An a posteriori test of the distribution using simulated systems confirms the correctness of this ansatz. The simulations led to distributions that indeed showed the skewed form of the ab initio distribution.

7   Parallelization of the Polymer System

The molecular dynamics part for the method outlined above is fairly general and can be considered on general grounds, except that we are dealing we long-chain molecules. If we were to consider only atoms, then we could apply the domain decomposition concept. For polymer systems, if we are to keep a single

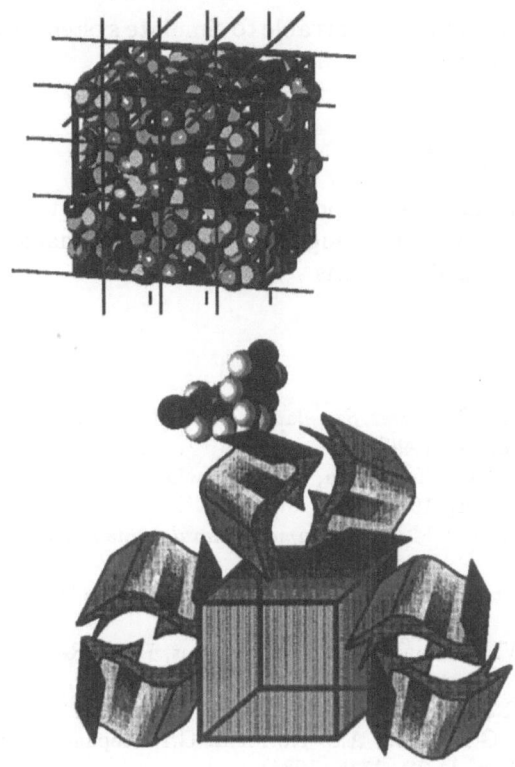

Fig. 2. Shown is a schematic picture for the parallelization of the polymer system. In this implementation polymers are not the units that are passed but atoms

chain as the "unit" then the computational box can not be partitioned into many subvolumes. The chain may stretch over many such subvolumes. From this point of view it is not advantageous to use a data structure and integration algorithms that work on the basis of the connectivity knowledge. Rather we should have the atom as the basic data structure. Then we could also apply the ideas developed for the domain decomposition to the polymer system. The price paid here is a higher administrative overhead. The algorithm can not be based on the connectivity and every "atomic data structure" must carry information on the chain connectivity. What is passed around is this data structure with the accumulated results for the force on that particular "atom". This also implies that there must be some instance that ensures that after a certain number of steps all information was gathered to guarantee a position update in terms of an integration step.

This approach also facilitates the application of the program to many different types of polymer systems. Since there is no reference to the chain connectivity with the program (all information of this nature is hidden in the program and build into the data structure) one is able to simulate systems ranging from simple linear chains to branches or even networks.

Acknowledgements

Partial support by the BMWF project 031240284 and the EU project CIPA-CT93-0105 is gratefully acknowledged. I would like to thank K. Zimmer and A. Linke for the numerous discussions.

References

[1] D.W. Heermann, *Computer Simulation Methods in Theoretical Physics*, 2nd ed. (Springer, Heidelberg 1990)

[2] K. Binder and D.W. Heermann, *Monte Carlo Simulation in Statistical Physics*, (Springer, Heidelberg 1988)

[3] M.P. Allen and D.J. Tildesley, *Computer Simulations of Liquids*, (Clarendon Press, Oxford 1987)

[4] S. Duane, A.D. Kennedy, B.J. Pendleton, and D. Roweth, Phys. Lett. B **195**, 216 (1987)

[5] B. Mehlig, D.W. Heermann, and B.M. Forrest, Phys. Rev. B, **45**, 679 (1992)

[6] B. Mehlig, D.W. Heermann, and B.M. Forrest, Mol. Phys. **76**, 1347 (1992)

[7] D. Rigby and R.J. Roe, J. Chem. Phys. **87**, 7285 (1987)

[8] G.G. Batrouni, G.R. Katz, A.S. Kornfeld, G.P. Lepage, B. Svetitsky, and K.G. Wilson, Phys. Rev. D **32**, 2735 (1985)

[9] M. Creutz and A. Goksch, Phys. Rev. Lett. **63**, 9 (1989)

[10] K. Binder, in *Molecular Level Calculation of the Structure and Properties of Non-Crystalline Polymers*, ed. J. Bicerano, (Dekker, New York 1991)

[11] D.N. Theodorou and U.W. Suter, Macromolcules, **18**, 1467 (1985)

[12] J. Batoulis, K. Binder, F.T. Gentile, D.W. Heermann, W. Jilge, K. Kremer, M. Laso, P.J. Ludovice, L. Morbitzer, W. Paul, B. Pittel, R. Plaetschke, K. Reuter, K. Sommer, U.W. Suter, R. Timmermann, and G. Weymans, Advanced Materials **3**, 590 (1991)

[13] K. Kremer and K. Binder, Comput. Phys. Rep. **7**, 259 (1988)

[14] W. Paul, K. Binder, D.W. Heermann, and K. Kremer, J. Phys. II **1**, 37 (1991)

[15] W. Paul, K. Binder, K. Kremer, and D.W. Heermann, Macromol., **24**, 6332 (1991)

[16] W. Paul, K. Binder, D.W. Heermann, and K. Kremer, J. Chem. Phys., **95**, 7726 (1991)

[17] W. Paul, K. Binder, D.W. Heermann, and K. Kremer, J. Non-Crystal. Solids **131**, 650 (1991)

[18] K. Binder, W. Paul, H.-P Wittmann, J. Baschnagel, K. Kremer, and D.W. Heermann, Progr. Colloid Polym. Sci. **91**, 5 (1991)

[19] A. Hartmann, D.W. Heermann, S. Ozawa, and K. Zimmer preprint (1994)

[20] W. Paul, K. Binder, D.W. Heermann, and K. Kremer, J. Phys. II **1**, 37 (1991)

[21] H. Sun, S.J. Mumby, J.R. Maple, and A.T. Hagler, J. Phys. Chem. **99**, 5873 (1995)

[22] J. Baschnagel, K. Binder, W. Paul, M. Laso, U.W. Suter, J. Batoulis, W. Jilge, and T. Bürger, J. Chem. Phys. **95**, 6014 (1991)

[23] K.M. Zimmer, A. Linke, D.W. Heermann, and J. Batoulis, preprint, (1995)

[24] D. Rigby and R.J. Roe, J. Chem. Phys. **87**, 7285 (1987)

[25] K.M. Zimmer and D.W. Heermann, J. Computer-Aided Materials Design **2**, 1 (1995)

Constraints in Molecular Dynamics, Nonequilibrium Processes in Fluids via Computer Simulations

Siegfried Hess

Institut für Theoretische Physik, Technische Universität Berlin, Hardenbergstr. 36, D-10623 Berlin, Germany, e-mail: `hess0433@w421zrz.physik.tu-berlin.de`

Abstract. The basic features of computer simulations for fluids are presented based on the molecular dynamics approach. Technical details, in particular the interaction potentials and the scaling of physical variables, the application of constraints to the dynamics, e.g., in order to simulate a thermostat, and the choice of integrators for the numerical solution of the equations of motion are discussed. Some examples which can be used as numerical exercises are outlined. The method of nonequilibrium molecular dynamics (NEMD) is introduced. Simulations of relaxation phenomena are mentioned briefly. The main emphasis is on simulations of a plane Couette flow as an example of a stationary transport process. Procedures to extract rheological properties, such as the (non-Newtonian) viscosity, normal pressure differences, and information on shear-flow-induced structural changes, are given, firstly for fluids composed of spherical particles. This comprises simple liquids and dense colloidal dispersions, in which the states far away from equilibrium studied in NEMD are accessible in experiments. Secondly, simulations for complex fluids, in particular polymeric melts, nematic and smectic liquid crystals, as well as ferrofluids and magnetorheological fluids are discussed.

1 Introduction

Transport and relaxation processes in fluids have been studied by nonequilibrium molecular-dynamics (NEMD) computer simulations for more than twenty years [1, 2]. By now the method is well established [3]-[12]. Physical phenomena in simple fluids far away from equilibrium and the material properties of complex fluids have been analyzed in recent years. The NEMD results provide information and insight into microscopic mechanisms of nonequilibrium phenomena similar to that available in real experiments, e.g. in light, x-ray or neutron scattering. However, even more detailed information is available in molecular dynamics, e.g. snapshots showing the positions of the particles or specific contributions to physical quantities of which only the sum can be measured in the real experiment. Thus computer simulations can often yield sharper tests of the assumptions underlying conventional theories than a real experiment where, in addition, some uncertainty regarding the particle–particle interaction always remains.

The maintenance of a stationary nonequilibrium state and, in particular, of a constant kinetic energy required to simulate a thermostat impose *constraints* on the dynamics. After some general remarks on molecular dynamics, potentials and the scaling of physical variables, methods for simulating a *thermostat* are presented. Some other technical details, including the use of various integrators for the numerical solution of the equations of motion are discussed.

Relaxation phenomena can be observed when a system is prepared in a nonequilibrium state and then allowed to relax to equilibrium. Examples are the decay of an initially crystalline positional order into a fluid state and stress relaxation after a sudden change of shape of the volume containing the particles. As an example of a *stationary transport process* a plane *Couette flow*, also referred to as "simple shear flow", is considered in some detail. Results are presented for "simple" and for "complex" fluids. The method of NEMD simulations is firstly discussed for fluids composed of spherical particles. The flow behavior of the so-called simple fluids and the shear-induced structural changes are not simple. A comparison with experimental results of (dense) colloidal dispersions of spherical particles can be made. The complex fluids studied are polymeric melts and anisotropic fluids such as nematic liquid crystals (LC) and ferrofluids, magneto- or electrorheological (MR or ER) fluids.

2 Basics of Molecular Dynamics

2.1 Equations of Motion

In a molecular dynamics computer simulation for a substance composed of N spherical particles Newtons equations of motion

$$m \frac{\mathrm{d}^2}{\mathrm{d}t^2} \mathbf{r}^i = \mathbf{F}^i = \sum_j \mathbf{F}^{ij} \ , \tag{1}$$

are integrated numerically. The particles are located at positions $\mathbf{r}^i$ in a volume V. The particle density is $n = N/V$. The particle i, $(i = 1, 2, ..., N)$, feels the force $\mathbf{F}^i = \sum_j \mathbf{F}^{ij}$ which is the sum of the forces $\mathbf{F}^{ij}$ exerted by all other particles $j \neq i$ on particle i. When one wants to avoid surface effects, periodic boundary conditions and the nearest-image convention are used. This means, particle i either feels the force caused by particle j or by one of its images depending on which one is closest to it.

The temperature T of the system is linked with the part of the kinetic energy K which is not associated with a macroscopic motion:

$$\frac{3}{2}(N - 1) k_{\mathrm{B}} T = K := \frac{1}{2} \sum_i m(\mathbf{c}^i)^2 \ , \tag{2}$$

where k_{B} is the Boltzmann constant and $\mathbf{c}^i$ is the "peculiar velocity", i.e., the velocity of a particle relative to the flow velocity $\mathbf{v} = \mathbf{v}(\mathbf{r}^i)$. The center of mass of the system of N particles is constant. Thus the number of degrees of freedom in three dimensions is $3(N - 1)$ rather than $3N$.

In order to simulate an isothermal system the temperature has to be kept constant. The simplest version of a "thermostat" consists of rescaling the peculiar velocity after each time step by the factor $(T_{\text{wanted}}/T_{\text{measured}})^{1/2}$. Other thermostats, e.g., those referred to as "Gaussian" and "Nose-Hoover" [7] are imposed as contraints and will be discussed in one of the following sections.

2.2   Extraction of Data from MD Simulations

The observables of interest, such as the internal energy and the components of the pressure or the stress tensor, can be calculated from the known positions and velocities of the particles as time averages according to the rules of statistical physics. Typically, in a stationary state, the required data are extracted after each tenth to hundredth time step. Similarly, more detailed information can be obtained from the simulation such as the velocity distribution function, the pair correlation function or the static structure factor, which can also be measured in scattering experiments.

In ordinary molecular dynamics (MD) simulations data are extracted when an equilibrium state with a specified density n (or pressure) and temperature T has been reached. In addition, dynamic quantities can be extracted from the temporal fluctuations, e.g., transport coefficients can be calculated from time correlation functions with the help of Green–Kubo relations. In nonequilibrium molecular dynamics (NEMD) simulations, on the other hand, relaxation and transport phenomena are investigated more directly and in close analogy to real experiments. States far away from equilibrium are also studied. In the following, some relaxation phenomena are mentioned and a stationary plane Couette flow is considered in more detail as a special case of a transport process.

Before NEMD simulations are discussed further, however, the presentation of some elementary material on potentials from which the forces are derived, on constraints and integrators is appropriate. Characteristic features of the dynamics can be studied in simple exercises involving one or two particles in one or two dimensions (without periodic boundary conditions).

3   Potentials, Constraints, and Integrators

3.1   Interaction Potential and Scaling

In the simulations dimensionless or "scaled" variables are used which are denoted by the same symbols as the physical variables when no danger of confusion exists. For a system of particles whose forces are derived from a binary interaction potential, the *Lennard–Jones* (LJ) *potential* is

$$\Phi = \Phi^{\text{LJ}} := 4\Phi_0 \left(\left(\frac{r_0}{r} \right)^{12} - \left(\frac{r_0}{r} \right)^6 \right) . \tag{3}$$

Lengths and energies are presented in units of the diameter r_0 and of the potential depth Φ_0. The units used for the particle density and for the temperature are

r_0^{-3} and $k_B^{-1} \Phi_0$. The time is scaled with the reference time $t_0 = r_0 m^{1/2} \Phi_0^{-1/2}$; m is the mass of a particle. The pressure, the shear rate and the viscosity of the LJ fluid are expressed in units of $r_0^{-3} \Phi_0$, t_0^{-1} and $r_0^{-3} \Phi_0 t_0 = r_0^{-2} m^{1/2} \Phi_0^{1/2}$. For many fluids composed of atoms or small molecules, the specific values of r_0 and Φ_0 that are needed to relate the theoretical results to the thermophysical properties of specific substances are available. In the simulations, the cutoff of the interaction at a finite distance r_c is often achieved just by putting the potential and the force equal to zero for $r > r_c$, e.g., with $r_c = 2.5 r_0$. A smoother cutoff whereby the force is (at least) continuous at r_c, however, is preferrable for the integration of the equations of motion. The "LJ-spline" potential Φ^{LJspl} of Evans and Holian is an example of a modified LJ potential with an improved cutoff. This potential is equal to the LJ potential for $r \le r_I$, where $r_I = (26/7)^{1/6} r_0 \approx 1.244 r_0$ is determined by the point of inflection, $\Phi''(r_I) = 0$, and by the third-order spline expression,

$$\Phi^{\mathrm{LJspl}} := s_2 \Phi_0 \left(\frac{r}{r_0} - \frac{r_c}{r_0} \right)^2 + s_3 \Phi_0 \left(\frac{r}{r_0} - \frac{r_c}{r_0} \right)^3 , \quad r_I = \left(\frac{26}{7} \right)^{1/6} r_0 \le r \le r_c$$

(4)

and, of course, $\Phi^{\mathrm{LJspl}} = 0$ for $r \ge r_c$. The coefficients s_2, s_3 and r_c, determined such that the interaction potential as well as its first and second derivatives are continuous at r_I, are given by

$$s_2 = -\frac{24192}{3211} \left(\frac{r_0}{r_I} \right)^2, \quad s_3 = -\frac{387072}{61009} \left(\frac{r_0}{r_I} \right)^3, \quad r_c = \frac{67}{48} r_I \approx 1.737 \, r_0 . \quad (5)$$

An even smoother cutoff where not only the force but also its first derivative vanish continuously at r_c is the potential Φ^{LJsmoc} which, between r_I and r_c is given by

$$\Phi^{\mathrm{LJsmoc}} := s_4 \Phi_0 \left(2 \left(\frac{r_c}{r_0} - \frac{r_I}{r_0} \right) \left(\frac{r}{r_0} - \frac{r_c}{r_0} \right)^3 + \left(\frac{r}{r_0} - \frac{r_c}{r_0} \right)^4 \right), \quad r_I \le r \le r_c$$

(6)

and, of course, also $\Phi^{\mathrm{LJsmoc}} = 0$ for $r \ge r_c$. Here one has

$$s_4 = \frac{133}{169} \left(\frac{36}{19} \right)^4 \left(\frac{r_0}{r_I} \right)^4, \quad r_c = \frac{55}{36} r_I \approx 1.901 \, r_0 . \quad (7)$$

A simpler short-range potential with a smooth cutoff and parameters chosen such that the force at $r = r_0$ and the well depth are equal to the corresponding values of the LJ potential is

$$\Phi^{\mathrm{shrat}} := 24 \Phi_0 \left(1 - \frac{r}{r_0} \right) \left(\frac{r_c - r}{r_c - r_0} \right)^3 , \quad r \le r_c = \frac{113}{81} r_0 \approx 1.395 \, r_0 , \quad (8)$$

and, of course, $\Phi^{\mathrm{shrat}} = 0$ for $r \ge r_c$. The minimum occurs at $r = r_{\min} := (89/81) r_0 \approx 1.099 r_0$, which is somewhat smaller than the corresponding value

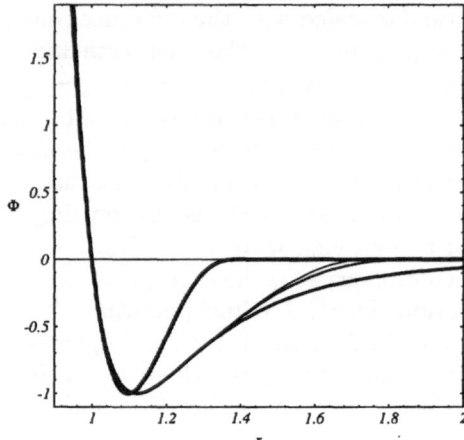

Fig. 1. Potentials Φ (in units of Φ_0) as functions of the distance r (in units of r_0). The Lennard–Jones (LJ) potential which extends farthest to the right is compared with the two smoothly cutoff versions "LJspline" (*second from left*) and "LJsmoc" (*second from right*). The *thick curve* is the short-range attractive "shrat" potential function

$2^{1/6} r_0 \approx 1.1225 \, r_0$ of the LJ potential. The LJ potential and the three other potential functions with a finite cutoff discussed here are displayed in Fig.1.

When only the repulsive "r^{-12}" part of the LJ interaction potential is taken into account one speaks of a "soft spheres" (SS) potential. The LJ-potential cutoff at its minimum $r \, r_0^{-1} = 2^{1/6} \approx 1.123$, which is also purely repulsive, is referred to as WCA potential. An interaction potential which, at $r = r_0$, has the same value and force as the WCA potential can be introduced in analogy to (8):

$$\Phi^{\text{shrep}} := 8\Phi_0 \frac{(r_c - r)^3}{r_0(r_c - r_0)^2} \, , \quad r \leq r_c = \frac{9}{8} r_0 = 1.125 r_0 \, , \tag{9}$$

and $\Phi^{\text{shrep}} = 0$ for $r \geq r_c$. The simple and smoothly cut off potentials (8) and (9) are recommended for test simulations with typical values of the kinetic energy per particle $\approx k_{\text{B}}T$ smaller than about $5\Phi_0$.

3.2   Thermostats

Gaussian Thermostat. A system with a vanishing average flow velocity, i.e., with $\mathbf{v} = 0$ is considered now; then $\mathbf{c}^i = \frac{\text{d}}{\text{d}t} \mathbf{r}^i$. The condition that the kinetic energy $K = (m/2) \sum_i \mathbf{c}^i \cdot \mathbf{c}^i$ be constant (isokinetic system) is a nonholonomic constraint which can be taken into account by supplementing the equations of motion for particle i with a constraint force $\mathbf{Z}^i$:

$$m \frac{\text{d}^2}{\text{d}t^2} \mathbf{r}^i = \mathbf{F}^i + \mathbf{Z}^i \, . \tag{10}$$

Gauss's principle which states that $\sum_i \mathbf{Z}^i \cdot \mathbf{Z}^i$ is extremal, hopefully minimal, leads to $\mathbf{Z}^i \sim \mathbf{c}^i$. The proportionality coefficient is put equal to $-m\zeta$ where the pseudo-friction coefficient ζ can be positive or negative, depending on the dynamic state of the system. More specifically, the *Gaussian* thermostat consists of replacing the equations of motion by:

$$
m \frac{\mathrm{d}^2}{\mathrm{d}t^2} \mathbf{r}^i = \mathbf{F}^i - m\zeta \, \mathbf{c}^i \, , \quad \mathbf{c}^i = \frac{\mathrm{d}}{\mathrm{d}t} \mathbf{r}^i \, , \tag{11}
$$

with

$$
\zeta = \zeta_{\mathrm{G}} := \frac{\sum_j \mathbf{c}^j \cdot \mathbf{F}^j}{m \sum_k \mathbf{c}^k \cdot \mathbf{c}^k} \, . \tag{12}
$$

The equations of motion remain time-reversal invariant provided that the force $\mathbf{F}^i$ guarantees this property. Notice that

$$
2 \, K \, \zeta_{\mathrm{G}} = -\frac{\mathrm{d}E^{\mathrm{pot}}}{\mathrm{d}t} \, , \tag{13}
$$

where E^{pot} is the potential energy. The Gaussian thermostat keeps the kinetic energy constant. Its desired value K_0 has to be initially assigned, e.g, by rescaling the initial velocities.

Velocity Rescaling. The frequently used method of rescaling the magnitude of the velocities after each integration time step is equivalent to the Gaussian thermostat. To show this, note that rescaling by the factor $(T_{\mathrm{wanted}}/T_{\mathrm{measured}})^{1/2} = (K_0/K)^{1/2}$, where K_0 and K are the wanted and the measured values of the kinetic energy, is equivalent to replacing, after each time step of length δt, the velocity $\mathbf{c}^i$ by $\mathbf{c}^i + \delta\mathbf{c}^i$ with

$$
\delta\mathbf{c}^i = \left(\left(\frac{K_0}{K} \right)^{1/2} - 1 \right) \mathbf{c}^i = \left(\left(\frac{K_0}{K_0 + \delta K} \right)^{1/2} - 1 \right) \mathbf{c}^i \approx -\frac{\delta K}{2K_0} \mathbf{c}^i = -\zeta_{\mathrm{G}} \mathbf{c}^i \delta t \, . \tag{14}
$$

Here, $\delta K = K - K_0$ is the change of the kinetic energy over one time step and $|\delta K| \ll 1$ is assumed. The last equality in (14) follows from the fact that the total energy $K + E^{\mathrm{pot}}$ remains constant, hence $\delta K = -\delta E^{\mathrm{pot}}$, and $dE^{\mathrm{pot}}/dt = \delta E^{\mathrm{pot}}/\delta t$ was used. For small enough time steps δt, the Gaussian constraint force occurring in (11) implies an extra change of the velocity given by (14). Thus the velocity rescaling is equivalent to the action of a Gaussian thermostat.

Nose–Hoover Thermostat. This thermostat consists of supplementing the equations of motion (11) with an equation of change for the coefficient ζ:

$$
\frac{\mathrm{d}\zeta}{\mathrm{d}t} = \frac{1}{\tau_{\mathrm{NH}}^2} \left(\frac{K}{K_0} - 1 \right) . \tag{15}
$$

As before, $K = (m/2) \sum_k \mathbf{c}^k \cdot \mathbf{c}^k$ and K_0 are the actual and the prescibed values of the kinetic energy. Furthermore, τ_{NH} is a relaxation time determining the speed of response of the thermostat. It is an extra parameter occurring in the simulation. Of course stationary and, in particular, equilibrium properties should not be affected by the choice of τ_{NH}. On the other hand, dynamic phenomena occurring on a time scale comparable with τ_{NH} are modified by the action of the thermostat. Again, time reversal invariance is not destroyed by the thermostat. The limiting cases $\tau_{NH} \rightarrow \infty$, $\zeta = 0$ and $\tau_{NH} = \delta t$, $\zeta = \zeta_G$ correspond to isoenergetic (adiabatic) and Gaussian isothermal simulations, respectively.

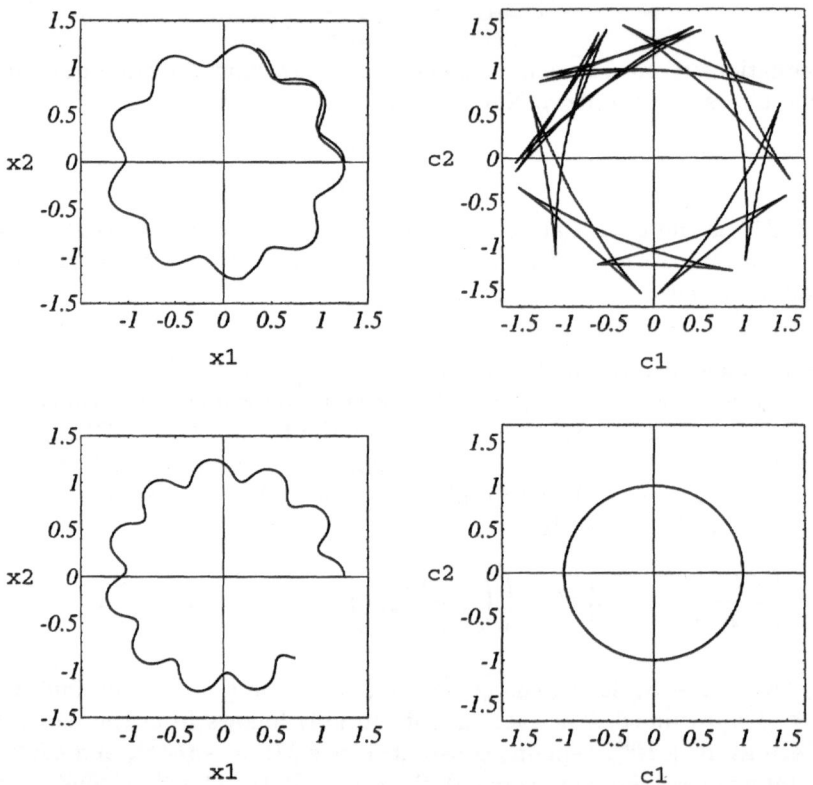

Fig. 2. The orbits $x2$ vs $x1$ and the velocity orbits $c2$ vs $c1$ for unconstrained (*top*) and Gaussian iso-kinetic (*bottom*) motions

Thermostats in Action. Next, as an exercise, the effects of the thermostats are studied by comparing contrained and unconstrained solutions for a simple two-dimensional one-particle problem. More specifically, a particle with mass m is subjected to an external force determined by the short-range attractive (shrat) potential (8). The cartesian components of the position vector and of the velocity are denoted by $x1$, $x2$ and $c1$, $c2$. In Fig.2, the orbits $x2$ versus $x1$ and the velocity orbits $c2$ versus $c1$, as calculated by *NDSolve* of *Mathematica*, are displayed for the unconstrained system and for the case of a Gaussian thermostat. The inititial conditions are, in terms of reduced, dimensionless variables: $x1 = 1.25$, $x2 = 0$, $c1 = 0$, $c2 = 1$, in both cases. The curves are shown for the reduced times t with $0 \leq t \leq 8$. In the absence of the constraint, the particle gains kinetic energy after the start. This is not allowed when the Gaussian thermostat is in action. For this reason, the particle does not travel as far as in the unconstrained case. The same holds true for the Nose–Hoover thermostat (Fig.3), where the corresponding orbits are presented for $\tau_{\mathrm{NH}} = 0.1$ (in reduced units). In this case, the instantaneous kinetic energy still varies and the velocity orbit shows a behavior intermediate between the cases compared in Fig.2.

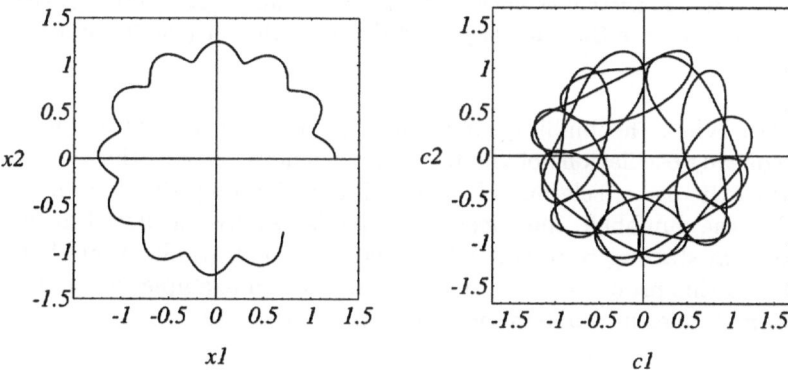

Fig. 3. The orbits $x2$ vs $x1$ and the velocity orbits $c2$ vs $c1$ for a Nose–Hoover thermostat

The time dependence of the pseudo-friction coefficient ζ corresponding to the Gaussian dynamics studied in the lower part of Fig.2 and the Nose-Hoover thermostat, cf. Fig.3, is shown in Fig.4. Note that ζ assumes positive and negative values of equal magnitude. The total energy E varies according to $\frac{\mathrm{d}E}{\mathrm{d}t} = -2K\zeta$. For comparison, in the case of an unconstrained motion, i.e., for $\zeta = 0$ corresponding to the top orbits of Fig.2, the maximum deviation of the energy from its initial value that is caused by inaccuracies of the numerical integration is less than 7×10^{-5}, in reduced units.

Other Constraints. Simulations in which the pressure or a specific component of the pressure tensor are constant can and have been performed by imposing an appropriate constraint on the dynamics of the N particle system. Various procedures analogous to the "rescaling", Gaussian and Nose–Hoover methods of thermostats were applied [7]-[9].

Different types of constraints are encountered when one wants to model molecules composed of atoms. Simple examples are dumbbells and chain

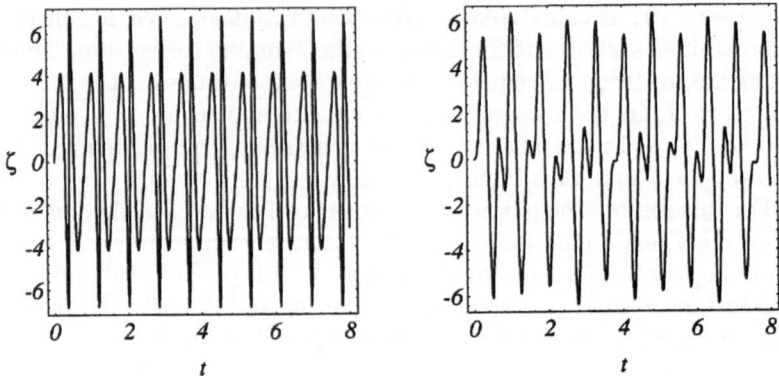

Fig. 4. The pseudo-friction coefficient ζ (in reduced units) as function of the (dimensionless) time t for the Gaussian (*left*) and the Nose–Hoover (*right*) thermostats

molecules. When the binding forces are derived from an extra potential, the high frequency oscillations of the bonds may require shorter time steps. This is avoided by the use of scleronomic constraints which keep the bond lengths constant. For the dumbbell, the constraint force is usually calculated in text books of Mechanics starting from the d'Alembert principle. Application of the Gauss principle to this problem yields the same result. An example for the simulation of chain molecules in a polymeric melt is presented later.

3.3 Integrators

Runge–Kutta and Predictor-Corrector Methods. The classical (second and higher order) Runge–Kutta methods used to integrate ordinary differential equations require more than one calculation of the forces per time step. For this reason, predictor-corrector methods which need only one of the time consuming force evaluations per time step became rather popular in MD and NEMD simulations [4, 16]. On the other hand, simple integrators referred to as "Stoermer–Verlet" and "velocity Verlet" methods used in the early days of MD simulations are still applied today. These and some "symplectic" integrators will be discussed in the following sections.

"Stoermer–Verlet" and "Velocity-Verlet" Methods. Consider the second order differential equation

$$\frac{d^2 x}{dt^2} = F(x) \,. \tag{16}$$

The Stoermer-Verlet method is based on approximating the second order time derivative by $(x(t + \delta t) - 2x(t) + x(t - \delta t))/(\delta t)^2$ where δt is the integration time step. The "force" F on the r.h.s. of (16) is taken as $F(x(t))$. With the identifications $x_{\text{new}} = x(t + \delta t)$, $x = x(t)$, $x_{\text{old}} = x(t - \delta t)$, the integrator corresponds to

$$x_{\text{new}} = 2x - x_{\text{old}} + F(x)(\delta t)^2. \tag{17}$$

Since x and the velocity v rather than x and x_{old} are given initially, x_{old} has to be calculated, e.g., by $x_{\text{old}} = x - v\,\delta t + 0.5F(x)(\delta t)^2$ before (17) can be applied for a sequence of time steps. Then velocity v, to be approximated by $v = (x_{\text{new}} - x_{\text{old}})/(2\,\delta t)$ is not involved in the integration procedure and thus can not be controlled by rescaling. The velocity-Verlet method, on the other hand, does involve the velocity. This procedure is based on

$$x_{\text{new}} = x + v\,\delta t + 0.5\,F(x)(\delta t)^2, \quad v_{\text{new}} = v + 0.5\,(F(x) + F(x_{\text{new}}))\,\delta t. \tag{18}$$

In Fortran notation, (18) is equivalent to the three consecutive statements ("leap-frog" scheme):

$$v = v + 0.5\,F(x)\,\delta t, \quad x = x + v\,\delta t + 0.5\,F(x)(\delta t)^2, \quad v = v + 0.5\,F(x)\,\delta t. \tag{19}$$

Symplectic Integrators. The Hamilton equations for the coordinates q and momenta p are:

$$\frac{dq}{dt} = \frac{\partial H}{\partial p} = G, \quad \frac{dp}{dt} = -\frac{\partial H}{\partial q} = F. \tag{20}$$

Since $\partial G/\partial q + \partial F/\partial p = 0$, the phase-space volume is conserved. The symplectic integrators guarantee this property. When constraints, for example, imposed by a stationary transport process lead to a decrease of the phase-space volume, one is sure that the integrator as such does not contribute to it. In the following, the special case $G = p$, $F = F(q)$ is considered. A sympletic integrator of order M determines the new values of q and p after one time step δt by a sequence of M consecutive changes δq and δp according to:

$$\delta p = b_1\,F\,\delta t, \quad \delta q = c_1\,p\,\delta t, \quad \delta p = b_2\,F\,\delta t, \quad ..., \quad \delta q = c_M\,p\,\delta t. \tag{21}$$

The coefficients b_i and c_i have the property $\sum_i b_i = 1$, $\sum_i c_i = 1$ and are chosen to optimize the conservation of energy. The simple case $b_1 = 0$, $b_2 = 1$, $c_1 = c_2 = 0.5$ corresponding to the scheme

$$\delta q = 0.5\,p\,\delta t, \quad \delta p = F\,\delta t, \quad \delta q = 0.5\,p\,\delta t, \tag{22}$$

is called "si2.a" in [17]. It is essentially equivalent to the Verlet integrator. An example of a third order integrator, denoted by "si3.b" [17], involves the coefficients:

$$b_1 = 0.2683301 = c_3, \quad b_2 = -0.1879916 = c_2, \quad b_3 = 0.9196615 = c_1. \tag{23}$$

Notice that some coefficients are negative. A fifth-order method, termed "si4.c" [17], uses the coefficients

$$b_1 = 0.06175885813563, \quad c_1 = 0.20517766154229,$$

$$b_2 = 0.33897802655364, \quad c_2 = 0.40302128160421,$$

$$b_3 = 0.61479130717558, \quad c_3 = -0.12092087633891, \quad (24)$$

$$b_4 = -0.14054801465937, c_4 = 0.51272193319241,$$

$$b_5 = 0.12501982279453, \quad c_5 = \qquad 0.$$

The difference $dE_{max} = E_{max} - E_{min}$ between the maximum and minimum values $E_{max} - E_{min}$ of the energy during an integration interval $0 \leq t \leq t_{end}$ can be used as a measure for the efficiency of an integrator. When comparing different integrators the number of force evaluations which is the main computational load in MD simulation, should be equal. For this reason, one may introduce a force time step dt_0 and put $\delta t = L dt_0$ where L is the number of force evaluations needed per integration time step δt. For the si2.a, as well as for the Verlet integrators, one has $L = 1$, whereas $L = 3$ and $L = 4$ for the si3.b and si4.c integrators. As an exercise, it is suggested that one determines dE_{max} for the harmonic oscillator with the simple Hamiltomian $H = 0.5(p^2 + q^2)$ integrated over one period corresponding to $t_{end} = 2\pi$, as function of dt_0, and compare the various integrators discussed here. For a slightly more complicated problem, viz. the one-dimensional motion of a particle with mass $m = 1$ in the short-range attractive shrat potential (8), such a comparison is presented in Fig.5. The initial position and velocity, expressed in variables analogous to those of the problem studied in previous numerical example, is $x0 = a/b$, $c0 = -1$. The particle starts outside the interaction region, flies towards the force center where it is first attracted, then feels the repulsion, is reflected and, at the time $t_{end} = 1$ is at a position close to where it started. This corresponds to a head-on collision. The curves displayed from top to bottom, are for the si2.a, si3.b schemes, a fifth order integrator proposed in [18], and the si4.c integrator. Notice that dE_{max} is plotted versus $1/dt_0$ and logarithmic scales are used; the value $dt_0 = 0.0025$, for example, corresponds to about 2.6 on the horizontal axis. Small force-time steps leading to smaller values for the energy inaccuracy are on the right. Clearly, the use of better integrators can improve the accuracy at no extra computational cost!

4   Nonequilibrium Phenomena

4.1   Relaxation Processes

Relaxation processes can be studied in MD simulations by starting from an intentionally prepared nonequilibrium state. Then the approach to equilibrium

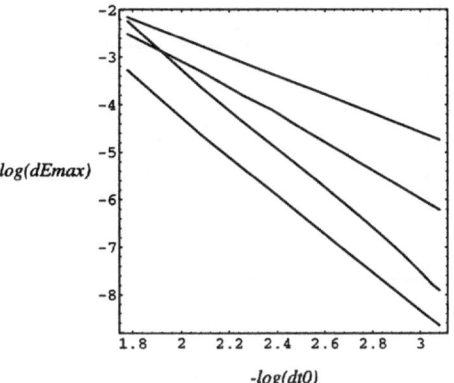

$-log(dt0)$

Fig. 5. The maximum energy deviation dE_{max} versus the reciprocal of the force time step dt_0 for a head-on collision with the "shrat" potential. The four curves, *from top to bottom*, are for the si2.a and si3.b schemes, a fifth order integrator proposed by Hoover et al. [18] and the si4.c integrator

is observed by analyzing the time dependence of (instantaneous) macroscopic variables such as the potential contributions to the internal energy and the pressure or quantities which are closer to a microscopic description of a fluid such as the pair-correlation function. Examples are the decay of an initial crystalline structure, e.g., of cubic type (bcc, fcc, sc), when the equilibrium state is a fluid. Variables sensitive to the bond-orientation (cubic) anisotropy show a much slower decay than scalar quantities like the energy or the pressure [19] when the equilibrium state point is close to the fluid-solid phase coexistence. From a simulation for $N = 1000$, LJ particles which started at simple cubic (sc) lattice sites with the values $T = 1$, $n = 0.9$ (reduced units) for the temperature and the density, Fig.6 shows the (high-frequency) shear modulus G and the cubic modulus G_c as a function of the (dimensionless) time t. These moduli are related to the Voigt elasticity coefficients c by $G = (c_{11} - c_{12} + 3\,c_{44})/5$ and $2\,G_c = c_{11} - c_{12} - 2\,c_{44}$. The latter coefficient vanishes in an isotropic state. These quantities are computed according to

$$VG = \sum_{ij} A(r^{ij}), \quad A(r) = \frac{1}{30} r^{-2} \left(r^4 \Phi(r)'\right)' \ , \tag{25}$$

$$VG_c = \sum_{ij} A_c(r^{ij}), \quad A_c(r) = \frac{5}{24} r^3 \left(r^{-1}\Phi(r)'\right)' K_4(\mathbf{r}), \tag{26}$$

where $\mathbf{r}^{ij} = \mathbf{r}^i - \mathbf{r}^j$ is the relative position vector of particles i, j, $r^{ij} = |\mathbf{r}^{ij}|$, and $\Phi(r^{ij})$ is their interaction potential. The prime denotes differentiation with respect to r, and V is the volume. The symbol $K_4 = (r_x^4 + r_y^4 + r_z^4 - (3/5)r^4)r^{-4}$

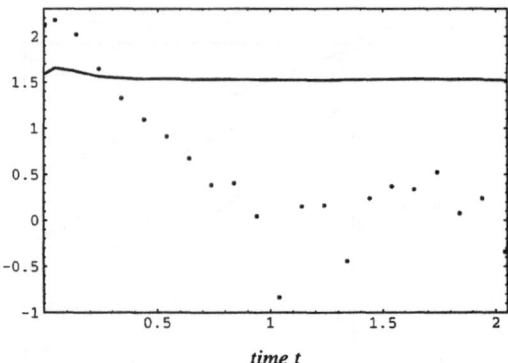

time t

Fig. 6. The logarithm of the shear modulus G (*full line*) and of the cubic modulus G_c (*data points*) as functions of the dimensionless time t during the relaxation of the simple cubic structure   •

is the fourth-order cubic harmonic with the full cubic symmetry. Except for the start values, the data presented in Fig.6 have been averaged over 10 time steps. Notice that G_c, for times $t < 1$, decays (approximately) exponential with a relaxation time ≈ 0.2. For $t > 1$, G_c shows strong fluctuations, G has already reached its stationary value.

Another example is stress relaxation which is usally just measured in solids. In simulations, it can also be studied for the fluid phase since it is possible to impose a deformation instantaneously. To be more specific, consider, at time $t = 0$, a rescaling of the x and y coordinates of all particles in an equilibrium fluid state, as well as the multiplication the pertaining sides of the basic periodicity box by the factors λ and $1/\lambda$, respectively ($\lambda > 0$). Then the normal presssure difference $p_{xx} - p_{yy}$ which flucutates about zero at equilibrium will suddenly become nonzero with a value from which one may, for small distortions ($\lambda \approx 1$), infer the high frequency shear modulus G. Expressions used for the calculation of the components of the pressure tensor are given in the following section. For times $t > 0$, the particles will rearrange themselves such that their spatial correlations, on average, become isotropic again. Consequently, $p_{xx} - p_{yy}$ will decay. For small distortions, the time dependence of the normal pressure difference is equal to that of the time correlation function which, via a Green–Kubo relation, yields the (shear) viscosity. An example for a stress relaxation simulation can be found in [20].

4.2   Plane Couette Flow

For a simple shear flow in the x direction with the gradient in the y direction, the shear rate γ is given by

$$\gamma = \frac{\partial v_x}{\partial y} \ .$$

(27)

Such a flow can be either be generated by moving boundaries or forces [1, 13], or as used here and indicated in Fig.7, by moving image particles undergoing an ideal Couette flow with the prescribed shear rate (homogeneous shear). Let the flow be switched on at $t = 0$. Then, at time t the image particles above (below) the basic (central) box have moved in the x direction to the right (left) by the distance $\gamma t L$ modulo(L) where L is the length of the periodicity box in the y direction. Of course, the periodic boundary conditions for the particles leaving and entering the basic box have to be modified (Lees–Edwards boundary conditions, [2]–[8]). For a system in a fluid state in equilibrium and for not-too-large shear rates, a linear velocity profile typical for a plane Couette flow is set up in the basic box (from which the data are extracted). At high shear rates where plug-like flow also occurs it is essential to use a velocity "profile unbiased thermostat" (PUT, [4, 14]). A shear flow can also be generated by modifying the equations of motion (SLLOD, [7, 8]). For a recent review of NEMD results for rheological properties simple and complex fluids see [21].

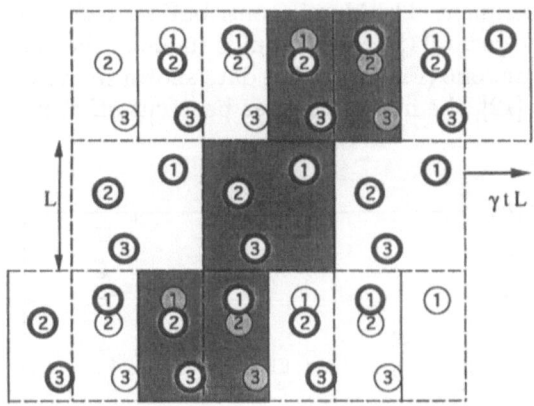

Fig. 7. Moving periodic images generate a plane Couette flow. The *thin circles* mark the positions the image particles would have without the shear flow

4.3   Viscosity

Rheological properties such as the (non-Newtonian) viscosity and normal pressure differences are obtained from the Cartesian components of the stress tensor $\sigma_{\mu\nu} = -p_{\mu\nu}$ or of the pressure tensor $p_{\mu\nu}$ which is the sum of "kinetic" und "potential" contributions:

$$p_{\mu\nu} = p_{\mu\nu}^{\mathrm{kin}} + p_{\mu\nu}^{\mathrm{pot}} \ , \tag{28}$$

$$V p_{\mu\nu}^{\mathrm{kin}} = \sum_i m_i c_\mu^i c_\nu^i \ , \tag{29}$$

$$Vp_{\mu\nu}^{\text{pot}} = \frac{1}{2}\sum_{ij} r_\mu^{ij} F_\nu^{ij} \ . \tag{30}$$

Here $\mathbf{c}^i$ is the peculiar velocity of particle i, i.e. its velocity relative to the flow velocity $\mathbf{v}(\mathbf{r}^i)$, $\mathbf{r}^{ij} = \mathbf{r}^i - \mathbf{r}^j$ is the relative position vector of particles i, j and $\mathbf{F}^{ij}$ is the force acting between them. The Greek subscripts μ, ν which assume the values $1, 2, 3$ stand for cartesian components associated with the x, y, z directions. In the simulations, the expression for the pressure tensor given is averaged over many (10^3 to 10^6) time steps.

For the present flow geometry, the (non-Newtonian) viscosity η is obtained by dividing the $yx(21)$ component of the stress or pressure tensor by the shear rate γ:

$$\eta = \sigma_{yx}/\gamma = -p_{yx}/\gamma . \tag{31}$$

The kinetic and potential contributions to the pressure tensor and to the viscosity can be computed separately from the simulation. Only the sum can be measured in a real experiment. The kinetic contribution to the viscosity is the dominating one in dilute gases [11]. In dense fluids (liquids) the potential contribution is the more important one (Figs.8,9). The data shown stem from simulations with $N = 512$ particles [12], the interaction has been cut off at $r = r_c = 2.5r_0$.

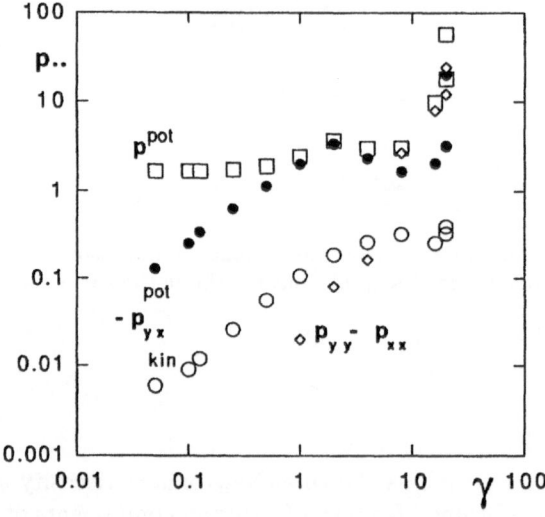

Fig. 8. The kinetic (*open circles*) and the potential (*closed circles*) contributions to the shear stress $\sigma_{yx} = -p_{yx}$, the potential contributions to the scalar pressure p^{pot} (*squares*) and to the normal pressure difference $p_{yy} - p_{xx}$ (*diamonds*) for an LJ fluid as functions of the shear rate γ. The density $n = 0.84r_0^{-3}$ corresponds to the triple point density, the temperature $T = \Phi_0/k_B$ is somewhat higher than the triple point temperature

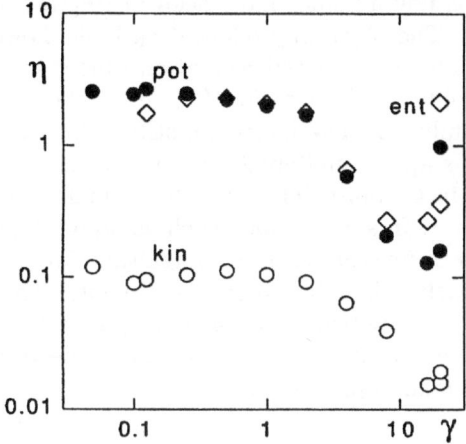

Fig. 9. The kinetic (*kin, open circles*) and potential (*pot, closed circles*) contributions to the viscosity $\eta = -p_{yx}/\gamma$, and the total viscosity inferred from the entropy production (*ent, diamonds*) for an LJ fluid as functions of the shear rate γ for the same density and temperature as in Fig.8

Normal stress or pressure differences, e.g., $\sigma_{xx} - \sigma_{yy} = p_{yy} - p_{xx}$ can be computed analagously, cf. Fig.8. At small shear rates one has $-p_{yx} \sim \gamma$ and $p_{yy} - p_{xx} \sim \gamma^2$, as well as $p_{xx} + p_{yy} - 2p_{zz} \sim \gamma^2$.

In Fig.9 the total vicosity is also shown as it follows from the entropy production, which is proportional to $\eta\gamma^2$ and is determined by the heat removed from the system by the thermostat.

4.4 Structural Changes

The shear-rate dependence of the viscosity as displayed in the "flow curve". Figure 9 shows four regimes: (*I*) the *Newtonian flow* regime where the shear viscosity η is independent of the shear rate γ and where normal pressure differences practically vanish; in the present case the Newtonian regime corresponds to $\gamma < 0.1$ (in LJ units); (*II*) a *weak shear thinning* for $0.2 < \gamma < 2$; (*III*) a *strong shear thinning* for $2 < \gamma < 20$; (*IV*) a *shear thickening* for $\gamma > 20$.

These qualitative differences of the flow behavior are linked with different flow-induced structural changes in the fluid. In regimes (*I*) and (*II*) these can be noticed in the pair-correlation function $g(\mathbf{r})$ or equivalently in its spatial Fourier transform, the static structure factor $S(\mathbf{k})$ which determines the scattering intensity. Both quantities become anisotropic in the presence of a viscous flow. The structure factor shows distorted Debye–Scherrer rings. In regime (*III*) a long-range partial positional ordering takes place which is apparent in real space and it is evident in snapshots [25, 26]. Of course, the long-range ordering is also seen in $g(\mathbf{r})$ and it leads to Bragg-like peaks in $S(\mathbf{k})$ [10–24].

Above, the various flow regimes have been distinguished by the shear rate expressed in LJ units. The physically relevant variable, however, is the product $\gamma\tau$ of the shear rate and the Maxwell relaxation time τ which, in turn, is given by the small shear rate limit of the ratio η/G, i.e., of the viscosity and the (high frequency) shear modulus G. The latter quantity, which can also be extracted from the simulation, is approximately 25 for the present system and the relaxation time is $\tau \approx 0.1$ in LJ units. Thus non-Newtonian flow phenomena can be observed for $\gamma > 0.1\tau^{-1}$. In simple fluids such as liquid Argon this corresponds to a shear rate which is several orders of magnitude larger than $10^6 s^{-1}$ which can reasonably be reached in laboratory experiments. The situation is different in (dense) colloidal dispersions of spherical particles. There, considerably shorter relaxation times occur and non-Newtonian effects can be noticed and are of importance for many applications.

4.5   Colloidal Dispersions

Dense colloidal dispersions of spherical particles exhibit flow curves which are qualitatively similar to those presented above. In the extreme shear thinning regime (III) where partial positional order is observed in the NEMD simulations, the static structure factor as measured in small-angle neutron scattering (SANS) experiments agrees very well with that computed from NEMD [24].

In addition to the direct interaction between the dispersed particles and the thermostating influence of the solvent already considered in the MD and NEMD simulations, particles in a dispersion feel a friction and the pertaining (Brownian) fluctuating forces. These additional forces are taken into account in the Brownian dynamics (BD) simulations. Results from such a BD simulation are presented in [21]. Furthermore, the particles experience hydrodynamic interactions and possess rotational degrees of freedom [22].

4.6   Mixtures

Recently, NEMD results were obtained [21] for a binary mixture of particles A and B interacting with an LJ potential cutoff in its minimum (WCA) with diameters $r_B = 0.58r_A$ for the $A - A$ and $B - B$ interactions, $r_{AB} = (r_A + r_B)/2$ for the $A - B$ interaction, masses $m_B = 0.195m_A$, and the energy parameters $\Phi_A = \Phi_B = \Phi_{AB}$. The size ratio corresponds to that occurring in opals, and mixtures of colloidal particles of this type have been studied experimentally [28]. The simulations were performed for the high number density $n = 1.26r_A^3$ corresponding to a packing fraction of 0.66 where the mixture is in a glassy state and where the viscosity, for small shear rates, decreases proportional to γ^{-1}. For the various mixtures studied, the total mass density (of the $N = 4000$ particles contained in the basic box) was kept constant. In Fig.10, the viscosity is shown as a function of the mass fraction of the smaller particles for the four shear rates $\gamma = 0.005, 0.05, 0.5, 5$. All quantities are in LJ units linked with the larger particles. At small shear rates, there is a region of concentrations where the mixture has a smaller viscosity than both pure components.

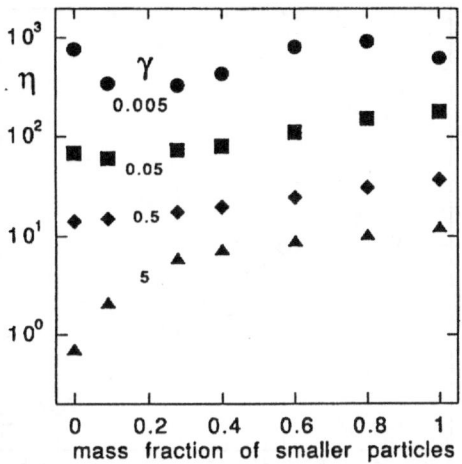

Fig. 10. The viscosity of mixtures as a function of the mass fraction of the smaller particles for the values of the shear rate γ indicated in the graph. The density and the temperature are $n = 1.26$ and $T = 1$ in LJ units (H. Voigt)

5   Complex Fluids

5.1   Polymer Melts

To model a polymer melt, one starts from a simple fluid of spherical particles and introduces extra binding forces or constraints [29, 36] in order to form molecular chains with a prescribed chain length of N_{ch} beads. Rheological studies for LJ fluids where the binding was achieved by increasing the energy parameter Φ_0 for neighbors in a chain by a factor showed many features of the nonlinear flow behavior typical for polymeric melts [27, 30].

The results to be presented here [32] follow from an extension of the previous simulations [31] for a system where all particles interact with the repulsive part of the LJ potential (WCA) and an attractive FENE potential with the maximal bond length R_0 is used for the binding within the chains. More specifically,

$$\Phi = \Phi^{WCA} := 4\Phi_0 \left[\left(\frac{r_0}{r}\right)^{12} - \left(\frac{r_0}{r}\right)^{6} + \frac{1}{4} \right] , \quad r \leq 2^{1/6} r_0 , \qquad (32)$$

and $\Phi^{WCA} = 0$ for $r > 2^{1/6} r_0$;

$$\Phi = \Phi^{FENE} := -0.5 \, k^* \, \Phi_0 \frac{R_0^2}{r_0^2} ln \left[1 - \frac{r^2}{R_0^2} \right] , \quad r \leq R_0 , \qquad (33)$$

and $\Phi^{FENE} = \infty$ for $r > R_0$. For this potential with $R_0 = 1.5$, $k^* = 30$, $T = \Phi_0/k_B$, and $n r_0^3 = 0.85$ extensive equilibrium MD studies have been made by Kremer and Grest [33]. In [31, 32] and for the data to be presented here, the same potential parameters and the same state point is used except

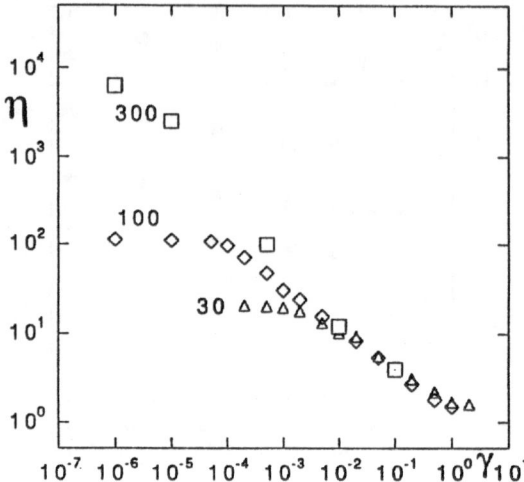

Fig. 11. The viscosity of melts consisting of polymer molecules with the chain lengths 30 (*triangles*), 100 (*diamonds*), and 300 (*squares*) as functions of the shear rate γ. The density and the temperature are $n = 0.84$ and $T = 1$ in LJ units (M. Kröger)

for a slightly smaller density of $n \, r_0^3 = 0.84$. Molecules with chain lengths $N_{\text{ch}} = 10, 30, 60, 100, 150, 200, 300$ and 400 were studied. The systems contained $N = 6000, 8400$ and 30000 monomers. In Fig.11, the viscosity η is displayed as function of the shear rate γ for $N_{\text{ch}} = 30, 100, 300$. Notice that the values of η and γ, both expressed in LJ units, span a much wider range than the data shown in Fig.9 for a simple LJ liquid. The Newtonian limit η_0 of the viscosity, for long molecules only reached at extremely small shear rates, is presented in Fig.12 as a function of the chain length N_{ch}. Two regimes, referred to as a Rouse regime where $\eta_0 \sim N_{\text{ch}}$ and as a reptation regime where $\eta_0 \sim N_{\text{ch}}^{3.5}$, can be distinguished. The transition between these regimes occures at $N_{\text{ch}} \approx 100$. This value is, as expected, about three times the entanglement lenght of ≈ 35 inferred from equilibrium studies [33]. A procedure to analyze and measure entanglements in MD simulations has recently been invented [34]. Other rheological properties, such as the first and second normal stress differences and the viscometric functions can and have been computed [31, 32]. The shear-induced bond orientation as it can be measured in flow birefringence, as well as the static structure factor of the whole melt or of marked chains and the shape of single polymer chains were analyzed and found to be in good agreement with experiments [35]. Other geometries, such as extensional flow of polymer melts (M. Kröger), the deformation of polymers in a glassy state, and systems with prescribed components of the pressure tensor (H. Voigt) have been simulated. Of course, dilute polymer solutions have also been studied under nonequilibrium conditions [36, 37]. Recently, "living polymers", i.e., chains which break and form again such as the worm-like micelles in surfactant solutions, can also be treated in NEMD simula-

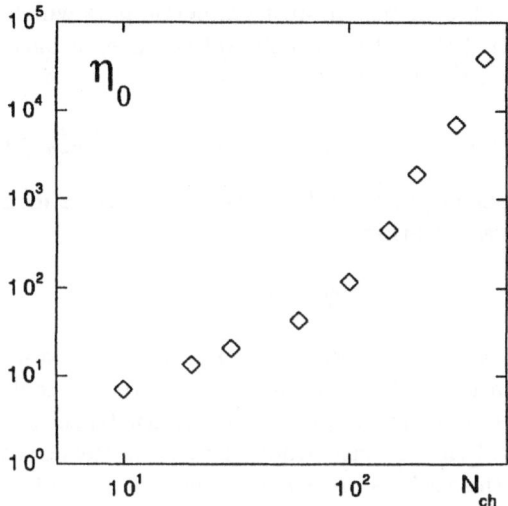

Fig. 12. The Newtonian viscosity η_0 of polymer melts as a function of the chain length N_{ch} (M. Kröger)

tions when the interaction potential is appropiately modified [38]. Furthermore, the investigation of equilibrium properties of chain molecules with stiff and flexible parts as in main-chain polymeric liquid crystals is under way [39], NEMD studies of the intriguingly complex rheological behavior of these substances are in preparation.

5.2 Nematic Liquid Crystals

Anisotropy of the Viscosity Coefficients. In nematic liquid crystals, the viscosity becomes anisotropic when the average direction of the molecules is fixed by an external magnetic (or electric) field. Four directions are needed to determine the full anisotropy of the shear viscosity. These cases, indicated by the labels $i = 1, 2, 3, 4$ for the pertaining shear viscosities η_i are the preferential direction chosen parallel to the flow velocity ($i = 1$), to its gradient ($i = 2$), to the vorticity which is perpendicular to both ($i = 3$), and to the bisector in the flow plane (xy plane) ($i = 4$). The first three viscosities are referred to as *Miesowicz* coefficients, the difference $\eta_{12} = 4\eta_4 - 2\eta_1 - 2\eta_2$ is called *Helfrich* viscosity. In the NEMD simulation, the viscosities are obtained according to

$$\eta_i = -p^i_{yx}/\gamma\,, \tag{34}$$

where p^i_{yx} is the yx component of the pressure tensor as given by (28), with (29,30) for the four above-mentioned cases. Of course, the interaction potential must now be modified appropriately in order to describe nonspherical particles. Special cases of nonspherical interaction potentials are given later. In the oriented system, the pressure tensor has an antisymmetric part which is associated with the torque acting on the particles. This antisymmetric part is used in NEMD simulations to obtain the *Leslie* viscosity coefficients γ_1 and γ_2 according to

$$\gamma_1 + \gamma_2 = 2(p^1_{xy} - p^1_{yx})/\gamma\,, \qquad \gamma_1 - \gamma_2 = 2(p^2_{xy} - p^2_{yx})/\gamma\,, \qquad (35)$$

where again the superscripts $1, 2$ refer to the orientations mentioned above. Due to the *Onsager–Parodi* relation ,

$$\gamma_2 = \eta_1 - \eta_2\,, \qquad (36)$$

only five of the six "nematic" viscosity coefficients used so far are linearly independent. In addition to the bulk viscosity, there are two coefficients, also linked by an Onsager relation, which couple the symmetric traceless and the trace parts of the pressure and of the velocity gradient tensors. Hence seven coefficients are needed to describe the viscous properties of nematic liquid crystals [40].

All coefficients, except for the bulk viscosity, have been calculated and the relation (36) has been tested for model fluids of perfectly oriented ellipsoidal particles where the nonspherical interaction potential is obtained from a spherical one, e.g., a LJ or SS interaction by an affine transformation [41, 42]. Both prolate and oblate ellipsoids of revolution with axis ratios $Q > 1$ and $Q < 1$,respectively, were studied. The inequalities

$$\eta_1 < \eta_3 < \eta_2\,, \qquad \gamma_2 < 0\,, \qquad (37)$$

typical for nematics $(Q > 1)$ are also found in the simulations. For nematic discotics $(Q < 1)$ one has

$$\eta_2 < \eta_3 < \eta_1\,, \qquad \gamma_2 > 0\,. \qquad (38)$$

For the affine transformation model, analytic results are available which compare favorably with simulations of perfectly oriented molecules. Motivated by the success in the comparison of MD data for the anisotropy of the diffusion in model fluids with variable degrees of orientation with a modified affine transformation model [43], similar considerations have been made for the viscosity coefficients [44].

A comparison of the nematic viscosities obtained from NEMD calculations with those based on Green–Kubo relations has been made for a fluid composed of particles interacting with a modified Gay–Berne potential [45].

Presmectic Behavior. The perfectly oriented ellipsoidal particles discussed above do not possess a smectic phase. In analogy to the ferrofluids and magnetorheological fluids discussed in the next section, a relatively simple model was introduced [46] in which the fluid does undergo a transition from the nematic to the smectic A phase. Incidentally, it possesses still another smectic phase (smB)

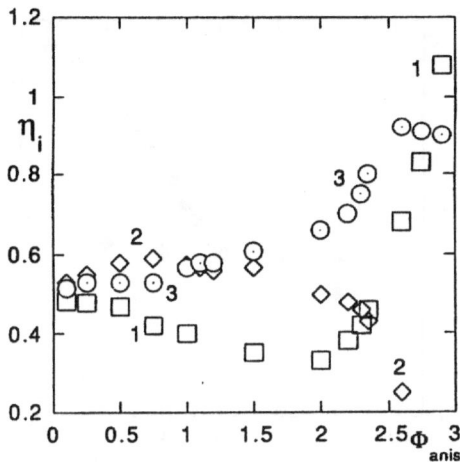

Fig. 13. The Miesowicz viscosity coefficients $\eta_i, i = 1, 2, 3$ as a function of the strength Φ_{anis} of anisotropic part of the interaction for $T = 0.25, n = 0.6, \gamma = 0.1$ in SS units (C. Pereira Borgmeyer)

in addition to the solid state. The potential is that of soft spheres and an extra anisotropic interaction of P_2 symmetry whose strength is determined by the parameter Φ_{anis}:

$$\Phi = \Phi_0^{\text{SS}}\left(\frac{r_0}{r}\right)^{12} + \Phi_{\text{anis}}\left(\frac{r_0}{r}\right)^6 \left(r^{-2}(\mathbf{r}\cdot\mathbf{n})^2 - \frac{1}{3}\right). \tag{39}$$

The unit vector $\mathbf{n}$ (director) specifies the preferential direction. For $\Phi_{\text{anis}} > 0$ this potential models elongated (prolate) particles. At the state point $T = 0.25, n = 0.6$, in SS units, and with the interaction cutoff at $r = 2.5r_0$, the transition nematic–smectic A occurs at $\Phi_{\text{anis}} \approx 2.3$ (in units of Φ_0^{SS}). The director $\mathbf{n}$ can be chosen parallel to the directions discussed above in connection with the anisotropy of the viscosity. The resulting Miesowicz viscosities $\eta_{1,2,3}$ are displayed in Fig.13 as functions of the anisotropy parameter Φ_{anis}. The shear rate is $\gamma = 0.1$ which, at least in the nematic phase, is in the Newtonian flow regime. The typical nematic order (37) in the magnitude of the viscosities is found for $0 < \Phi_{\text{anis}} < 1$. For $1 < \Phi_{\text{anis}} < 2.3$, presmectic effects change this behavior. In the smectic phase, for $\Phi_{\text{anis}} > 2.8$, the order of the viscosities corresponds to that of a nematic discotic system (38). The shear flow breaks the smectic layers but disk-like correlated clusters remain. Of course, the shear-induced structural changes can and have been analyzed with standard methods. Model fluids of short-chain molecules with a stiff central part and flexible ends possess broad smectic A phase [39]. The study of their rheological properties and of the shear-induced structural changes is in preparation.

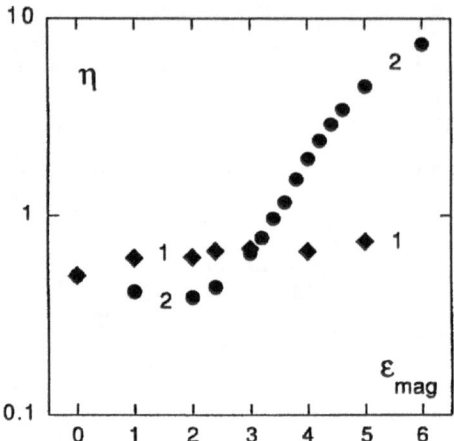

Fig. 14. The viscosity coefficients η_1, η_2 as a function of the strength ϵ_{mag} of the dipole–dipole interaction for $T = 0.25, n = 0.6, \gamma = 0.06$ in SS-units (T. Weider)

5.3   Ferrofluids and Magneto-Rheological Fluids

Spherical colloidal particles with a magnetic core, such as occur in ferrofluids in the presence of an applied magnetic field, have been modelled by soft spheres plus a dipole–dipole interaction [47]:

$$\Phi = \Phi_0^{SS} \left[\left(\frac{r_0}{r}\right)^{12} - \epsilon_{mag} \left(\frac{r_0}{r}\right)^3 \left(r^{-2}(\mathbf{r} \cdot \mathbf{n})^2 - \frac{1}{3} \right) \right] . \tag{40}$$

The parameter $\epsilon_{\mathrm{mag}} > 0$ is proportional to the square of the (induced) magnetic moment of the particles which are parallel to $\mathbf{n}$. The angular dependence of the nonspherical part of the interaction potential is the same as in (39), the sign of the prefactor and the r dependence, however, are different. Pairs of particles feel a disk-like interaction since, for fixed relative kinetic energy, they can approach each other more closely in the direction parallel to $\mathbf{n}$ than in the perpendicular directions. Thus it is not surprising that ferrofluids show an anisotropy analogous to nematic discotic liquid crystals [40, 42]. When the dipole–dipole interaction is stronger, however, chains are formed which, at higher densities, are arranged in partially ordered spatial structures. This affects the viscous behavior in a dramatic way. An example is shown in Fig.14 in which the viscosities η_1 (magnetic field parallel to the flow velocity) and η_2 (magnetic field parallel to the gradient of the flow velocity) are plotted as functions of the anisotropy parameter ϵ_{mag}. The state point is $T = 0.25, n = 0.6$, in SS units and the shear rate is $\gamma = 0.06$. The interaction is again cut off at $r = 2.5 r_0$, $N = 1000$ particles are used [48]. For $0 < \epsilon_{\mathrm{mag}} < 3$, the discotic behavior $\eta_1 > \eta_2$ is observed. For $\epsilon_{\mathrm{mag}} > 3$, the viscosity η_2 for the field parallel to the gradient direction increases strongly with increasing ϵ_{mag}. Notice that a logarithmic scale is used for the viscosity. A yield stress occurs for the higher values of the dipole–dipole interaction. This is typical for the magnetorheological (MR) fluids which are similar to the ferrofluids but are composed of particles with stronger dipole–dipole interactions and usually contain a higher volume fraction of colloidal particles. Electrorheological (ER) fluids can, with appropriate modifications, also be treated theoretically by the model.

Acknowledgements

Scientific interactions with H. Hanley (Boulder, Colorado) were of crucial importance for getting me started on NEMD simulations. Helpful discussions with him, as well as with M.P. Allen, G. Ciccotti, D.J. Evans, W.G. Hoover and D. Frenkel are gratefully acknowledged. I thank my former and present (graduate) students and postdocs F. Affouard, C. Aust, L. Bennet, D. Baalss, O. Hess, M. Kröger, W. Loose, N. Herdegen, C. Pereira Borgmeyer, R. Schramek, J. Schwarzl, H. Sollich, U. Stottut, H. Voigt, and T. Weider for their efficient NEMD investigations on the rheological and structural properties of simple and of complex fluids. Financial support was provided by the Deutsche Forschungsgemeinschaft (DFG) via the Sonderforschungsbereich (SFB) 335 "Anisotrope Fluide" and the Graduiertenkolleg "Polymerwerkstoffe" and by the BASF AG (Ludwigshafen). Many enlightning discussions with R. Bung, H.M. Laun and C. Kormann (BASF), as well as the generous donation of computer time by the Konrad-Zuse-Zentrum für Informationstechnik (Berlin) and by the Höchstleistungsrechenzentrum der KFA Jülich GmbH are gratefully acknowledged.

References

[1] W.T. Ashurst and W.G. Hoover, Am. Phys. Soc. **17**, 1196 (1972); Phys. Rev. Lett. **31**, 206 (1972)

[2] A.W. Lees and S.F. Edwards, J. Phys. C **5**, 1921 (1972)

[3] G. Ciccotti, G. Jacucci, and I.R. McDonald, Phys. Rev. A **13**, 426 (1975); J. Stat. Phys. **21**, 1 (1979); C. Trozzi and G. Ciccotti, Phys. Rev. A **29**, 916 (1984)

[4] W.G. Hoover, Annu. Rev. Phys. Chem. **34**, 103 (1983); D.J. Evans and G.P. Morriss, Comp. Phys. Rep. **1**, 287 (1984); D.J. Evans and W.G. Hoover, Ann. Rev. Fluid Mech. **18**, 243 (1986)

[5] B.D. Holian and D.J. Evans, J. Chem. Phys. **78**, 5157 (1983)

[6] D.M. Heyes, J. Chem. Soc. Faraday II, **82**, 1365 (1986)

[7] W.G. Hoover, *Molecular Dynamics*, (Springer, Berlin 1986); *Computational Statistical Mechanics*, (Elsevier, Amsterdam 1991); Physica A **194**, 450 (1993)

[8] M.P. Allen and D.J. Tildesley, *Computer Simulation of Liquids*, (Clarendon, Oxford 1987)

[9] R. Haberlandt, S. Fritzsche, G. Peinel, and K. Heinzinger, *Molekular-Dynamik*, (Vieweg, Braunschweig 1995)

[10] S. Hess and W. Loose, in *Constitutive Laws and Microstructure*, eds. D. Axelrad and W. Muschik, (Springer, Berlin 1988), p. 92

[11] W. Loose and S. Hess, Phys. Rev. Lett. **58**, 2443 (1988); Phys. Rev. A **37**, 2099 (1988)

[12] S. Hess, in *Rheological Modelling: Thermodynamical and Statistical Approaches*, eds. J. Casas-Vázques and D. Jou, Lecture Notes in Physics 381, (Springer, Berlin 1991), p. 51; Physikal. Blätter **44**, 325 (1988)

[13] S. Hess and W. Loose, Physica A **162**, 138 (1989)

[14] W. Loose and S. Hess, Rheol. Acta **28**, 91 (1989); W. Loose and S. Hess, in *Microscopic Simulation of Complex Flows*, ed. M. Mareschal, NATO ASI series, (Plenum, New York 1990); S. Hess and W. Loose, Ber. Bunsenges. Phys. Chem. **94**, 216 (1990)

[15] W. Loose and G. Ciccotti, Phys. Rev. **45**, 3859 (1992)

[16] W.C. Gear, *Numerical Initial Value Problems in Ordinary Differential Equations*, (Prentice Hall, London 1971)

[17] S.K. Gray, D.W. Noid, and B.G. Sumpter, J. Chem. Phys. **101**, 4062 (1994)

[18] W.G. Hoover, O. Kum, and N.E. Owens, J. Chem. Phys. **103**, 1530 (1995)

[19] S. Hess, Physica A **127**, 509 (1984); J.de Physique **46**, C3-191 (1985)

[20] S. Hess, Phys. Lett. **105** A, 113 (1984)

[21] S. Hess, M. Kröger, W. Loose, C. Pereira Borgmeyer, R. Schramek, H. Voigt, and T. Weider, Simple and Complex Fluids Under Shear, in eds. K. Binder and G. Ciccotti in press

[22] G. Bossis, Y Grasselli, E. Lemaire, A. Meunier, J.F. Brady, and T. Phung, Physica Scripta T **49**, 89 (1993)

[23] O. Hess, W. Loose, T. Weider, and S. Hess, Physica B **156/157**, 505 (1989); T. Weider, U. Stottut, W. Loose, and S. Hess, Physica A **174**, 1 (1991); S. Hess, D. Baalss, O. Hess, W. Loose, J. F. Schwarzl, U. Stottut and T. Weider in *Continuum Models and Discrete Systems*, ed. G. A. Maugin, (Longman, Essex 1990), p. 18

[24] H.M. Laun, R. Bung, S. Hess, W. Loose, O. Hess, K. Hahn, E. Hädicke, R. Hingmann, F. Schmidt, and P. Lindner, J. Rheol. **36**, 743 (1992)

[25] J.J. Erpenbeck, Phys. Rev. Lett. **52**, 1333 (1984)

[26] S. Hess, J. Mécanique Théor. Appl., Numéro spécial, 1 (1985); Int. J. Thermophys **6**, 657 (1985);

[27] S. Hess, J. Non-Newtonian Fluid Mech. **23**, 187 (1987)

[28] P. Bartlett and P.N. Pusey, Physica A **194**, 415 (1993)

[29] J.P. Ryckaert, Mol. Phys. **55**, 549 (1985)

[30] M. Kröger and S. Hess, Physica A **195**, 336 (1993)

[31] M. Kröger, W. Loose, and S. Hess, J. Rheol. **37**, 1057 (1993); M. Kröger, *Rheologie und Struktur von Polymerschmelzen*, (W&T, Berlin 1995)

[32] M. Kröger, Rheology **95**, 66 (1995)

[33] K. Kremer and G.S. Grest, J. Chem. Phys. **92**, 5057 (1990)

[34] M. Kröger and H. Voigt, Macromol. Theory Simul. **3**, 639 (1994)

[35] R. Muller, J.J. Pesce, and C. Picot, Macromol. **26**, 4356 (1993)

[36] C. Pierleoni and J.P. Ryckaert, Phys. Rev. Lett. **71**, 1724 (1993); Macromolecules **28**, 5087 (1995)

[37] C. Aust, *Molekulardynamik-Untersuchungen von Polymerketten in strömenden Lösungen*, Thesis, TU-Berlin, (1995)

[38] M. Kröger and R. Makhloufi, Phys. Rev. E **53**, 2531 (1996)

[39] A. Affouard, M. Kröger, and S. Hess, in preparation

[40] S. Hess, J. Non-Equilibr. Thermodyn. **11**, 176 (1986)

[41] D. Baalss and S. Hess, Phys. Rev. Lett. **57**, 86 (1986); Z. Naturforsch. **43** a, 662 (1988); H. Sollich, D. Baalss, and S. Hess, Mol. Cryst. Liq. Cryst. **168**, 189 (1989)
[42] S. Hess, J. Schwarzl, and D. Baalss, J. Phys. Condens. Matter **2**, 279 (1990)
[43] S. Hess, D. Frenkel, and M.P. Allen, Mol. Phys. **74**, 765 (1991)
[44] H. Ehrentraut and S. Hess, Phys. Rev. E **51**, 2203 (1995)
[45] S. Sarman and D.J. Evans, J. Chem. Phys. **99**, 9021 (1993)
[46] C. Pereira Borgmeyer, Thesis, TU-Berlin, unpublished, (1995)
[47] S. Hess, J.B. Hayter, and R. Pynn, Mol. Phys. **53**, 1527 (1984)
[48] T. Weider, *Molekulardynamik-Simulation kristalliner Strukturen unter dem Einfluss einer Scherströmung*, Dissertation, TU-Berlin (1995)

Molecular-Dynamic Simulations of Structure Formation in Complex Materials

Thomas Frauenheim, Dirk Porezag, Thomas Köhler, and Frank Weich

Technische Universität, Institut für Physik, D–09107 Chemnitz, Germany,
e-mail: frauenheim@physik.tu-chemnitz.de

Abstract. We are describing fundamental principles for molecular-dynamic simulations of structure formation in real materials at finite temperature. Various concepts for the calculation of total energies and interatomic forces are reviewed: Classical concepts based on the construction of empirical potentials and quantum-mechanical concepts combining the atom dynamics with a simultaneous solution of the electron problem of the many-atom configuration within density-functional theory. Out of these concepts we introduce in more detail a parameter-free density-functional-based nonorthogonal tight-binding scheme. This method combines the advantages of the simplicity and efficiency of semiempirical tight-binding approaches with the accuracy and transferability of ab initio calculations. After describing the simulation geometries and regimes for clusters, bulk structures and surface modifications the accuracy and high transferability of the interatomic potentials to the simulations of all-scale systems including also heteronuclear interactions are verified. Various successful applications of the method to the study of C_{60}-polymerization, the stability of highly tetrahedral amorphous carbon and the characterization of diamond surface reconstructions are summarized.

1 Introduction

There is currently a growing interest in materials science at an atomic level of understanding of physical and chemical properties of real structures. While common experimental scattering techniques only provide a one-dimensional structurally averaged picture of the atomic arrangement, spectroscopic data are more sensitive to the local chemical bonding environment and local defect configurations. However, this information is also averaged over a considerable sample volume. Real local probes have become available only recently. Well-known examples are atomic-scale imaging of surfaces by scanning tunneling microscopy (STM) [1] or single molecule spectroscopy (SMS) [2] in fluids or solids. These techniques are now modified for very different applications. However, the very high complexity of the measured signal in relation to energy-dependent charge density distributions, vibrational excitations, and electronic transitions makes a further theoretical treatment necessary.

Going beyond phenomenological considerations, an atomic-scale interpretation of the experimental data has to be based on realistic structure models (stable and metastable minimal-energy configurations) of various-scale complex systems ranging from clusters and molecules to crystalline and amorphous solids and solid surfaces. Starting from the atomic coordinates of these systems, a detailed theoretical analysis of material properties becomes possible. In combination with the experimentally derived mechanical, vibrational, electronic, and optical data, these results may lead to a fundamental understanding of the material physics and chemistry.

In a further step, even more interesting questions can be raised about how particular structures with desirable properties are going to be formed in optimized technological processes. For instance, there are various applications of molecular-dynamics modeling to study thin-film growth on solid substrate surfaces [3]. Applying rather different approaches ranging from very crude binary collision modeling [4] through empirical potentials [5] up to highly sophisticated density-functional methods [6], the thin-film formation by energetic particles [7], the nucleation of amorphous bulk materials [8], and elementary mechanisms in homoepitaxial and heteroepitaxial growth [9] are currently under investigation. In such simulations, a rapidly increasing number of new hypothetical structural configurations of (possibly metastable) clusters, fullerenes, nanotubes, and solids with interesting properties has been predicted. In this context, the computer simulations have been developed into a very nice tool for the structural design of real materials and complex systems.

2   Simulation Methods

In order to compare theoretical results to experimental data, the ultimately investigated model structures should represent stable or metastable minimal-energy configurations which have to be obtained by finite-temperature structure optimization. The two main techniques which can be applied are the Monte Carlo (MC) and the molecular-dynamics (MD) simulated annealing (SA) techniques [10, 11]. Both methods have to rely on a mathematical description of the total energy of the system as a function of all atomic coordinates, $E_{\text{tot}}(\mathbf{R}_i)$. Whereas the MC method follows a stochastic path to search for the minimal-energy configuration by repeatedly generating random atom configurations that are weighted by the total energy itself, the MD scheme determines the atom trajectories $\{\mathbf{R}_i(t), \mathbf{P}_i(t)\}$ that are driven by the interatomic forces.

In both methods, simulated annealing means the consideration of finite temperature during the minimal-energy search. If a MC optimization step results in a configuration with lower energy, it will be accepted as a new reference system for the following steps. However, in case an energetically less-favorable configuration is produced, the new structure is only accepted if the *Boltzmann probability* for this configuration, defined as $P_{\text{B}} = \exp\{-(E_{\text{tot}}^{\text{new}} - E_{\text{tot}}^{\text{old}})/k_{\text{B}}T\}$, is larger than some random number out of the interval (0,1). In this way, a finite-temperature noise is introduced into the search path, which allows the system to overcome energy barriers in favor of establishing the global rather than a local minimum.

During the MD, the incorporation of finite temperature is achieved more naturally by coupling the kinetic energy of the particles (mass M_l, velocity v_l) with the thermal energy of the system, (*equipartition theorem*),

$$\sum_l \frac{M_l}{2} v_l^2 = \frac{3}{2} N k_B T . \tag{1}$$

Focusing further on the MD method, the *Newton equations of motion* for all atoms in the structure have to be solved,

$$\mathbf{F}_l = M_l \ddot{\mathbf{R}}_l = \frac{-\partial E_{tot}(\{\mathbf{R}_i\})}{\partial \mathbf{R}_l} . \tag{2}$$

To realize this on a computer, one makes use of time-discretization techniques [12] using finite difference methods. From the difference equations one derives recursion relations for the positions and/or velocities (momenta). These algorithms proceed in the time direction yielding the phase-space trajectories of all atoms.

The most straightforward discretization of the differential equation stems from the Taylor series expansion of the atom positions $\mathbf{R}_l$ at an infinitesimal time step $t + h$,

$$\mathbf{R}_l(t + h) = \mathbf{R}_l(t) + \sum_i^{n-1} \frac{h^i}{i!} \mathbf{R}_l^{(i)}(t) + \mathbf{X}_n , \tag{3}$$

where $\mathbf{X}_n$ gives the error involved in the approximation and $\mathbf{R}_l^{(i)}$ is the i-th time derivative of the particle position. Using this equation in two steps at $n = 3$ a very simple, accurate and efficient algorithm is derived:

$$\mathbf{R}_l(t + h) = \mathbf{R}_l(t) + h \frac{d\mathbf{R}_l(t)}{dt} + \frac{1}{2} h^2 \frac{d^2\mathbf{R}_l(t)}{dt^2} + \mathbf{X}_3 \tag{4}$$

$$\mathbf{R}_l(t - h) = \mathbf{R}_l(t) - h \frac{d\mathbf{R}_l(t)}{dt} + \frac{1}{2} h^2 \frac{d^2\mathbf{R}_l(t)}{dt^2} + \mathbf{X}_3^* . \tag{5}$$

Note, that $\mathbf{X}_3 \neq \mathbf{X}_3^*$. Adding one to the other equation and neglecting higher order corrections, we derive the *Verlet algorithm* [13],

$$\mathbf{R}_l(t + h) = 2\mathbf{R}_l(t) - \mathbf{R}_l(t - h) + \frac{h^2}{M_l} \mathbf{F}_l(t) . \tag{6}$$

In this form the recursion relations produce only the positions. However, it is important to know the velocities to controle the kinetic energy or to study transport properties via the velocity autocorrelation function, for example. The velocity can be approximated as follows:

$$\mathbf{v}_l(t) = \frac{\mathbf{R}_l(t + h) - \mathbf{R}_l(t - h)}{2h} . \tag{7}$$

Notice, that at time $t + h$ the computed velocities are those of the previous time! Hence, the kinetic energy is one step behind the computed potential energy. In order to achieve control of the energy conservation it would be desirable to determine the atomic positions and velocities at the same time. By using $\mathbf{v}_l(t) = \{\mathbf{R}_l(t - h) - \mathbf{R}_l(t)\}/h$, we finally arrive at

$$\mathbf{R}_l(t) = \mathbf{R}_l(t - h) + h\mathbf{v}_l(t - h) \tag{8}$$

$$\mathbf{v}_l(t) = \mathbf{v}_l(t - h) + \frac{h}{M_l}\mathbf{F}_l(t) , \tag{9}$$

thus representing together with (6, 7) two simple and efficient algorithms.

However, before going into any MD simulation, we have to turn back to the basic principles of the calculation of total energies and interatomic forces.

3   Total Energies and Interatomic Forces

Currently, rather different theoretical approaches are applied in performing total energy calculations on clusters and extended systems.

3.1   Classical Concepts

These are fast, computationally efficient and based on the construction of parameterized empirical potentials. The potential parameters are typically fitted to data of equilibrium crystalline structures provided by experiments or ab initio methods which will be discussed later. The total energy of the system may be described by many-body potentials expanding the interatomic force terms in increasingly collective interactions (binary, three-body, ...),

$$E_{\text{tot}}(\{\mathbf{R}_l\}) = \sum_i^N \sum_{j>i}^N V_2(\mathbf{R}_i, \mathbf{R}_j) + \sum_i^N \sum_{j>i}^N \sum_{k>j}^N V_3(\mathbf{R}_i, \mathbf{R}_j, \mathbf{R}_k) + \ldots \tag{10}$$

As the simplest pair potential neglecting higher-order terms, the Lennard–Jones-type potential (LJP) [14],

$$V_2(\mathbf{R}_i, \mathbf{R}_j) = 4\epsilon \left[\left(\frac{\sigma}{r_{ij}} \right)^{12} - \left(\frac{\sigma}{r_{ij}} \right)^6 \right], \tag{11}$$

has been used to accurately model noble gases to yield reliable bond energies and bond lengths. r_{ij} is the interatomic distance and ϵ, σ are parameters determined by fitting to known properties of the gas phase such as viscosity, virial coefficients, etc. The potential is suitable for atoms which are only slightly distorted from their stable closed-shell configurations. The attractive forces between the atoms are due to fluctuating dipole interactions, which vary as the inverse sixth power of the interatomic distance. However, when the atoms get too close to each other, the repulsion of the ionic cores and, in part, the repulsion of the filled electron shells (Pauli's exclusion principle) become dominant. The balance between both parts finally defines the equilibrium interatomic distance $\sqrt[6]{2}\,\sigma$ and the binding energy $-\epsilon$. Also, the curvature of the potential well determines the force constant for vibrational motions.

Although the parameterization has been developed by fitting to gas-phase data, the LJP is found to accurately describe the bond energies and equilibrium bond length of the solid state of noble gases as well. Due to it's mathematical simplicity, the LJP has also been widely used for modeling covalent and especially metallic systems in the past. However, it can be quite incorrect for those materials, because it does not include the nature of covalent bonding.

The determination of energy and characterization of chemical bonding in covalent systems is much more difficult because the bond energies and bonding neighbor arrangements are strongly determined by the local coordination around each atom. For example, a bond between two carbon atoms can be a single, double, or triple bond, depending on how many atoms are in the nearest neighbor sphere at particular bond angles. Typical many-body semiconductor potentials such as the Stillinger–Weber (SW) [15] or Biswas–Hamann (BH) [16] potentials are terminated with the three-body terms. Considering for example the SW potential, it is preferentially fixed for a certain hybridization type of the atom-bonding configurations and does not allow for any hybridization change forced by the neighboring atom arrangement. To deal more successfully with this problem, in 1987 Tersoff suggested a potential which may be written in terms of effective two-particle contributions [17],

$$E_{\text{tot}} = \sum_i E_i = \frac{1}{2} \sum_{i \neq j} V_{ij} , \qquad (12)$$

$$V_{ij} = f_c(r_{ij}) \left[a_{ij} f_R(r_{ij}) + b_{ij} f_A(r_{ij}) \right] . \qquad (13)$$

The effective-pair potential V_{ij} consists of a Morse-type potential,

$$f_R + f_A = A e^{-\lambda_1 r} - B e^{-\lambda_2 r}, \qquad (14)$$

modeling a similar potential well as in the LJP parameterization by the superposition of an attractive and a repulsive contribution, which are modified by introducing the coefficients a_{ij} and b_{ij}. $f_c(r_{ij})$ is a simple cutoff function limiting the range of the interatomic interaction. Whereas in most cases considered up to now (Si and C) the coefficient of the repulsive term a_{ij} is chosen to be equal to 1, the coefficient b_{ij} of the attractive term represents an angular-sensitive bonding strength between atoms i and j depending on how many additional binding neighbors these atoms have. By this very intelligent, generally coordination-dependent control of the attractive forces, it becomes possible to model many-atom potential contributions which are able to describe coordination-dependent hybridization changes between sp, sp^2 and sp^3 as needed in the carbon bonding cases. For more details we refer the reader to [17].

In application to carbon, a total number of 12 parameters have been fitted to properly describe the energetics and structural properties of the zero-pressure graphite and diamond phase, as well as the higher-coordinated high-pressure sc and bcc lattices. The potential has been shown to be very efficient for use in MD simulations and to work well for describing the properties of crystalline systems. But as all empirical potentials, also the *Tersoff* potential may only hardly be transferred to bonding situations which have not been included into the parameterization. Consequently, small clusters, defects, amorphous structures and atom configurations on surfaces are described incorrectly in many cases. For example, the dihedral angular interactions orienting the bonding planes of paired undercoordinated sp^2 atoms in an amorphous matrix favoring their π-bonding are completely missing. Additional effort made by D. Brenner to generalize the *Tersoff* potential [18] has been successful by partly removing these problems and including interactions with hydrogen. However, by including even more parameters, such a scheme is hardly applicable for describing additional heteronuclear interactions, needed for doping and alloying studies.

3.2   Density-Functional Theory, Car–Parrinello MD

The problem of transferability, is solved in general by using ab initio MD concepts on the basis of density-functional theory (DFT). Hohenberg and Kohn proved that the ground-state total energy of an electron gas, including exchange and correlation may be expressed as a unique functional of the electron density [19], even in the presence of an external static potential. Kohn and Sham further showed how the many-electron problem may formally be replaced by an exactly equivalent set of self-consistent one-electron equations [20]. Following these guidelines, the total energy of a system with an electronic density $n(\mathbf{r})$ can be written as:

$$E[n(\mathbf{r})] = T_0[n(\mathbf{r})] + \frac{1}{2} \int \frac{n(\mathbf{r})n(\mathbf{r}')}{|\mathbf{r} - \mathbf{r}'|} \, dV dV'$$
$$+ \int n(\mathbf{r}) V_{\text{nucl}}(\mathbf{r}) \, dV + \int n(\mathbf{r}) \varepsilon_{\text{xc}}[n(\mathbf{r})] \, dV \ . \tag{15}$$

In the above equation, T_0 is the kinetic energy functional for a *non-interacting* electron gas, the second and third terms represent the electron–electron and electron–nuclear Coulomb energies, respectively. The last term is referred to as exchange-correlation energy. For the ground-state density, $E[n(\mathbf{r})]$ has a minimum, thus including particle conservation:

$$\frac{\delta \left[E[n(\mathbf{r})] + \mu(N - \int n(\mathbf{r}) \, dV] \right]}{\delta n(\mathbf{r})} = 0 \ . \tag{16}$$

The additional Lagrange parameter μ guarantees the correct normalization of $n(\mathbf{r})$ (N is the number of electrons). Using (16) and the assumption that the electron density can be expressed as a sum of single-particle wavefunction densities,

$$n(\mathbf{r}) = \sum_{i=1}^{N} |\Psi_i(\mathbf{r})|^2 \qquad \int |\Psi_i(\mathbf{r})|^2 \, dV = 1 \ \forall \ i \tag{17}$$

leads to the Kohn–Sham equation:

$$\left[-\frac{\nabla^2}{2} + \int \frac{n(\mathbf{r}')}{|\mathbf{r} - \mathbf{r}'|}\, dV' + V_{\text{nucl}}(\mathbf{r}) + V_{\text{xc}}[n(\mathbf{r})] \right] \Psi_i(\mathbf{r}) = \varepsilon_i \Psi_i(\mathbf{r}) \ , \qquad (18)$$

$$V_{\text{xc}}[n(\mathbf{r})] = \frac{\delta\left(n(\mathbf{r})\varepsilon_{\text{xc}}[n(\mathbf{r})]\right)}{\delta n(\mathbf{r})} \ . \qquad (19)$$

This equation is similar to the Schrödinger equation; however, the Coulomb potential due to the nuclear charges V_{nucl} is replaced by an effective potential V_{eff},

$$V_{\text{eff}}(\mathbf{r}) = \int \frac{n(\mathbf{r}')}{|\mathbf{r} - \mathbf{r}'|}\, dV' + V_{\text{nucl}}(\mathbf{r}) + V_{\text{xc}}[n(\mathbf{r})] \ , \qquad (20)$$

which depends also on the electron density and single-particle wavefunctions. Consequently, one has to deal with a self-consistent (scf) problem which is usually solved by an iterative procedure. In practical applications, the $\Psi_i(\mathbf{r})$ are expanded in terms of different basis functions. Consequently, the solution of the Kohn–Sham equations is transformed into the solution of a matrix eigenvalue problem.

The strength of density-functional theory lies in the fact that reasonable approximations have been found to express the exchange-correlation energy density $\varepsilon_{\text{xc}}[n(\mathbf{r})]$. Widely used is the local-density approximation (LDA), wherby $\varepsilon_{\text{xc}}[n(\mathbf{r})]$ depends only on the local electron density:

$$\varepsilon_{\text{xc}}[n(\mathbf{r})] = \varepsilon_{\text{xc}}(n(\mathbf{r})) \ . \qquad (21)$$

The best available LDA parameterizations for ε_{xc} are based on Quantum-Monte-Carlo calculations performed for the uniform electron gas. They have been proven to adequately describe equilibrium properties of clusters, molecules, and solids, such as electronic and geometric structures, vibrational frequencies, and relative energies of different phases. However, there are also serious problems. For instance, the electronic gap in semiconductors is significantly underestimated.

If one has to perform density-functional-based dynamic simulations in atomic systems, one can take advantage of the fact that the electron mass is much smaller than the mass of any nucleus and hence the motion of the nuclei can be treated in a classical way whereas the electrons will follow the nuclei adiabatically. This approximation is called the Born–Oppenheimer approximation. Consequently, one can calculate the ground-state electron density for a given atomic configuration and then determine the forces as derivatives of the total energy with respect to the nuclear coordinates. These forces can be used in molecular-dynamics (MD) simulations. However, the determination of the ground-state density is very time consuming due to the necessary scf procedure.

For that reason, Car and Parinello [21] developed a different scheme to use in MD simulations. They introduced a fictitious Lagrangian for the two sets of independent degrees of freedom $\mathbf{R}_l$ (nuclear coordinates) and Ψ_i (electronic wavefunctions):

$$L = \sum_i \mu_i \int |\dot{\Psi}_i^* \dot{\Psi}_i| \, dV + \frac{1}{2} \sum_l M_l \dot{\mathbf{R}}_l^2 - E[\{\Psi_i\}, \mathbf{R}_l]$$
$$+ \sum_{ij} \Lambda_{ij} \left(\int \Psi_i \Psi_j^* \, dV - \delta_{ij} \right) , \tag{22}$$

leading to the equations of motion:

$$\mu_i \ddot{\Psi}_i(\mathbf{r}, t) = -\frac{\delta E}{\delta \Psi_i^*(\mathbf{r}, t)} + \sum_k \Lambda_{ik} \Psi_k(\mathbf{r}, t) , \tag{23}$$

$$M_l \ddot{\mathbf{R}}_l = -\nabla E . \tag{24}$$

M_l are the nuclear masses and μ_i are fictitious masses related to the electronic degrees of freedom. The Lagrange parameters Λ_{ij} have been introduced to guarantee an orthonormal behavior of the Ψ_i. The Car–Parinello scheme has the advantage that electronic and nuclear degrees of freedom can be treated simultaneously, thus avoiding the explicit determination of the electronic ground-state for each geometry.

Although there has been much success in applying these methods to ever larger systems, they are still too slow for the investigation of many interesting problems. On the other hand, the accurate ab initio calculations based on scf density-functional (DF) [21, 22] or Hartree–Fock (HF) [23] theory represent without any doubt very reliable benchmarks for all other methods.

Due to the limitations in the transferability of empirical potentials to all scale systems including also heteronuclear interactions and the use of time-consuming ab initio methods, *semiempirical* techniques have been developed to simulate extended systems with reasonable computational costs. In addition to numerous traditional quantum chemical methods, tight-binding (TB) schemes have been very successful [26, 27, 28, 29]. These methods can be seen as simplified ab initio methods which still use the formalism of quantum mechanics to determine the electronic properties of the system. However, instead of actually calculating the Hamiltonian matrix, the elements of this matrix are fitted to reproduce an arbitrary set of input data. In general, only two-center contributions to the Hamiltonian matrix are considered: that means that terms which include wavefunctions and potentials on three different atoms are neglected. In many cases, the results of these schemes deviate only slightly from those of more sophisticated methods. However, the usual way of fitting the electronic Hamiltonian (matrix elements) to some input data is rather complicated and not very straight forward. That is why there is considerable interest in almost parameter-free tight-binding methods such as the one developed by our group in collaboration with Gotthard Seifert from the Technical University Dresden, which has almost the accuracy of scf LDA calculations but does not suffer from the drawback of exploding computational costs with growing system size. Our method is a *nonorthogonal* TB scheme which means that the basis functions used in the calculations are allowed to overlap.

Our method tries to avoid the difficulties arising from an empirical parameter-ization by calculating the elements of Hamiltonian and overlap matrices ab initio out of a local orbital basis with the help of DFT LDA and some integral ap-proximations. For this reason, it can be seen as an approximate LCAO DFT scheme yielding exactly the same energy expression as common nonorthogonal TB schemes. The only, but important, difference is that there is a well-defined procedure for how to determine the desired matrix elements. We will refer to our method as a nonorthogonal density-functional-based tight-binding (DF TB) scheme. As in usual TB formulations, only two-center Hamiltonian matrix ele-ments are considered. Despite the extreme simplicity of this approach compared to scf [21, 22] and ab initio calculations using the *Harris functional* [24, 25], the method has proven to be transferable to complex carbon and hydrocarbon sys-tems and recently to Si(H), Ge(H), SiC, BN, CN systems, GaAs and SiO_2. In this way we support discussions in the literature [25, 28, 29] that nonorthogo-nality is a key to transferability, but in our opinion this has to be combined with a *first-principle* based derivation of all interaction and overlap matrix elements.

4   Density-Functional Based Tight-Binding Method

Our method [30], based on the work of *Seifert, Eschrig* and *Bieger* [31, 32], ap-plies the formalism of optimized linear combination of atomic orbitals (O-LCAO) as introduced by *Eschrig* and *Bergert* for band structure calculations [33]. In this approximation, the Kohn–Sham orbitals ψ_i of the system are expanded in terms of atom-centered localized basis functions ϕ_μ,

$$\psi_i(\mathbf{r}) = \sum_\nu C_{\nu i}\phi_\nu(\mathbf{r} - \mathbf{R}_k) \,, \tag{25}$$

solving the Kohn–Sham equations in an effective one-particle potential $V_{\text{eff}}(\mathbf{r})$:

$$\hat{H}\psi_i(\mathbf{r}) = \varepsilon_i\psi_i(\mathbf{r}) \,, \quad \hat{H} = \hat{T} + V_{\text{eff}}(\mathbf{r}) \,. \tag{26}$$

As a result, the Kohn–Sham equations are transformed into a set of algebraic equations:

$$\sum_\nu C_{\nu i}(H_{\mu\nu} - \varepsilon_i S_{\mu\nu}) = 0, \quad \forall\mu, i \,, \tag{27}$$

where

$$H_{\mu\nu} = \langle\phi_\mu|\hat{H}|\phi_\nu\rangle, \qquad S_{\mu\nu} = \langle\phi_\mu|\phi_\nu\rangle \,. \tag{28}$$

It has already been shown by a number of authors [34, 35, 36, 37] that the total energy of the system can be approximated as a sum over the band structure energy (sum of the eigenvalues of all occupied Kohn–Sham orbitals) and a short-range repulsive two-body potential:

$$E_{\text{tot}}(\{\mathbf{R}_k\}) = E_{\text{BS}}(\{\mathbf{R}_k\}) + E_{\text{rep}}(\{|\mathbf{R}_k - \mathbf{R}_l|\})$$

$$= \sum_i n_i \varepsilon_i(\{\mathbf{R}_k\}) + \sum_{k<l} V_{\text{rep}}(|\mathbf{R}_l - \mathbf{R}_k|) , \tag{29}$$

where n_i is the occupation number of orbital i.

Interatomic forces for MD applications can easily be derived from an exact calculation of the gradients of the total energy at the considered atom sites,

$$\mathbf{F}_l = -\frac{\partial E_{\text{tot}}(\{\mathbf{R}_l\})}{\partial \mathbf{R}_l}$$

$$= \sum_i n_i \sum_\mu \sum_\nu C_{\mu i} C_{\nu i} \left[-\frac{\partial H_{\mu\nu}}{\partial \mathbf{R}_l} + \varepsilon_i \frac{\partial S_{\mu\nu}}{\partial \mathbf{R}_l} \right] - \frac{\partial E_{\text{rep}}}{\partial \mathbf{R}_l}. \tag{30}$$

In order to get the necessary matrix elements and the repulsive contributions V_{rep}, we perform the construction of our potential in three steps which are discussed in detail below:

1. Creation of (spin-unpolarized) pseudoatoms by solving a modified atomic Kohn–Sham equation,
2. Calculation of all Hamiltonian and overlap matrix elements,
3. Fitting of the short-range repulsive potential V_{rep}.

4.1   Creation of the Pseudoatoms

We write the pseudoatomic wavefunctions in terms of Slater-type orbitals and spherical harmonics:

$$\phi_\nu(\mathbf{r}) = \sum_{n,i,l_\nu,m_\nu} a_{n\alpha_i} r^{l_\nu+n} e^{-\alpha_i r} Y_{l_\nu m_\nu}\left(\frac{\mathbf{r}}{r}\right) . \tag{31}$$

As many tests have shown [38], five different values of α_i and $n = 0, 1, 2, 3$ form a sufficiently accurate basis set for all elements up to the third row of the periodic table of elements. functions does not yield any significant changes, this basis can be considered converged.

Using (31), we perform a self-consistent solution of modified atomic Kohn–Sham equations:

$$[\hat{T} + V^{\text{psat}}(r)]\phi_\nu(\mathbf{r}) = \varepsilon_\nu^{\text{psat}} \phi_\nu(\mathbf{r}) , \tag{32}$$

$$V^{\text{psat}}(r) = V_{\text{nucleus}}(r) + V_{\text{Hartree}}[n(r)] + V_{\text{xc}}^{\text{LDA}}[n(r)] + \left(\frac{r}{r_o}\right)^N . \tag{33}$$

V_{xc} is expressed in terms of the local density approximation as parameterized by Perdew and Zunger [39]. The additional term $(r/r_o)^N$ appearing in $V(r)$

in (33) was first introduced by Eschrig et al. [33, 38] in order to improve band-structure calculations performed within LCAO. It forces the wavefunctions to avoid areas far away from the nucleus, thus resulting in an electron density that is compressed in comparison to the free atom. The parameter N has only a rather small influence on the results; we choose $N = 2$ for all types of atoms. The radius r_o may be optimized to yield best results; however, we have found that $r_o \approx 2r_{cov}$ is usually a good choice, where r_{cov} is the covalent radius of the element.

4.2   Calculation of Matrix Elements

We use the solutions ϕ_ν of (32) as basis functions for the LCAO treatment of the system. Within a minimal basis description only valence orbitals are considered. As an approximation, we write the one-electron potential of the many-atom structure as a sum of spherical atomic contributions:

$$V_{\text{eff}}(\mathbf{r}) = \sum_k V_0^k(|\mathbf{r} - \mathbf{R_k}|) , \tag{34}$$

where V_0 is the Kohn–Sham potential of a neutral pseudoatom due to its *compressed* electron density, but not containing the additional term $(r/r_o)^N$ any more. This equation differs from the one used in older studies [31, 34, 35] where the potentials of free neutral atoms were used to evaluate the matrix elements. Using the potentials of compressed pseudoatoms for the evaluation of the matrix elements has two advantages:

– Numerous self-consistent calculations on molecules and solids have shown that the electron densities in these structures can be roughly approximated as a superposition of *compressed* atomic densities. Thus, by using this information, we anticipate the results of a more sophisticated calculation up to a certain extent. In addition to that, as has already been shown by Seifert et al. [32], the densities due to superposed pseudoatomic potentials are even more realistic than a simple superposition of pseudoatomic densities.
– The necessary integral approximations work better if one uses basis functions that decay more rapidly than those of the free atom. Furthermore, Eschrig [33] has shown that the modified wavefunctions form a better basis set in condensed-matter applications. Similar ideas on confined orbitals have been discussed more recently by Jansen and Sankey [40] and Chetty et al. [41].

The overlap matrix consists only of two-center elements and can be calculated in a straightforward way. Consistent with (34), one can neglect several contributions to the Hamiltonian matrix elements $H_{\mu\nu}$ [31] yielding:

$$H_{\mu\nu} = \begin{cases} \varepsilon_\mu^{\text{free atom}} & \text{if } \mu = \nu \\ \langle\phi_\mu^A|\hat{T} + V_0^A + V_0^B|\phi_\nu^B\rangle & \text{if } A \neq B \\ 0 & \text{otherwise} \end{cases} . \tag{35}$$

The indices A and B indicate the atom on which the wavefunctions and the potentials are centered. As can be seen easily, only two-center Hamiltonian matrix elements are dealt with. Approximation (35) may be seen as an LCAO variant of a cellular Wigner-Seitz method as applied for instance by Inglesfield [42]. As follows from (35), the eigenvalues of the free atom serve as diagonal elements of the Hamiltonian, thus guaranteeing the correct limit for isolated atoms.

Due to the fact that all matrix elements depend only on interatomic distances, we need to calculate them only once for each pair of atom types. For the two-center integral evaluation, the analytic formula of Eschrig [43] is applied. Matrix elements corresponding to a given interatomic distance can easily be obtained by interpolating between the stored values. Therefore, the creation of the Hamiltonian requires about the same time as common TB models. The calculation time is mainly determined by the efficiency of the diagonalization routines. We are still using Householder and QL algorithms but the implementation of recently developed linear-scaling methods [44] is in progress.

4.3   Fitting of Short-Range Repulsive Part

The construction of the Hamiltonian as described in the previous subsection allows us to calculate the band structure energy E_{BS} non-self-consistently. Thus, the short-range repulsive part $V_{rep}(R)$ can easily be determined as the difference of the total energy resulting from a self-consistent calculation and E_{BS} for different values of interatomic distances R:

$$V_{rep}(R) = E_{LDA}^{sc}(R) - E_{BS}(R) .$$
(36)

We write $V_{rep}(R)$ as a sum of polynomials:

$$V_{rep}(R) = \begin{cases} \sum\limits_{n=2}^{N_P} d_n(R_c - R)^n & \text{for } R < R_c \\ 0 & \text{otherwise} \end{cases} .$$
(37)

Equation (37) guarantees $V_{rep}(R)$ to be zero for $R \geq R_c$ and a smooth behavior at the cutoff radius R_c. In many cases, this expression is sufficient enough to fit the points given by (36) at $N_P = 5$.

In most cases, diatomic molecules can be used to fit $V_{rep}(R)$. However, these small systems sometimes tend to show level crossings causing sudden changes of orbital occupation numbers (as long as occupation numbers are restricted to integers) and thus discontinuities in the first derivatives of the energies. This behavior makes a reasonable fit in the vicinity of the level crossing almost impossible. Fortunately, one is not restricted to diatomic molecules. Other information on systems with uniform bond length available from self-consistent calculations can be included in the fit too, e.g., equilibrium crystalline phases.

5   Vibrational Properties

For a reliable vibrational analysis one has to use fully relaxed model structures (amorphous bulk supercells, surface slabs or finite clusters/molecules), in which the forces do not exceed a minimal critical value.

The vibrational properties of an atomic arrangement may be derived within the harmonic approximation by a calculation of eigenvectors and eigenvalues of the dynamical matrix. This matrix is the matrix of the total energy second derivatives with respect to the nuclear coordinates and may be constructed in the following way. If one displaces each atom i ($i = 1, \ldots, N$) by Δr from its equilibrium position into the directions of the three basis vectors $\mathbf{e}_\alpha$ ($\alpha = 1, 2, 3$) of the Cartesian coordinate system and into the corresponding opposite directions, one can calculate the elements $H_{ij}^{\alpha\beta}$ of the dynamical matrix using the forces $F_{j,\|\mathbf{e}_\beta}$ acting on each atom j in the directions $\mathbf{e}_\beta$ ($\beta = 1, 2, 3$):

$$H_{ij}^{\alpha\beta} = \frac{F_{j,\|\mathbf{e}_\beta}^- - F_{j,\|\mathbf{e}_\beta}^+}{2\Delta r} . \tag{38}$$

The signs refer to the two possible displacements of atom i in the direction $\pm \mathbf{e}_\alpha$. As can be seen from a Taylor expansion of the total energy, (38) eliminates errors of $\mathcal{O}(\Delta r)$ in the elements of H. To find a value for δr in practical applications, one has to consider two sources of errors, higher order terms in the total energy expansion favoring a very small Δr and the numeric instability of expression (38) for very small displacements. We found $\Delta r = 0.02\,\mathrm{a_B}$ (Bohr radii) to be a reasonable choice.

After symmetrizing the dynamical matrix and projecting out translational and rotational modes, we solve the general eigenvalue problem

$$H\mathbf{y} = \omega^2 M\mathbf{y}, \tag{39}$$

where M denotes a matrix with the atomic masses on the main diagonal whereas ω and $\mathbf{y}$ are the eigenvalues and their corresponding eigenvectors.

From this we obtain the total vibrational density of states (VDOS)

$$g(\omega) = \frac{1}{3N} \sum_{i=1}^{3N} \delta(\omega - \omega_i) . \tag{40}$$

For any practical evaluation of such a type of expression we have to replace the *Dirac's function* by an approximation $F(\lambda, \omega - \omega_i)$ (Gaussian, Lorentzian, ...) where the parameter λ controls the "width" of the eigenfrequencies within the spectra. Going beyond this, we can convolute the spectra for the purpose of comparison with experimental results by the resolution function of the specified detector. From the mathematical point of view we are also able to define the following projection technique for the calculation of the partial VDOS $g_\mathcal{H}(\omega)$ for any given arbitrary index set of coordinates $\mathcal{H} = \{\nu_i\}_{i=1}^{M \leq 3N}$. Restricting one-self to normalized eigenvectors, which guarantees the strict additivity of various partial VDOS to the resulting total VDOS, we get

$$g_{\mathcal{H}}(\omega) = \frac{1}{3N} \sum_{s \in \mathcal{H}} \sum_{i=1}^{3N} \delta(\omega - \omega_i) |\langle \mathbf{p}_s | \mathbf{y}_i \rangle|^2 \,, \tag{41}$$

where we have to sum over all elements of the index set and $\langle \mathbf{p}_s | \mathbf{y}_i \rangle$ is the scalar product of a vector $\mathbf{p}_s = (\underbrace{0, \ldots, 0}_{s-1}, 1, \underbrace{0, \ldots, 0}_{3N-s})$ with the ith eigenvector $\mathbf{y}_i$. This turns out as a helpful tool to characterize changing vibrational contributions of chemically different atom groups within the systems. In an amorphous carbon system, for example, we project out the partial VDOS of atoms of equal hybridization type. We can do the same to select out defect atoms in a crystal or cluster. Another important example is the description of surface related phonons of two-dimensional periodic slab models. Here we have to use the method to separate the surface excitations from bulk-like modes. To avoid errors connected with the finite depth of such slab models one has to saturate the dangling bonds of the bottom crystalline layer with hydrogen and keep these atoms fixed at their equilibrium positions by giving them infinite masses. This simulates the continuous transition from bulk-like to surface-like behavior.

Furthermore, the determination of the degree of localization of the vibrations gives a lot of useful information about the topological defect structure of the systems. The definition of a localization measure, such as an inverse participation ratio of the jth mode,

$$P_j^{-1} = \sum_{i=1}^{3N} |\langle \mathbf{p}_s | \mathbf{y}_i \rangle|^4 \,, \tag{42}$$

enables the discussion of a lot of interesting physics around structural defects in amorphous and crystalline systems.

6   Simulation Geometries and Regimes

In addressing applications to various scale systems from clusters to solids and solid surfaces we have to describe the geometries and regimes used for the simulation.

6.1   Clusters, Molecules

In the case of small clusters and molecules we are dealing with finite systems in which, depending on the interaction radius, each atom interacts with all other atoms. Consequently, the potential energy surface $E_{\text{tot}}(\{\mathbf{R}_i\})$ in the $(3N\text{-}6)$ dimensional configuration space $\{\mathbf{R}_i\}$ is highly complex, developing a large number of extrema and saddle points even for a few atoms. This fact is already manifested in systems with only simple two-particle interactions and becomes even more obvious if the atoms interact in a more complicated way.

To theoretically determine the minimal-energy configurations of most stable structures, complex search strategies based on chemical intuition or simulated annealing (SA) techniques have to be used. In the former case, the starting structures are constructed from the experience of investigating similar systems that are believed to be close to the configurations that we are looking for. The atom configurations may be relaxed into the next lower-lying local energy minimum by well-established gradient search methods. Much more efficient and less restricted are the annealing MD simulations because a larger area of configuration space is scanned. Thus, it is more likely that global minima will be found with these methods [45]. In the SA regime, the clusters may be heated to high enough temperatures leading to high atom mobilities and rearrangements. About every 500 time steps, an output cluster is generated which is gradually cooled down by simply rescaling the velocities followed by a final relaxation using conjugate gradient techniques. The resulting set of clusters or molecules consisting of an equal number of atoms is then analyzed by comparing the total energies of the different structures to determine the energetic order of the isomers and to select the most stable ones for a more detailed investigation of their physico-chemical properties.

6.2   Bulk-Crystalline and Amorphous Solids

To simulate bulk-solid structures and their properties, one has to remove all surfaces from the calculations by constructing three-dimensional periodic supercells (for example, simple cubic) incorporating an appropriate large number of atoms ($N \geq 100$) for the problem to be studied. The artificially introduced translational symmetry is described by supercell translation vectors, generating the atom positions in the surrounding 26 cells from the atom coordinates in the center cell. Using such a construction, the investigation of structural properties or effects of interatomic order as in amorphous systems are restricted to the range of the supercell size. To properly describe such systems, the interaction of all atoms in one cell with all other atoms falling into the interaction range has to be considered. That also implies that interactions between atoms in neighboring cells constructed via a supercell translation will be included if the interaction radius propagates into a neighboring cell. In this simple version of a model with periodic boundary conditions, the interaction radius by itself has to be chosen less than half the side-length of the cubic supercell, otherwise we would have a multiple counting of interactions.

For the generation of a metastable bulk amorphous model structure, we use a simulation regime in which the system is propagated along a path of constant energy in phase space by the algorithms (6–9). The initial positions fix the contribution of the potential energy to the total energy, and the velocities determine the kinetic contribution. The common technique for searching for amorphous equilibrium structures by computer simulations is then to quench or cool a liquid phase to room temperature over a certain time interval in the picosecond to nanosecond range. If we choose, as the most simple case, the volume of the supercell to be constant during the search, and adjust the system at each temperature to a given constant energy, we maintain a microcanonical ensemble during the MD run, N, V, E = constant.

In the first step of the simulation, we equilibrate the liquid phase by adjusting the system at a corresponding high energy. This is accomplished by scaling the velocities until the temperature of the liquid phase has been established by equilibration. This can be controlled by checking the validity of the *virial theorem* (equipartition of averaged potential and kinetic energy) and following the temperature which should fluctuate around the mean value according the thermal energy defined by the *equipartition theorem*. Proceeding in steps of 500 K, the structure is then cooled down until the final energy corresponding to an amorphous phase at room temperature is determined. At each temperature step, the system has to be partly equilibrated. Algorithmically, the equilibration procedure is: (i) integrate the equation of motion, (ii) compute the kinetic and potential energies, (iii) if the kinetic energy is not equal to the desired, then scale the velocities, (iv) repeat until the system has reached equilibrium.

6.3   Surfaces and Adsorbates

Considering surface configurations, the annealing MD simulations for determining (meta)stable surface structures make use of conventional two-dimensional boundary conditions in a three-dimensional formulation. The periodicity in the direction normal to the surface is strongly oversized, making the system practically finite in this direction. The surface slabs themselves are composed of several (10 to 20) atomic monolayers, each one consisting of a sufficiently large number of atoms. The two bottom layers of the slab should be held fixed to simulate the infinite crystalline substrate. Additionally, all occuring dangling orbitals have to be removed by saturation with hydrogen. The stability and dynamical restructuring are then studied on the remaining top layers, including possible adsorbates and finite temperatures.

In performing MD simulations of periodic supercells, one generally has to calculate the energies and forces by a sampling of values at different k-points in the *Brillouin zone*. However, by making the supercells much larger than the primitive ones, e.g., including a large number of atoms in the supercell, the evaluation of physical quantities at only the Γ-point ($\mathbf{k} = 0$) is practically equivalent to a summation over a collection of different k-points in the primitive cell. For that reason, the calculation of interatomic forces at only the Γ-point is a valid approach in MD applications if the supercell is chosen large enough.

7   Accuracy and Transferability

To prove the accuracy and transferability of the described DF-TB MD method we now present results on small clusters, molecules and crystalline solid modifications, and compare with experimental data as well as with results of more sophisticated calculations on the basis of scf LDA and HF (MP) theory treating correlation effects in Møller–Plesset perturbation theory.

7.1   Small Silicon Clusters, Si_n

As one prominent example for small clusters which has been broadly investigated during the past decade by experiments [46] and theory [47, 48], we discuss the equilibrium configurations of silicon clusters Si_{2-10} [49]. While for small carbon clusters the minimal-energy configurations are either linear chains or cyclic structures, the situation in the silicon case is more complex. In contrast to carbon, silicon does not form strong π–bonds, but can be found more than fourfold coordinated. Therefore, many metastable isomers have to be considered even for small clusters.

Because of this variety of possible equilibrium structures, the determination of the lowest-energy geometry and the corresponding ground-state properties is a good test for any method. Whereas empirical potentials fitted to bulk crystalline structures usually fail in correctly describing small finite atom configurations, we obtain the same minimal-energy clusters as Raghavachari et al. within *ab initio* calculations (HF/MP4) [47], see Fig. 1. There are two exceptions: for Si_8 we find a distorted octahedron with two neighboring faces capped (C_2 symmetry) at almost the same energy as the distorted octahedron with two opposite faces capped (C_{2h} symmetry) reported by ab initio calculations. To clarify the question as to which of these two clusters is more stable, we have performed additionally a self-consistent calculation with a generalized gradient functional [50] and found our new cluster to be only 0.05 eV/atom higher in energy. In the case of Si_9, we confirm a distorted tricapped trigonal prism (C_{2v} symmetry), which was proposed recently by Ordejón [51], to be the most stable configuration.

Figure 2 shows the (zero-point corrected) cohesive energies of the minimal-energy clusters compared to ab initio and experimental results. To account for the spin polarization of the isolated atoms, we have corrected our energies by 0.64 eV/atom, which can be determined within a self-consistent local-spin density calculation. The deviations of our results from the experimental values are very small and comparable to that of the ab initio calculations at 6-31G* level [47].

7.2   Molecules, Hydrocarbons

Carbon shows many different types of bonding. All of them can be found in the huge class of hydrocarbon molecules known to play an important role in the structure formation of many systems. For that reason, it is important to

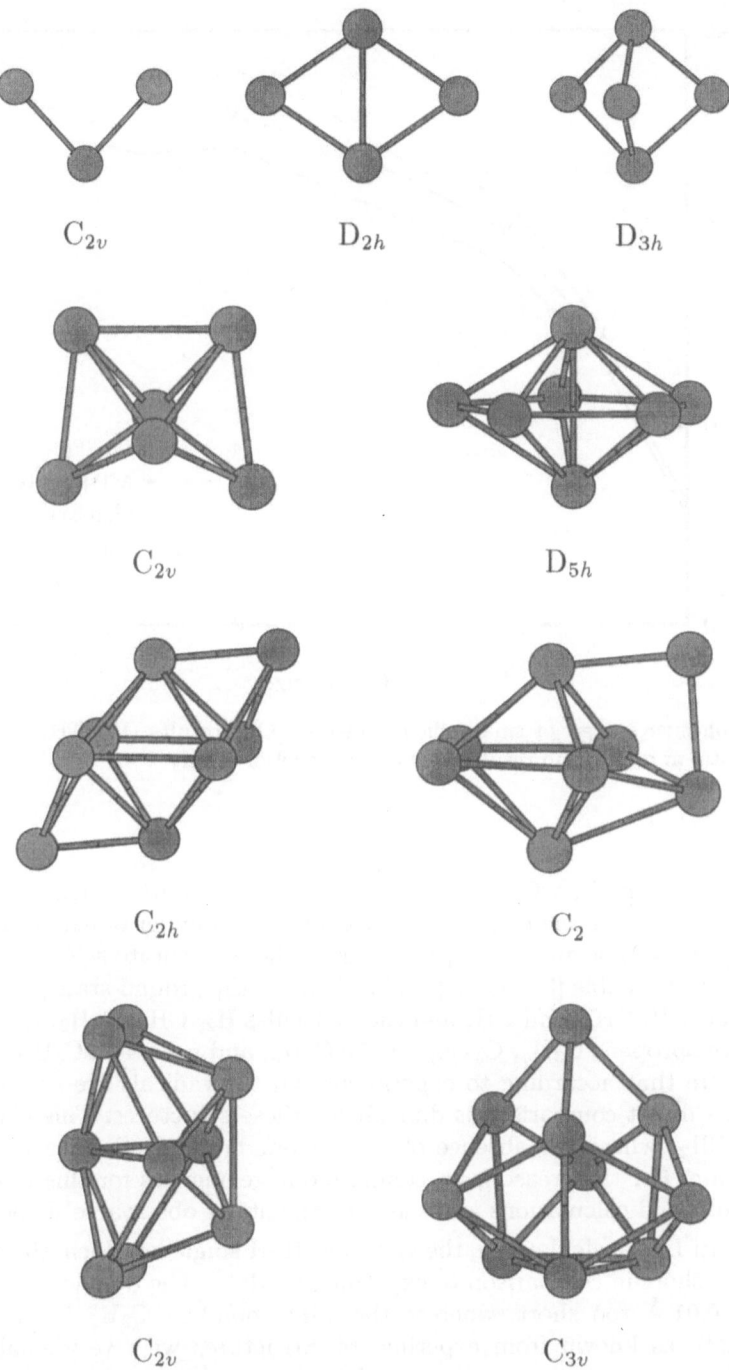

C_{2v} D_{2h} D_{3h}

C_{2v} D_{5h}

C_{2h} C_{2}

C_{2v} C_{3v}

Fig. 1. Minimal-energy configuration of small silicon clusters Si_n (n=3-10). For Si_8 both discussed structures are displayed

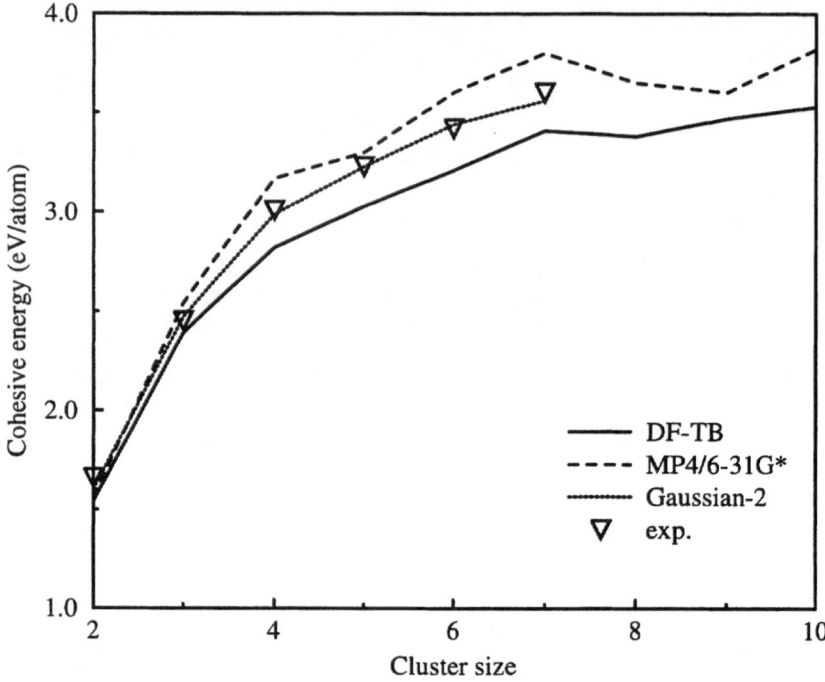

Fig. 2. Cohesive energy of small silicon clusters. Our results (DF-TB) compared to experimental and ab initio results at different level of theory

know how a method performs on these systems. In addition, hydrocarbons are well understood and one can refer to an abundant number of experimental and theoretical data. For all the properties tested here, accurate self-consistent calculations are available [52, 53, 54]. Table 1 shows the ground-state geometries of the radicals CH, CH_2, and CH_3 and the molecules H_2, CH_4, C_2H_2, C_2H_4, C_2H_6, C_6H_6, Cyclopropene C_3H_4, Cyclopropane C_3H_6, and n-Butane C_4H_{10}. We want to note here that according to experiments all the radicals are spin-polarized, therefore a direct comparison is difficult for these structures. This is especially true for CH_2, where the absence of spin in our model leads to a more stable singlet state. For that reason, we compare our geometries for this radical with spin-unpolarized calculations and the experimentally observable singlet state.

As in scf LDA calculations, the C–C and H–H single bond lengths are about 0.02 Å too short in comparison to experimental data. The double bond in C_2H_4 is about 0.01 Å too short, whereas the triple bond in C_2H_2 has almost the same length as known from experiments. Structures with very small C–C–C bond angles, such as cyclopropane and cyclopropene are also well described. C–H bonds are systematically overestimated by about 0.03 Å, a little bit more than the overestimation in scf LDA calculations. Bond angles agree within 2°

Table 1. Geometric properties (bond lengths XY in Å and bond angles XYZ in degree) obtained for selected radicals and molecules. The SCF and experimental values have been taken from [54] (H_2 through ethane) and [53] (cyclopropene through benzene). GGA values refer to calculations using generalized gradient approximations for the exchange-correlation functional as described in [54] and [53]

Molecule	Variable	DF-TB	LSD	GGA	Exp.
H_2	HH	0.765	0.765	0.748	0.741
CH	CH	1.138	1.152	1.108	1.120
CH_2 (singlet)	CH	1.134	1.135	1.117	1.111
	HCH	98.6	99.1	99.1	102.4
CH_3	CH	1.114	1.093	1.090	1.079
	HCH	116.8	120.0	120.0	120.0
CH_4	CH	1.116	1.101	1.100	1.086
C_2H_2 (acetylene)	CC	1.206	1.212	1.215	1.203
	CH	1.099	1.078	1.073	1.061
C_2H_4 (ethene)	CC	1.503	1.513	1.541	1.526
	CH	1.119	1.105	1.104	1.088
	CCH	116.3	116.4	116.2	117.8
C_2H_6 (ethane)	CC	1.503	1.513	1.541	1.526
	CH	1.119	1.105	1.104	1.088
	HCH	108.0	107.2	107.5	107.4
C_3H_4 (cyclopropene)	C_1C_2	1.318	1.305		1.296
	C_2C_3	1.509	1.510		1.509
	C_1H	1.109	1.091		1.072
	HC_1C_2	148.4	149.5		149.9
C_3H_6 (cyclopropane)	CC	1.503	1.504		1.510
	CH	1.114	1.095		1.089
C_4H_{10} (n-butane)	C_1C_2	1.511	1.517		1.533
	C_2C_3	1.520	1.532		1.533
C_6H_6 (benzene)	CC	1.389	1.396		1.399
	CH	1.114	1.095		1.089

or even better, exceptions are the radicals where the H-C-H angles are clearly underestimated by about 4°.

Table 2 shows atomization energies (including zero-point corrections) and reaction energies for some typical hydrocarbon reactions. The atomization energies are with respect to free, *spin-polarized* atoms: that means the spin polarization energies of free carbon and hydrogen atoms (1.13 eV and 0.90 eV, respectively, calculated within LSDA) have been subtracted from the actual atomization energies determined by the DF-TB method. The atomization energies as taken from [52] are already corrected for zero-point vibrations. For the selected reaction energies presented here, accurate calculations and experimental values are available from [53] and [54]. Comparing with our results, the atomization energies are found to be almost excellent, whereas the reaction energies calculated self-consistently show a better error cancelation.

Table 2. Atomization and reaction energies for some typical reactions of organic chemistry. Atomization energies are with respect to the free spin-polarized atoms (see text). Calculated energies are not corrected for zero-point vibrations, but experimental values are extrapolated to zero and corrected for zero-point vibrations. All reference energies were taken from [52] (atomization) and [53] (reactions)

Atomization energies (kcal/mol)					
Molecule	DF-TB	HF	LSD	GGA	Exp.
H_2	113	84	113	105	109
CH_4	425	332	463	422	424
C_2H_2	422	300	461	417	408
C_2H_4	577	431	634	574	568
C_2H_6	735	557	795	720	719
C_6H_6	1438	1041	1577	1413	1375
Rms err/bond	3.5	27.4	12.9	2.5	

Reaction energies (kcal/mol)					
Reaction	DF-TB	HF	LSD	GGA	Exp.
$C_2H_6 + H_2 \rightarrow 2\ CH_4$	2	21	18	19	19
$C_2H_4 + 2\ H_2 \rightarrow 2\ CH_4$	47	64	67	60	57
$C_2H_2 + 3\ H_2 \rightarrow 2\ CH_4$	89	118	131	114	89
Rms error	14.3	8.6	16.1	5.5	
$C_2H_4 + 2\ CH_4 \rightarrow 2\ C_2H_6$	43	22	32	22	20
$C_2H_2 + 4\ CH_4 \rightarrow 3\ C_2H_6$	84	54	77	58	49
$C_3H_4 + 3\ CH_4 \rightarrow 2\ C_2H_6 + C_2H_4$	87	50	50	44	45
$C_3H_6 + 3\ CH_4 \rightarrow 3\ C_2H_6$	63	26	26	23	25
Rms error	35.2	3.7	15.4	4.7	

7.3   Solid Crystalline Modifications, Silicon

During the past few years, there has been progress in the application of parameterized TB models to solid-state modifications of silicon [28, 55] and other group-IV semiconductors [27, 56]. For comparison, the results of accurate self-consistent methods are available for the group-IV equilibrium structures and some higher-coordinated high-pressure modifications [57, 58]. Considering the results of all empirical TB schemes, no general transferability is obtained for all systems. Either the solids are described with high accuracy or small clusters, but never both at similar quality as obtained with more sophisticated methods.

In addressing possible applications of the DF-TB to bulk systems and crystalline surfaces, we have also performed calculations for the total energy as a function of nearest-neighbor distance for the diamond silicon lattice, and for the high-pressure crystalline phases sc, bcc and fcc. The results are displayed in Fig. 3 compared to the ab initio data of Yin and Cohen [57, 58]. The equilibrium distance of the diamond crystal (2.346 Å) is close to the reference value

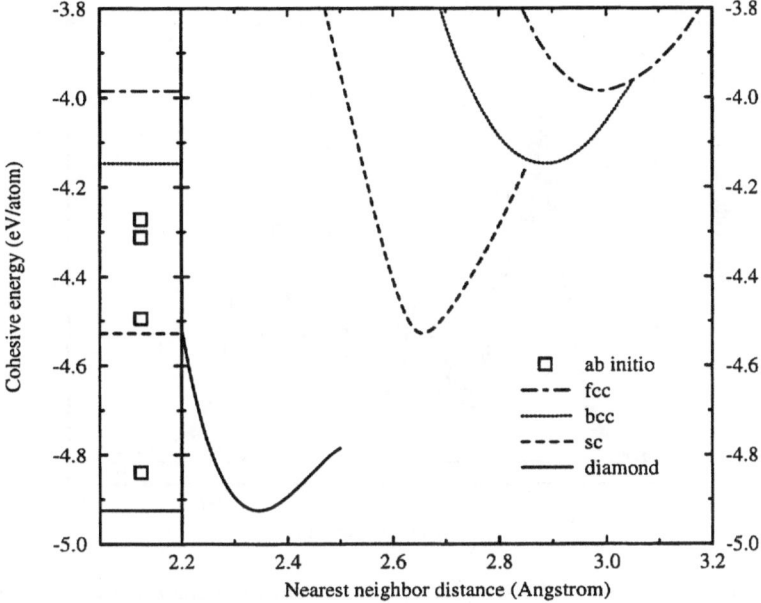

Fig. 3. Cohesive energies per atom for different lattice types versus nearest-neighbor distance obtained with respect to the spin-unpolarized silicon atom. On the left the energy values at the equilibrium positions are compared to ab initio values [57]

(2.360 Å) and the experimental result (2.351 Å). The energy decrease for the simple cubic phase is close to the SCF data whereas the energy differences to the other high-pressure phases are overestimated by about 30%.

As an additional benchmark, we have calculated the vibrational density of states of a *diamond*-Si 216-atom supercell which has been plotted in Fig. 4 in comparison to inelastic neutron-scattering data obtained on a polycrystalline sample [59]. The calculated spectrum has been convoluted by the experimental resolution function. We find the most characteristic modes with appropriate intensities at the correct wave numbers and the overall width of the vibrational spectrum to be in good agreement with the experimental data.

8 Applications

8.1 Structure and Stability of Polymerized C_{60}

Since the discovery of the C_{60} molecule [60], there has been a growing interest in fullerene-based carbon systems. At room temperature, the only weakly interacting molecules in solid C_{60} form a face centered cubic lattice (fcc). Recent experiments [61, 62] provide convincing evidence that ultraviolet (UV) and visible light causes the molecules in solid C_{60} to polymerize. The phototransformed

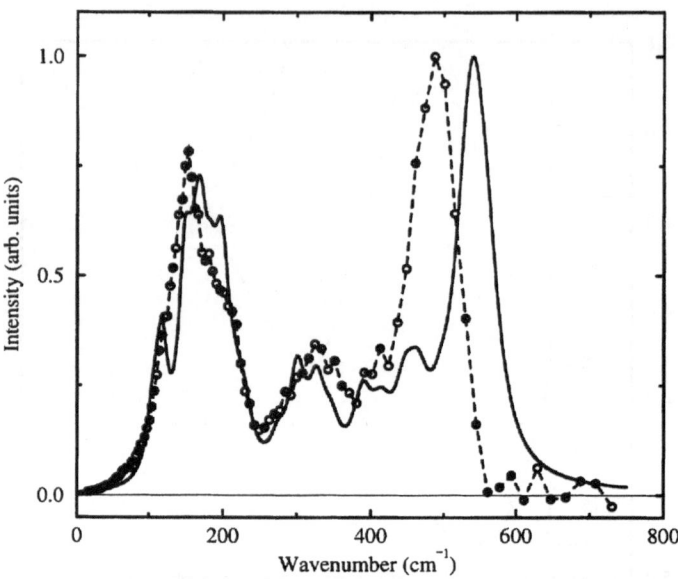

Fig. 4. Vibrational density of states for crystalline silicon using a convolution with the experimental resolution function (*solid line*) compared to inelastic neutron scattering data [59] (*dashed line*). The intensities have been scaled to a maximum value of one

material is no longer soluble in toluene and the associated mass spectra consists of integer multiples of the C_{60} mass. Further, the characteristic infrared and Raman modes of pristine C_{60} show significant shifts and splittings when the material is phototransformed and new infrared and Raman modes appear in the spectra. In particular, the high energy $A_g(2)$ pentagonal pinch mode shifts by about 10 cm^{-1} from 1469 cm^{-1} in the pristine solid to 1459 cm^{-1} in the photo-transformed structure. In addition, a new low-energy Raman peak appears at 118 cm^{-1}. With respect to the stability of this structure, experiments show that the phototransformed material returns to pristine C_{60} at approximately 200° C. Further analysis of the experimental data leads to the conclusion that the energy barrier for the thermal dissociation of the polymerized material is about 1.25 eV [61].

A number of preceding theoretical works [63, 64, 65] have also investigated the phenomenon of C_{60} dimerization. In these publications, it has been established that the energetically most favorable bonding between two C_{60} molecules is accomplished by a 66/66 2+2 cycloaddition reaction which leads to nearly rigid molecules that are bound together by a four-membered square ring as displayed in Fig.5. However, although empirical and generalized tight-binding (TB) methods [63] determine the polymerized structure to be less stable than isolated C_{60}, the Harris Functional approach [64], Hartree-Fock [63] and an all-electron self-consistent LDA calculation [65] come to the opposite conclusion. One major problem is that due to the rather large size of the dimer, accurate self-consistent methods relied on partially relaxed geometries. Since the expected energy difference will be close to zero, this effect may alter the qualitative result of the calculation. As such, it is desirable to derive the dimer binding energy using an accurate, self-consistent scheme *and* geometries that are fully relaxed.

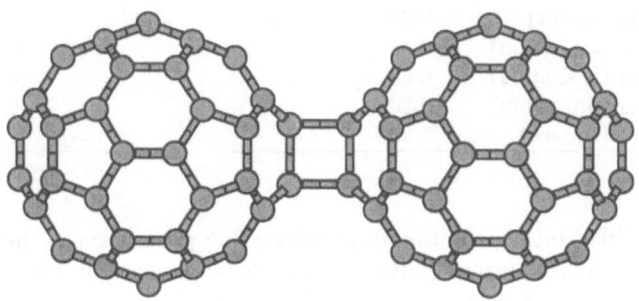

Fig. 5. Ground-state geometry of the C_{60} dimer as calculated with the DF-TB method

To investigate the problems discussed above, we have applied both DF-TB and *first principle* DF methods. Geometry optimizations were performed by a conjugate gradient algorithm using the DF-TB energy functional as described in [30].

In our calculations we have considered different oligomers of C_{60}, among them linear structures and a ring formed by four bucky balls. However, the basic physics can be understood if one only considers the $[C_{60}]_2$ dimer. The monomers of all polymeric compounds are bound by a 66/66 2+2 cycloaddition as suggested in previous studies [63, 64, 65]. For the structures investigated here, it is found that the shape of the resulting four-membered rings is almost independent of the number of polymeric connections between a particular C_{60} monomer and other monomers. We determine a length of 1.583 Å for the intermolecular bonds and a length of 1.590 Å for the intramolecular bonds. Further, the bond length between the four fourfold coordinated atoms and their threefold coordinated neighbors is 1.514 Å. These values are in excellent agreement with those of [64] who found 1.588 Å, 1.578 Å and 1.511 Å, respectively.

Table 3 displays the cohesive energies of the five polymeric compounds with respect to isolated C_{60}. Except for minor differences, DF-TB leads to an energy gain of about 0.3 eV per polymer bond. This is in accord with the findings of Adams et al. [64] who found an energy gain of 0.47 eV per polymer bond for the dimer and 0.44 eV per polymer bond for the infinite chain. Based on these results, it should be possible to create stable at least two-dimensional polymeric solids by photopolymerizing the material.

While energetic stability is one figure of merit for potential applications of

Table 3. Cohesive energies ΔE and zero-point corrected cohesive energies ΔE_{zpc} for the investigated structures with respect to isolated C_{60} as determined by the DF-TB, SCF-LDA, and SCF-GGA methods. Note that ΔE is defined as $\Delta E = E((C_{60})_N) - N \times E(C_{60})$

Structure	ΔE [eV]			ΔE_{zpc} [eV]
	DF-TB	SCF-LDA	SCF-GGA	DF-TB
C_{120} (dimer)	0.30	1.20	0.32	0.32
C_{180} (lin. chain)	0.58			0.61
C_{180} (L-shaped)	0.60			0.63
C_{240} (lin. chain)	0.86			0.90
C_{240} (square)	1.20			1.26

fullerene assembled materials, the more interesting parameter is the energy that is required to break a fullerene polymer apart. Since a reaction barrier is encountered when the dimer forms, the observed dissociation energy should be larger than the binding energy. An estimate of this barrier for the C_{60} dimer dissociation has been determined within the DF-TB approach. We have looked at two different dissociation pathways for the 120 atom complex. For the first path, both interball bonds were fixed in each step at a certain length while the rest of the cluster was allowed to relax. We find that the balls dissociate as the interball bond length becomes larger than 2.16 Å. The activation energy for this path is 1.9 eV. For the second path, we have fixed only one of the two interball bonds and allowed the rest of the structure to relax. Finally, when the length of the fixed bond exceeds 2.62 Å, the other bond breaks spontaneously and the balls separate. The activation energy for this path is only 1.6 eV which is in reasonable agreement to the experimental estimate of 1.25 eV [61]. Adams et al. [64] found an upper limit of 2.4 eV for this energy. Further, the fact that the second bond would break spontaneously when the first one is already broken is in accord with the results presented in [64] where a single bond connection between two balls was found to be unstable.

Finally, we had a look at the changes in the vibrational density of states occurring as a result of polymerization. In excellent agreement with experiments, we find the $A_g(2)$ pentagonal pinch mode shift downward by 10 cm^{-1}. A look at the vibrational behavior of the larger oligomers shows peaks at downward shifts by the same magnitude and by 16 cm^{-1} and 20 cm^{-1}. Further analysis of this phenomenon leads to a very simple model: balls with ring connections to only one neighbor shift down by 10 cm^{-1}, balls with two connections at an angle of 90° (such as the balls in the $[C_{60}]_4$ square) by 16 cm^{-1} and those with two connections at opposite sides of the balls (such as the ones in the middle of a chain) by 20 cm^{-1}. Consequently, the vibrational behavior in the high-frequency region is mainly determined by the local bonding environment of the balls. From this fact we can draw the conclusion that the solids on which the experiments have been performed consist mainly of dimer-like structures.

8.2   Stability of Highly Tetrahedral Amorphous Carbon, ta-C

Carbon in combination with hydrogen is one of the most promising chemical elements for molecular structure design in nature. An almost infinite richness of possible structures with a wide variety of physical properties can be produced. Even the two crystalline inorganic modifications, graphite and diamond, show diametrically opposite physical properties. Whereas graphite with is typical layered sp^2 bonding planes is black, soft, lubricating, electrically conducting and adsorbing, the fully three-dimensionally connected sp^3 hybridized brightly sparkling diamond is light, hard, brittle, electrically insulating and transparent. Graphite behaves as a semimetal due to a nonvanishing density of states (DOS) near the Fermi energy produced by delocalized π-states. In contrast, a large σ-band gap makes diamond behave like an insulator, which is also known as one of the best heat conductors.

Successful handling of different deposition methods allows the preparation of metastable amorphous carbon structures with interesting physical properties between diamond and graphite [66]. "Tetrahedral amorphous carbon" (ta-C) [67], which was recently grown by McKenzie et al. using ion-beam techniques, is promising for future technological applications due to its diamond-like properties. The material is extremely hard, chemically inert, develops large band gaps [68] and may be n-type doped by phosphorus [69] and nitrogen [68].

The growing interest in a fundamental investigation of structure-property relations and mechanisms of structure formation in carbon-based amorphous materials has lead to recent applications of molecular-dynamical simulations [25, 70, 71, 72]. As the result of our simulations, we have obtained final metastable amorphous carbon modifications at different mass densities. The relaxation of 128 atoms containing supercell structures has been realized by applying a "rapid" dynamical cooling of a partly equilibrated liquid at a cooling rate 10^{15}K/s over 2 ps under constant volume conditions. In all finally obtained structures there is a clear tendency for the different hybrids to separate from each other and to form small interconnected subclusters. Owing to the fixed composition and constant atom number in the supercells, the cohesive energies at different densities have been compared to determine 3.0 g/cm^3 as a magic density at which the most stable amorphous carbon modifications are formed. In Fig. 6 we have plotted the cohesive energies of all a-C models versus density and as a reference we note that the diamond cohesive energy per atom within the present DF-TB scheme is -8.02 eV. For a comprehensive description of how the structures, the chemical bonding properties, and the global band gap properties change with the simulation regime, we refer the reader to [72]. Here we only focus on some important facts. For illustration, we show in Fig. 7 an image of the structure and in Fig. 8 the electronic density of states (DOS), completed by the vibrational density of states (VDOS). The latter spectra has been split into the different hybrid contributions.

The results confirm the stability of high-density tetrahedral amorphous carbon ta-C [68], which has been deposited in different laboratories by var-

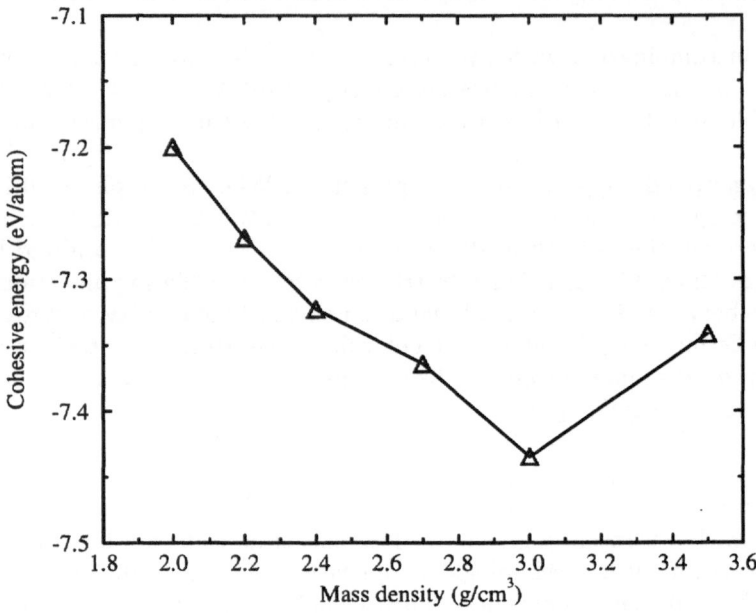

Fig. 6. Cohesive energies of amorphous carbon models versus mass density

ious techniques. This is also supported independently by another ab initio MD method [25]. At this density, the internal strain is maximally reduced from the network by the separation of small, favorably even-membered π clusters between undercoordinated sites yielding minimal defect concentration. The fraction of fourfold coordinated atoms reaches about two thirds, which is very close to the experimentally obtained values. The increase in the π–π^* splitting up to 3.0 eV compared to low-density materials is mainly determined by the changing size distribution of π clusters with respect to smaller ones and by the ability of the π bonds to relax to a mean value for the p–p–π overlap of 0.6 – 0.7 times the overlap in C_2H_4. There is a balance between the gain of π-bonding energy due to sp^2 clustering and the residual stress in the amorphous network, favoring the high stability at the considered density.

Comparing the simulated diffraction data of our 3.0 g/cm³ model in both momentum and real space with neutron-diffraction experiments, the agreement is very good. Additionally, we have calculated the vibrational spectra to allow for further comparison with Raman- and infrared-spectroscopical data. For the total VDOS of our magic density model we find a characteristic half-sphere shape in a frequency range ω between 250 and 1500 cm^{-1} in complete support of similar-shaped spectra obtained by Drabold et al. [25] and Wang et al. [71] using different methods. The most surprising property of the spectra is the complete loss of all reminiscence of the split graphite and diamond behavior. This is a remarkable difference to the comparable situation of a-Si near the crystalline density, where

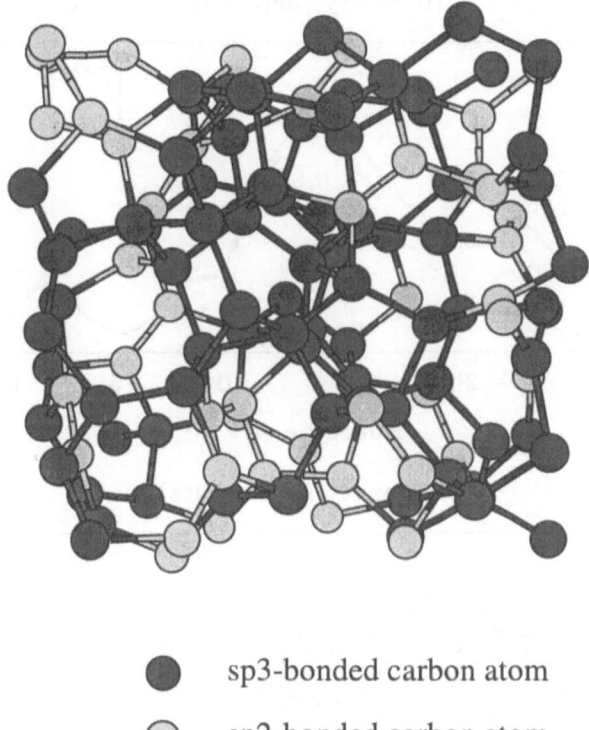

sp3-bonded carbon atom

sp2-bonded carbon atom

Fig. 7. Structural image of the 3.0 g/cm^3 model

the VDOS shows a typical diamond-lattice vibrational behavior. Responsibility for this different situation might lie with the dominating π-interaction in carbon even at the considerable density of 3.0 g/cm^3. The small band features above 1500 cm^{-1} are due to strongly localized stretch-type modes caused by embedding of undercoordinated sp^2-units (pairs, ...) in a rigid sp^3 bonding environment.

8.3   Diamond Surface Reconstructions

An atomistic understanding of growth-related properties of various diamond surfaces is becoming important in order to achieve an optimization of deposition conditions and the control of the surface chemistry for the production of high-quality diamond films. Regarding this, the combination of surface-sensitive experimental techniques such as high resolution electron-energy-loss spectroscopy (HREELS) [73, 74], scanning tunneling microscopy (STM) [75] and theoretical modeling [76, 77, 78, 79, 80, 81] becomes particularly helpful to develop ideas about growth mechanisms at various metastable surface modifications.

Intensity (arb. units)

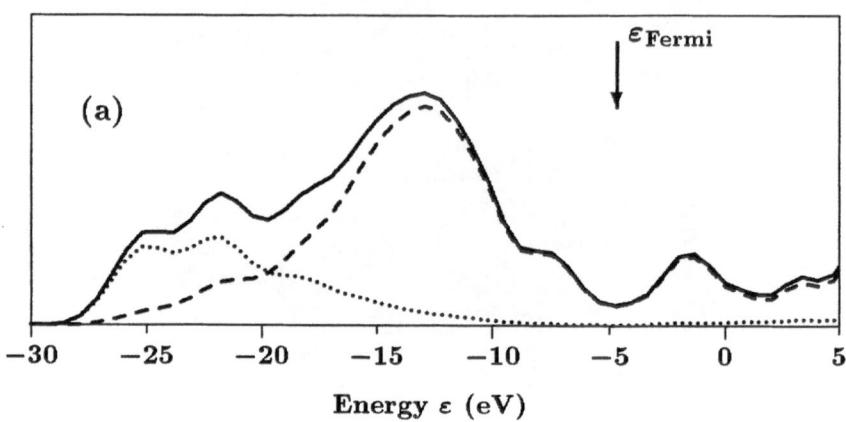

Energy ε (eV)

Intensity (arb. units)

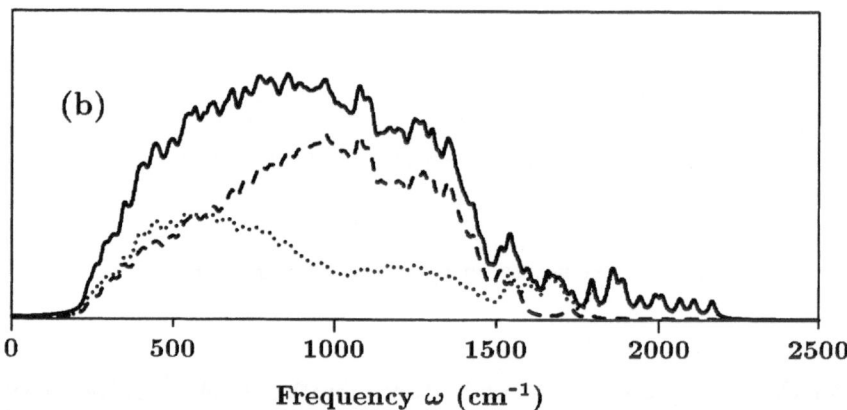

Frequency ω (cm^{-1})

Fig. 8. Electronic and vibrational data for the high-density *ta*-C model. (**a**) Total (*solid line*) electronic density of states (DOS) split into orbital contributions (s-DOS *dotted line*, p-DOS *dashed line*). (**b**) Total (*solid line*) and hybrid-fractional (sp^2 *dotted line*, sp^3 *dashed line*) vibrational density of states (VDOS)

For reasons of limited space we only focus on some important results about stability and vibrational properties of diamond (111) reconstructions. For more detailed information about the (100) and (111) diamond surface models we refer the reader to [79, 82, 83]. More recent studies about growth mechanisms forced by CH$_x$-radical adsorption are discussed elsewhere [84, 85].

After termination of growth, three hydrogen terminated (111) surface reconstructions, shown in Fig. 9 with their related surface vibrational densities of states (SVDOS), are found to be stable on (111) facets. A typical observation of hydrogen terminated (1×1) bulk-like structures on as-grown (111) facets by STM is confirmed by the C(111)(1×1):H surface model. In the absence of hydrogen, the (2×1) reconstructed π-bonded *Pandey-chain* model has been established as the most stable configuration [76, 79], giving an energy gain of 0.7 eV/surface atom relative to the unreconstructed bulk situation. Hydrogenating the chains maintains the (2×1) *Pandey-chain* reconstruction, now forming a metastable surface state shown in the C(111)(2×1)PC:H model. This geometry is less stable than the hydrogenated bulk C(111)(1×1):H structure by 0.6 eV/surface atom. In recent MD adsorption studies of CH$_3$ radicals onto (111) diamond, another stable surface layer reconstruction has been found. Whereas a (1×1) adsorption of CH$_3$ radicals on the clean (111) diamond surface will lead to steric interference between the hydrogens [86], an alternating adsorption of CH$_3$ at every second surface site as discussed by Sasaki and Kawarada [87] is found to be highly stable. In this case, the adsorbed CH$_3$ species may form a (2×2) reconstruction, compare Fig. 9. Considering this surface configuration, the subsurface C atoms that bind the adsorbed CH$_3$ radicals are sp^3 bonded, whereas the remaining C atoms of the subsurface have one dangling bond favoring a behavior towards local graphitization. If these dangling bonds are saturated by hydrogen, as shown in the C(111)(2×2):CH$_3$, the diamond surface is again completely stabilized. The theoretical confirmation of various hydrogenated reconstruction types that were observed in experiments on as-grown (111) facets may provide important information about successive growth steps.

For reasons of comparison with recent HREELS studies of the (111) diamond surface we have calculated and discussed the VDOS of various surface reconstructions [83]. In the following, we only briefly list some of the most characteristic properties.

C(111)(1×1):H The distinct and sharp peak centered at about 3000 cm^{-1} is exclusively due to C–H stretching vibrations. Towards lower frequencies, the band region of $\omega \sim 1000$–1500 cm^{-1} is occupied by various types of C–H bending modes. Most of them are coupled to sublattice excitations. Within a lower frequency band region, 500–900 cm^{-1}, we additionally assign various (C–H) complex vibrations. We find pure translational motions of the (C–H) complexes parallel to the surface as well as pure bouncing vibrations. All theoretically obtained features may be related to their measured counterparts from HREELS experiments [73].

C(111)(2×2):CH$_3$ Comparing with the surface VDOS of the H-terminated C(111) $-$ (1×1) we find two characteristic differences. Instead of an intense and sharp peak in the high-frequency region, the feature located around 3000 cm^{-1} is clearly split into two peaks centered at 2990 and 3060 cm^{-1}, respectively. As origin for the two distinct features we determine the C–H stretching vibrations of the CH$_3$ adsorbates. While asymmetric C–H stretching of the CH$_3$ groups occur at the highest frequencies around 3060 cm^{-1}, the symmetric stretching modes

Intensity (arb. units)

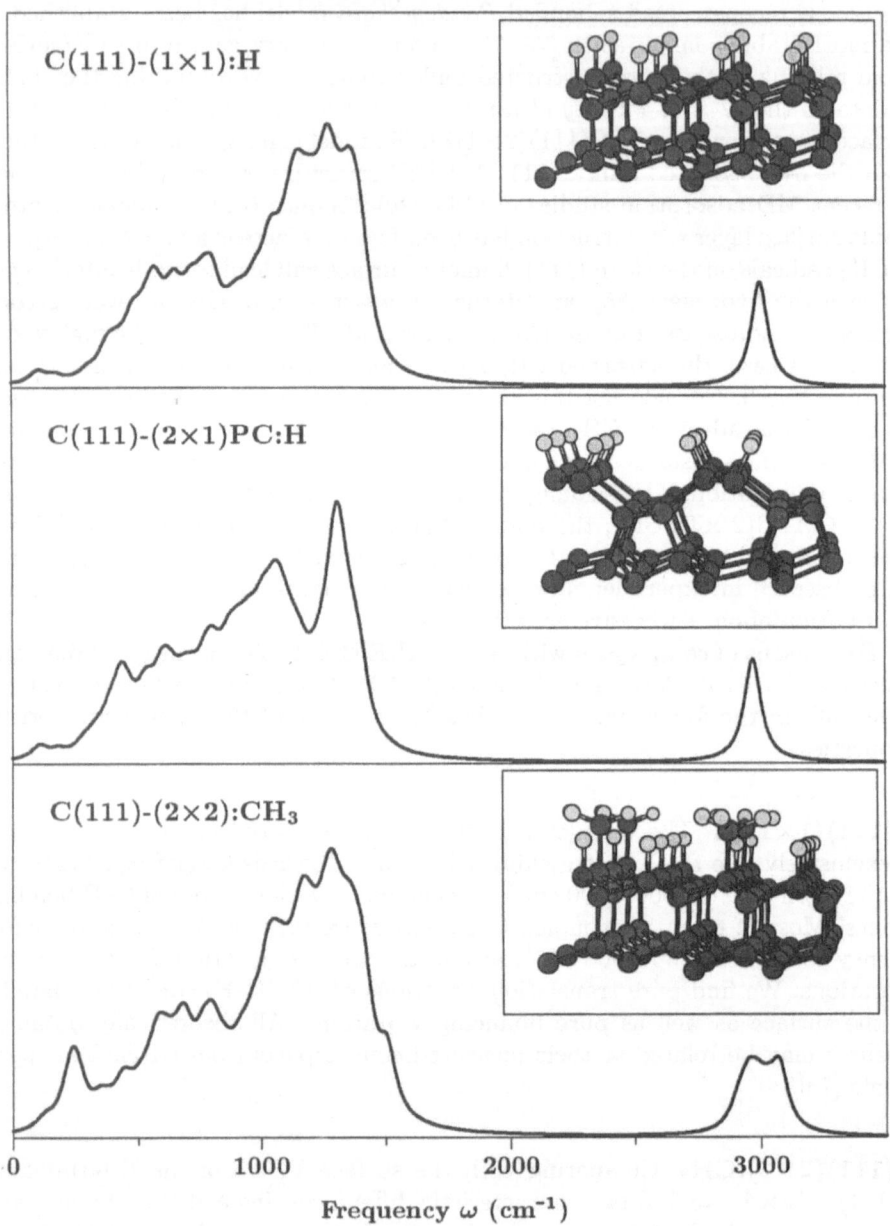

Fig. 9. Surface projected vibrational density of states (SVDOS) and structural images of the C(111)(1 × 1):H, C(111)(2 × 1)PC:H and C(111)(2 × 2):CH₃ reconstruction models

cause the lower-frequency peak at 2990 cm^{-1}. Another interesting feature caused by very soft translational modes of the entire CH$_3$ radical parallel to the surface is visible at about 250 cm^{-1}.

The results are in good agreement with recent HREELS studies of epitaxially grown diamond (111) by Aizawa et al. [88]. These authors conclude from their off-specular HREELS spectra of as-grown (111) diamond that the high-frequency peak splitting is clear evidence for the presence of CH$_3$ groups. Additionally, they report an acoustic surface mode in the lower-frequency region which may be discussed in correlation to our CH$_3$ translational modes parallel to the surface. On the basis of the experimental data, the authors have predicted a coexistence of C–CH$_3$ and C–H on the surface. The appearance of the acoustic phonon mode may serve as an important indication for possible high symmetry ordering of the adsorbed CH$_3$, as predicted by the C(111)(2×2):CH$_3$.

C(111)(2×1)PC:H The existence of σ-bonded chains in the top surface layer are responsible for most of the observable phonon modes. The single distinct peak centered at 2970 cm^{-1} is due to symmetric and asymmetric C–H stretching and compressing modes, in very good agreement to experiments and theoretical predictions [81, 89, 90]. The broad band region $\omega \sim 1200$–1500 cm^{-1} is dominated by hindered rotational vibrations of C–H complexes within the chains. Towards lower frequencies, we find a transition from rocking- and bending- to scissoring-like behavior. The most intense modes dominating this band, e.g., the peak maximum at 1318 cm^{-1}, are due to C–H complex bending vibrations. For example, a symmetric in-phase shearing translation of C–H complexes within the chains may be responsible for the experimentally observed peaks between 1240 and 1300 cm^{-1}, see [89, 90, 91]. The most significant modes belonging to the lower frequencies, $\omega \leq 1200$ cm^{-1}, again represent vibrations in which the C–H complexes of the chains move as a whole.

9 Summary

Throughout this review we have described basic principles for total energy and interatomic force calculations in molecular-dynamics simulations of structure formation of real materials. Discussing the advantages and shortcomings in the use of classical empirical potentials and fully self-consistent density-functional calculations, we outline a third *"hybrid"* method – a density-functional based tight-binding molecular dynamics, which combines the efficiency of empirical concepts with the accuracy of scf calculations.

Considering the simple and straightforward ab initio concept for the construction of the electronic Hamiltonian within a nonorthogonal two-center approach, all tests and applications on various-scale homo- and heteronuclear systems confirm a transferability which is almost comparable to scf-LDA calculations. This opens the possibility for performing highly predictive computer simulations in molecular and solid structure design.

References

[1] G. Binning, H. Rohrer, Ch. Gerber, and E. Weibel, Phys. Rev. Lett. **49**, 57 (1982)
[2] W.E. Moerner and Th. Basché, Angew. Chemie **105**, 537 (1993)
[3] B.A. Pailthorpe, J. Appl. Phys. **70**, 543 (1991)
[4] W. Eckstein, *Computer Simulation of Ion-Solid Interaction*, (Springer, Heidelberg 1991)
[5] H. Balamane, T. Halicioglu, and W.A. Tiller, Phys. Rev. B **46**, 2250 (1992)
[6] R.G. Parr, W. Yang, *Density Functional Theory of Atoms and Molecules*, (Oxford University Press, Oxford 1989)
[7] H.-P. Kaukonen and R.M. Nieminen, Phys. Rev. Lett. **68**, 620 (1992)
[8] Th. Frauenheim, G. Jungnickel, Th. Köhler, and U. Stephan, J. Non-Cryst. Solids **182**, 186 (1995)
[9] S. Skokov, B. Weiner, M. Frenklach, Th. Frauenheim, and M. Sternberg, Phys. Rev. B **52**, 5426 (1995)
[10] K. Binder, ed., *Monte Carlo Methods in Statistical Physics*, 2nd ed., (Springer, Berlin 1984)
[11] G. Ciccotti, D. Frenkel, and I.R. McDonald, eds., *Simulation of Liquids and Solids*, (North-Holland, Amsterdam 1990)
[12] D.W. Heermann, *Computational Simulation Methods in Theoretical Physics*, 2nd ed., (Springer, Berlin 1990)
[13] L. Verlet, Phys. Rev. **159**, 98 (1967)
[14] J.E. Lennard-Jones, Proc. Royal Soc. A **106**, 463 (1924)
[15] F.H. Stillinger and T.A. Weber, Phys. Rev. B **31**, 5262 (1985)
[16] R. Biswas and T.R. Hamann, Phys. Rev. B **36**, 6434 (1987)
[17] J. Tersoff, Phys. Rev. Lett. **56**, 632 (1986) 632; Phys. Rev. B **37**, 6991 (1988); Phys. Rev. Lett. **61**, 2879 (1988)
[18] D.W. Brenner, Phys. Rev. B **42**, 9458 (1990)
[19] P. Hohenberg and W. Kohn, Phys. Rev. **136**, 864B (1964)
[20] W. Kohn and L.J. Sham, Phys. Rev. **140**, A1133 (1965)
[21] R. Car and M. Parinello, Phys. Rev. Lett. **55**, 2471 (1985)
[22] M.R. Pederson and K.A. Jackson, Phys. Rev. B **41**, 7453 (1990)
[23] K. Raghavachari and J.S. Binkley, J. Chem. Phys. **87**, 2191 (1987)
[24] O.F. Sankey and D.J. Niklewski, Phys. Rev. B **40**, 3979 (1989)
[25] D.A. Drabold, P.A. Fedders, and P. Stumm, Phys. Rev. B **49**, 16415 (1994)
[26] K. Laasonen and R.M. Nieminen, J. Phys. Condens. Matter **2**, 1509 (1990)
[27] C.H. Xu, C.Z. Wang, C.T. Chan, and K.M. Ho, J. Phys. Condens. Matter **4**, 6047 (1992)
[28] M. Menon, and K.R. Subbaswamy, Phys. Rev. B **47**, 12754 (1993)
[29] L.M. Canel, A.E. Carlsson, and P.A. Fedders, Phys. Rev. B **48**, 10739 (1993)
[30] D. Porezag, Th. Frauenheim, Th. Köhler, G. Seifert, and R. Kaschner, Phys. Rev. B **51**, 12947 (1995)
[31] G. Seifert and H. Eschrig, Phys. Stat. Sol. (b) **127**, 573 (1985)
[32] G. Seifert, H. Eschrig, and W. Bieger, Z. Phys. Chem. (Leipzig) **267**, 529 (1986)
[33] H. Eschrig and I. Bergert, Phys. Stat. Sol. (b) **90**, 621 (1978)
[34] G. Seifert and R.O. Jones, Z. Phys. D **20**, 77 (1991)
[35] P. Blaudeck, T. Frauenheim, D. Porezag, G. Seifert, and E. Fromm, J. Phys. Condens. Matter **4**, 6389 (1992)
[36] W.M.C. Foulkes, R. Haydock, Phys. Rev. B **39**, 12521 (1989)

[37] D. Tomanek and M.A. Schluter, Phys. Rev. B **36**, 1208 (1987)

[38] H. Eschrig, *Optimized LCAO Method and the Electronic Structure of Extended Systems*, (Akademie, Berlin 1988)

[39] J.P. Perdew and A. Zunger, Phys. Rev. B **23**, 5048 (1981)

[40] R. Jansen and O.F. Sankey, Phys. Rev. B **36**, 6520 (1987)

[41] N. Chetty, K. Stokbro, K.W. Jakobsen and J.K. Norskov, Phys. Rev. B **46**, 3798 (1992)

[42] J.E. Inglesfield, Mol. Phys. **37**, 873 (1979)

[43] H. Eschrig, Phys. Stat. Sol. (b) **96**, 329 (1979)

[44] P. Ordejon, D.A. Drabold, R.M. Martin, and M.P. Grumbach, Phys. Rev. B **48**, 14646 (1993); Phys. Rev. B **51**, 1456 (1995)

[45] D. Hohl, R.O. Jones, R. Car, and M. Parrinello, Chem. Phys. Lett. **139**, 540 (1987)

[46] Q. Ran, R.W. Schmude, M. Miller, and K.A. Gingerich, Chem. Phys. Lett. **230**, 337 (1994)

[47] K. Raghavachari and C.M. Rohlfing, J. Chem. Phys. **89**, 2219 (1988)

[48] L.A. Curtiss, P.W. Deutsch, and K. Raghavachari, J. Chem. Phys. **96**, 6868 (1992)

[49] Th. Frauenheim, F. Weich, Th. Köhler, D. Porezag, and G. Seifert, Phys. Rev. B, **52**, 11492 (1995)

[50] M.R. Pederson and K.A. Jackson, Phys. Rev. B **43**, 7312 (1991)

[51] P.J. Ordejon, D. Lebedenko, and M. Menon, Phys. Rev. B **50**, 5645 (1994)

[52] J.P. Perdew, J.A. Chevary, S.H. Vosko, K.A. Jackson, M.R. Pederson, D.J. Singh, and C. Fiolhais, Phys. Rev. B **46**, 6671 (1992)

[53] J. Andzelm and E. Wimmer, J. Chem. Phys. **96**, 1280 (1992)

[54] B.G. Johnson, P.M.W. Gill, and J.A. Pople, J. Chem. Phys. **98**, 5621 (1993)

[55] I. Kwon, R. Biswas, C.Z. Wang, K.M. Ho, and C.M. Soukoulis, Phys. Rev. B **49**, 7242 (1994)

[56] J.L. Mercer and M.Y. Chou, Phys. Rev. B **49**, 8506 (1994)

[57] M.T. Yin and M.L. Cohen, Phys. Rev. B **24**, 6121 (1981)

[58] M.T. Yin and M.L. Cohen, Phys. Rev. B **29**, 6996 (1984)

[59] W.I. Kamitakahara, NIST, Reactor Radiation Division, Gaithersburg, MD 20899, private communication

[60] H.W. Kroto, J.R. Heath, S.C. O'Brian, R.F. Curl, and R.E. Smalley, Nature (London) **318**, 162 (1985)

[61] Y. Wang, J.M. Holden, X. Bi, and P.C. Eklund, Chem. Phys. Lett. **217**, 3 (1993)

[62] Y. Wang, J.M. Holden, Z. Dong, X. Bi, and P.C. Eklund, Chem. Phys. Lett. **211**, 341 (1993)

[63] D.L. Strout, R.L. Murry, C. Xu, W.C. Eckhoff, G.K. Odom, Y. Wang, Chem. Phys. Lett. **214**, 576 (1993)

[64] G.B. Adams, J.B. Page, O.F. Sankey, M. O'Keeffe, Phys. Rev. B **50**, 17471 (1994)

[65] M.R. Pederson and A.A. Quong, Phys. Rev. Lett. **74**, 2319 (1995)

[66] J. Robertson, Diamond Related Materials **2**, 984 (1993)

[67] D.R. McKenzie, D. Muller, and P.A. Pailthorpe, Phys. Rev. Lett. **67**, 773 (1991)

[68] V.S. Veerasamy, J. Yuan, G.A.J. Amaratunga, W.I. Milne, K.W.R. Gilkes, M. Weiler, and L.M. Brown, Phys. Rev. B **48**, 17954 (1994)

[69] V.S. Veerasamy, G.A.J. Amaratunga, C.A. Davis, A.E. Tins, W.I. Milne, and D.R. McKenzie, J. Phys. Condens. Matter **5**, L169 (1993)

[70] G. Galli, R.M. Martin, R. Car, and M. Parinello, Phys. Rev. B **42**, 7470 (1990)

[71] C.Z. Wang, K.M. Ho, and C.T. Chan, Phys. Rev. Lett. **70**, 611 (1993)

[72] Th. Frauenheim, G. Jungnickel, Th. Koehler, and U. Stephan, J. Non-Cryst. Solids **182**, 186 (1995)

[73] B.J. Waclawski, D.T. Pierce, N. Swanson, and R.J. Celotta, J. Vac. Sci. Technol. **21**, 368 (1982)

[74] S.-T. Lee and G. Apai, Phys. Rev. B **48**, 2684 (1993)

[75] H.-G. Busmann, W. Zimmermann-Edling, S. Lauer, H. Hertel, Th. Frauenheim, P. Blaudeck, and D. Porezag, Surf. Sci. **295**, 340 (1993)

[76] S. Iarlori, G. Galli, F. Gygi, M. Parinello, and E. Tosatti, Phys. Rev. Lett. **69**, 2947 (1992)

[77] S.H. Yang, D.A. Drabold, and J.B. Adams, Phys. Rev. B **48**, 5261 (1993)

[78] F. Bechstedt and D. Reichardt, Surf. Sci. **202**, 83 (1988)

[79] Th. Frauenheim, U. Stephan, P. Blaudeck, D. Porezag, H.-G. Busmann, W. Zimmermann-Edling, and S. Lauer, Phys. Rev. B **48**, 18189 (1993)

[80] J. Furthmüller, J. Hafner, and G. Kresse, Europhys. Lett. **28**, 659 (1994)

[81] D.R. Alfonso, D.A. Drabold, and S.E. Ulloa, Phys. Rev. B **51**, 14669 (1995)

[82] Th. Frauenheim, Th. Köhler, M. Sternberg, D. Porezag, and M. Pederson, Thin Solid Films, (in print)

[83] Th. Köhler, M. Sternberg, D. Porezag, and Th. Frauenheim, Phys. Stat. Sol. special issue March 1996

[84] B.J. Garrison, E.J. Dawnkaski, D. Sriavastava, and D.W. Brenner, Science **255**, 835 (1992)

[85] S.J. Harris and D.G. Goodwin, J. Phys. Chem. **97**, 23 (1993)

[86] Th. Frauenheim and P. Blaudeck, Appl. Surf. Sci. **60/61**, 182 (1992)

[87] H. Sasaki and H. Kawarada, Trans. Mat. Res. Soc. Jpn. **14B**, 1475 (1994)

[88] T. Aizawa, T. Ando, M. Kamo, and Y. Sato, Phys. Rev. B **48**, 18348 (1993)

[89] B.D. Thoms, P.E. Pehrsson, and J.E. Butler, J. Appl. Phys. **75**, 1804 (1994)

[90] B. Sun, X. Zhang, Q. Zhang, and Z. Lin, Appl. Phys. Lett. **62**, 31 (1993)

[91] N.J. DiNardo, W.A. Thompson, A.J. Schell-Sorokin, and J.E. Dermuth, Phys. Rev. B **34**, 3007 (1986)

Finite Element Methods for the Stokes Equation

Jochen Reichenbach and Nuri Aksel

Lehrstuhl für Strömungsmechanik, Technische Universität, D-09107 Chemnitz, Germany, e-mail: `jochen.reichenbach@mb1.tu-chemnitz.de`

Abstract. The description of the finite element method (FEM) for the Stokes problem is considered. Boundary conditions that are important in fluid mechanics are discussed. The condition of incompressibility leads to a saddle-point problem. The approximation by a mixed finite element method requires the choice of suitable finite elements. Otherwise the computation suffers from instability and useless results are produced.

1 Introduction

In fluid mechanics the basic conservation equations of mass, momentum, and energy are usually made tractable by neglecting terms that are assumed to be small. In this way examples of the standard types of partial differential equations, i.e., elliptic, parabolic, and hyperbolic, are obtained. These three types of differential equations have different characteristics. Hence different initial and/or boundary conditions are necessary to formulate a well-posed problem.

An analytical solution can be found in only very few cases, even for simplified physical problems. The application of different numerical approximation methods has therefore a long tradition in fluid mechanics.

In recent years the finite element method has found an increasing use and a wider acceptance for the solution of the equations governing viscous incompressible fluid flows. The advantages are a well-established mathematical foundation, great geometrical flexibility, and the possibility to design general computer implementations. Nowadays, there are computer codes having the capability of simulating complex industrial processes. On the other hand the FEM applications in fluid mechanics are subject to further research investigations.

The different types of partial differential equations have special properties, which have to be considered in the numerical treatment. Here we consider the elliptic Stokes equation, which plays a fundamental role in approximation problems in the field of fluid mechanics. Furthermore, the theory of FEM for this problem is very well developed. In particular, the problem of the condition of incompressibility has been studied.

In the following, we shall give an outline of FEM for the Stokes problem. A strongly mathematical study can be found in the literature. We especially refer to the comprehensive books of [1, 4, 5, 8].

2   Stokes Equation

2.1   Conservation Equations

The domain Ω is occupied by an incompressible fluid with density ρ, and we consider a stationary process. If we denote the velocity by $\underline{u}$, the conservation of mass gives

$$\underline{\nabla} \cdot \underline{u} = 0 \,, \tag{1}$$

and when we neglect convective terms (creeping flow) the conservation of momentum reads

$$-\underline{\nabla} \cdot \underline{\underline{\sigma}} = \rho \underline{f} \,. \tag{2}$$

There $\underline{f}$ is an external force on a unit volume and $\underline{\underline{\sigma}}$ denotes the total or Cauchy stress tensor. This tensor can be split into a pressure term p and the extra stress tensor $\underline{\underline{\tau}}$ due to viscosity,

$$\sigma_{ij} = -p\delta_{ij} + \tau_{ij} \,. \tag{3}$$

The Newtonian constitutive equation relates the stresses with the viscosity η by

$$\tau_{ij} = 2\,\eta\,\frac{1}{2}(u_{i,j} + u_{j,i}) \,. \tag{4}$$

The abbreviation $u_{i,j} = \partial u_i / \partial x_j$ denotes the partial derivative of u_i with respect to the coordinate x_j. From (3), (1), and (2) we derive the Stokes equation

$$-\eta\,\Delta\underline{u} + \underline{\nabla}p = \rho\underline{f} \,, \tag{5}$$

where the symbols $\underline{\nabla}$ and Δ denote the classical grad and Laplacian operators. In (5) the pressure is determined only up to an additive constant. Usually, in the theory it is normalized by

$$\int_{\Omega} p\,\mathrm{d}x = 0 \,; \tag{6}$$

in practice p is prescribed in one point.

2.2   Function Spaces and Variational Formulation

In the FEM the fundamental relation is the variational or weak formulation of the differential problem. The unknown functions are looked for in some function spaces and the equations are (scalar) multiplied by some test functions.

Let Ω be a bounded open set in $\mathrm{I\!R}^n$, $n = 2$ or 3, with a sufficiently regular boundary Γ. Usually we denote by

$$H^1(\Omega) = \{q \in L^2(\Omega) : \mathrm{D}q \in L^2(\Omega)\} \tag{7}$$

the space of functions with integrable derivatives of order $k = 0, 1$ on Ω. Here $L^2(\Omega)$ is the space of square integrable functions in Ω and $\mathrm{D}q$ symbolizes a (generalized) derivative of q. Consider the boundary-value problem

$$-\eta\,\Delta\underline{u} + \nabla p = \rho\underline{f} \,, \quad \underline{\nabla}\cdot\underline{u} = 0 \,, \quad \underline{u}|_\Gamma = \underline{g} \,, \quad \text{with} \quad \int_\Gamma \underline{g}\cdot\underline{n}\,\mathrm{d}\Gamma = 0 \,, \tag{8}$$

where the function g describes the velocity on the boundary Γ. We look for $\underline{u}$ in

$$V_g = \{\underline{v} \in (H^1(\Omega))^n : \underline{\nabla} \cdot \underline{v} = 0, \ \underline{v}|_\Gamma = \underline{g}\} \tag{9}$$

and p in

$$L_0^2 = \left\{ q \in L^2(\Omega) : \int_\Omega q \, d\boldsymbol{x} = 0 \right\} . \tag{10}$$

The last condition for g in (8) stems from application of the Gauss theorem to the continuity equation, with $\underline{n}$ outer normal on Γ:

$$0 = \int \underline{\nabla} \cdot \underline{u} \, d\boldsymbol{x} = \int_\Gamma \underline{u} \cdot \underline{n} \, d\Gamma = \int_\Gamma \underline{g} \cdot \underline{n} \, d\Gamma . \tag{11}$$

This is a necessary condition for the boundary velocities for the existence of a divergencefree velocity field in Ω.

Multiplying the first equation in (8) by the test functions $\underline{v} \in V_0$, defined by (9) with $\underline{g} = 0$, and applying the Gauss theorem, we obtain the weak formulation

$$\int (-\eta u_{i,jj} + p_{,i} - \rho f_i) v_i \, d\boldsymbol{x} = \int (-\eta(u_{i,j}v_i)_{,j} + (pv_i)_{,i} + \eta u_{i,j}v_{i,j} - pv_{i,i} - \rho f_i v_i) \, d\boldsymbol{x}$$

$$= \int_\Gamma (-\eta u_{i,j} n_j v_i + p v_i n_i) \, d\Gamma + \int (\eta u_{i,j} v_{i,j} - pv_{i,i} - \rho f_i v_i) \, d\boldsymbol{x}$$

$$= \int (\eta u_{i,j} v_{i,j} - \rho f_i v_i) d\boldsymbol{x} = 0 . \tag{12}$$

Here, and subsequently, we employ the usual summation convention, i.e., if an index is repeated, summation about this index is implied. The surface integral is zero due to homogeneous boundary values and the pressure term vanishes because the functions $\underline{v} \in V_0$ are divergence free. Given

$$a(\underline{u}, \underline{v}) := \eta \int u_{i,j} v_{i,j} \, d\boldsymbol{x} \quad , \quad \langle \underline{f}, \underline{v} \rangle := \int f_i v_i \, d\boldsymbol{x} , \tag{13}$$

the weak formulation of (8) reads:

$$\text{find} \quad \underline{u} \in V_g , \quad \text{with} \quad a(\underline{u}, \underline{v}) = \rho \langle \underline{f}, \underline{v} \rangle \quad \forall \underline{v} \in V_0 . \tag{14}$$

If $\underline{u}_g \in V_g$ is a known extension of the boundary function $\underline{g}$ to Ω, we have $\underline{u} = \underline{u}_0 + \underline{u}_g$, $\underline{u}_0 \in V_0$. With the right-hand side

$$l(\underline{v}) = \rho \langle \underline{f}, \underline{v} \rangle - a(\underline{u}_g, \underline{v}) , \tag{15}$$

problem (14) reads as a problem with homogeneous boundary conditions:

$$\text{find} \quad \underline{u}_0 \in V_0 , \quad \text{with} \quad a(\underline{u}_0, \underline{v}) = l(\underline{v}) \quad \forall \underline{v} \in V_0 . \tag{16}$$

The form $a(\underline{u}, \underline{v})$ is bilinear and symmetrical. Therefore the problem is equivalent to the variational problem

$$J(\underline{v}) := \frac{1}{2} a(\underline{v}, \underline{v}) - l(\underline{v}) \longrightarrow \inf_{V_0} . \tag{17}$$

This will be seen by writing the Euler equations, i.e., for vanishing variation of $J(\underline{v})$,

$$\delta J(\underline{v})|_{\underline{u}} := \lim_{t \to 0} \frac{1}{t} \{ J(\underline{u} + t\underline{v}) - J(\underline{u}) \} = 0 , \qquad (18)$$

we find (16). This equivalence gives the name 'variational formulation' to (16). More mathematical details can be found in the standard literature cited above.

We note that such a relation to a variational problem exists only in the case of a bilinear and symmetrical functional $a(\ ,\ )$, i.e., the differential equation problem has to be linear and symmetrical. This is not fulfilled for example for the Navier Stokes equations due to the convective terms, which are nonlinear.

However, the derivation of the weak formulation is always possible. The name results from the fact that, in general, the solution of (17) or (16) is not sufficiently regular to satisfy (8) in the classical sense (pointwise).

2.3   Saddle Point Problem

We have assumed the functions $\underline{v} \in V_0$ are (pointwise) divergence free, $\underline{\nabla} \cdot \underline{v} = 0$. This can be weakened:

$$b(\underline{v}, q) := \int \underline{\nabla} \cdot \underline{v} \, q \, dx = 0 , \quad \forall q \in L_0^2(\Omega) . \qquad (19)$$

Replacing V_g by

$$\tilde{V}_g = \left\{ \underline{v} \in (H^1(\Omega))^n : b(\underline{v}, q) = 0, \ \forall q \in L_0^2(\Omega); \ \underline{v}|_\Gamma = \underline{g} \right\} \qquad (20)$$

and repeating the procedure concerning (12) with $\underline{v} \in \tilde{V}_g$ we find the results (13) – (17), i.e., the variational problem

$$J(\underline{v}) := \frac{1}{2} a(\underline{v}, \underline{v}) - l(\underline{v}) \longrightarrow \inf_{\tilde{V}_0} . \qquad (21)$$

Substituting $\tilde{V}_0$ by its definition, we can rewrite this as a constrained minimisation problem:

$$J(\underline{v}) = \frac{1}{2} a(\underline{v}, \underline{v}) - l(\underline{v}) \longrightarrow \inf_{X_0} , \quad \text{with} \quad b(\underline{v}, q) = 0 , \quad \forall q \in L_0^2(\Omega) . \qquad (22)$$

Here

$$X_0 = \{ \underline{v} \in (H^1(\Omega))^n : \underline{v}|_\Gamma = 0 \} . \qquad (23)$$

Using the Lagrangian multiplier, we obtain a minimisation problem without constraints:

$$\mathcal{L}(\underline{v}, q) := J(\underline{v}) + b(\underline{v}, q) = \frac{1}{2} a(\underline{v}, \underline{v}) - l(\underline{v}) + b(\underline{v}, q) \longrightarrow \inf_{\underline{v} \in X_0} \sup_{q \in L_0^2} . \qquad (24)$$

We remark that the usual Lagrangian multiplier λ is implied in the function q. Such a problem is called a saddle point problem because the solution is a minimum with respect to $\underline{v}$ and a maximum with respect to q. For vanishing variation,

$$\delta\mathcal{L}(\underline{v},q)|_{(\underline{u},p)} := \lim_{t\to 0}\frac{1}{t}\{\mathcal{L}(\underline{u}+t\underline{v},p+tq)-\mathcal{L}(\underline{u},p)\} = 0 , \qquad (25)$$

we find

$$a(\underline{u},\underline{v})+b(\underline{v},p) = l(\underline{v}) , \qquad \forall \underline{v}\in X_0 ,$$
$$b(\underline{u},q) = 0 , \qquad \forall q\in L_0^2 . \qquad (26)$$

It can be seen that the pressure p plays the role of the Lagrangian multiplier of the constraint that the velocity has to be divergence free. Sommerfeld obtained this relation by applying the Hamilton method [9].

The first component of the solution $(\underline{u},p)$ of (26) is always a solution of (21), but the unicity of the Lagrangian parameter in (24) or (26), i.e., the equivalence to (22), can be guaranteed only if the following inf-sup or LBB condition (Ladyshenskaja–Babuška–Brezzi) is satisfied:

$$\inf_{q\in L_0^2}\sup_{\underline{v}\in X_0}\frac{b(\underline{v},q)}{\|\underline{v}\|\,\|q\|} \geq \beta > 0 . \qquad (27)$$

Here $\|.\|$ denotes the norm in the spaces X_0 and L_0^2, respectively. This condition is fulfilled by the continuous Stokes problem but is not automatically satisfied by the discretized equations. It is a condition for the approximation of the spaces X_0 and L_0^2, which cannot be arbitrarily chosen.

It is interesting to note that the problem (26) will also be obtained if we multiply the first equation in (8) by $\underline{v}\in X_0$ and the second one by $q\in L_0^2$. In such a simple manner, FEM had been realized long before the relation to the saddle point problem and the inf-sup condition were revealed.

2.4   General Boundary Conditions

In fluid mechanics we often find boundary conditions that are more complicated than Dirichlet conditions for the velocity assumed above. To solve the Stokes problem with general boundary conditions we have to reconsider (2). Let us multiply this equation by an arbitrary but sufficiently regular function $\underline{v}$:

$$\int(-\sigma_{ij,j}-\rho f_i)v_i\,\mathrm{d}x = \int(-(\sigma_{ij}v_i)_{,j}+\sigma_{ij}v_{i,j}-\rho f_i v_i)\,\mathrm{d}x$$
$$= \int_\Gamma -\sigma_{ij}n_j v_i\,\mathrm{d}\Gamma + \int(\sigma_{ij}\frac{1}{2}(v_{i,j}+v_{j,i})-\rho f_i v_i)\,\mathrm{d}x . \qquad (28)$$

In manipulating the second term on the right-hand side, we have used the symmetry of the stress tensor σ_{ij}. For a Newtonian fluid we obtain, with (3) and (4),

$$\int\sigma_{ij}\frac{1}{2}(v_{i,j}+v_{j,i})\,\mathrm{d}x = \int\frac{1}{2}\eta(u_{i,j}+u_{j,i})(v_{i,j}+v_{j,i})\,\mathrm{d}x - \int pv_{i,i}\,\mathrm{d}x . \qquad (29)$$

In the surface integral in (28) we find the stress vector $\underline{\sigma} = \underline{\underline{\sigma}} \cdot \underline{n}$. This vector can be split into a normal and a tangential component: $\underline{\sigma} = \sigma_n \underline{n} + \sigma_t \underline{t}$. This holds in $\mathbb{R}^2$. In $\mathbb{R}^3$ there are two tangential vectors $\underline{t}_1$, $\underline{t}_2$. Thus, we have

$$\int_\Gamma \sigma_{ij} n_j v_i \, d\Gamma = \int_\Gamma (\sigma_n \underline{n} + \sigma_t \underline{t}) \cdot \underline{v} \, d\Gamma . \tag{30}$$

We now can formulate the boundary conditions that allow the computation of this integral [7]:

1. The Dirichlet conditions on Γ_0 are

$$\Gamma_0 : \quad \underline{u} = \underline{g} . \tag{31}$$

The test function $\underline{v}$ has to vanish on Γ_0, $\Gamma_0 : \underline{v} = 0$. The line integral on Γ_0 is then zero.

2. On the boundary Γ_t the tangential component of the velocity is known,

$$\Gamma_t : \underline{u} \cdot \underline{t} = g_t . \tag{32}$$

Then the tangential component of $\underline{v}$ on Γ_t has to vanish on $\Gamma_t : \underline{v} \cdot \underline{t} = 0$. Knowing the line integral on Γ_t, we find the natural boundary condition

$$\Gamma_t : \quad \sigma_n = (\underline{\underline{\sigma}} \cdot \underline{n}) \cdot \underline{n} = -p + (\underline{\underline{\tau}} \cdot \underline{n}) \cdot \underline{n} = \sigma_n^g , \tag{33}$$

i.e., the normal component of the stress vector $\underline{\sigma}$ is prescribed. This boundary condition is useful to model entry and exit flows.

The equation (33) shows that in this case the pressure is completely determined.

3. On the boundary Γ_n the normal component of the velocity is known,

$$\Gamma_n : \underline{u} \cdot \underline{n} = g_n . \tag{34}$$

The normal component of $\underline{v}$ has to be zero on Γ_n, $\Gamma_n : \underline{v} \cdot \underline{n} = 0$. The natural boundary condition is then

$$\Gamma_n : \quad \sigma_t = (\underline{\underline{\sigma}} \cdot \underline{n}) \cdot \underline{t} = (\underline{\underline{\tau}} \cdot \underline{n}) \cdot \underline{t} = \sigma_t^g . \tag{35}$$

This can be used to formulate a line of symmetry.

4. On Γ_f both σ_n and σ_t are prescribed. Then the line integral on Γ_f is known without any assumption for $\underline{v}$. This formulation is useful to model a free surface.

The notion of a natural boundary condition stems from the fact that the trial functions do not fulfill these conditions. However, the weak solution satisfies them (in a weak sense) in a natural way.

The definite form of the natural boundary conditions depends on the formulation of the differential operator. Starting with (5) we arrive at other natural boundary conditions which do not have a simple physical interpretation.

2.5   Example

We illustrate the boundary conditions and the weak formulation by considering a branching flow (Fig.1).

The line Γ_n is a line of symmetry:

$$\Gamma_n: \qquad \underline{u} \cdot \underline{n} = 0 \quad, \sigma_t = 0 \,. \tag{36}$$

Across the boundaries $\Gamma_{kt}, k = 1, 2$, there is an exit flow in the normal direction:

$$\Gamma_{kt} : \underline{u} \cdot \underline{t} = 0 \,, \sigma_n = -p + (\underline{\underline{\tau}} \cdot \underline{n}) \cdot \underline{n} = -p_k^g \,. \tag{37}$$

The prescribed entry flow and the no-slip condition belong to Γ_0 with Dirichlet conditions for $\underline{u}$:

$$\Gamma_0: \qquad \underline{u} = \underline{g} \,. \tag{38}$$

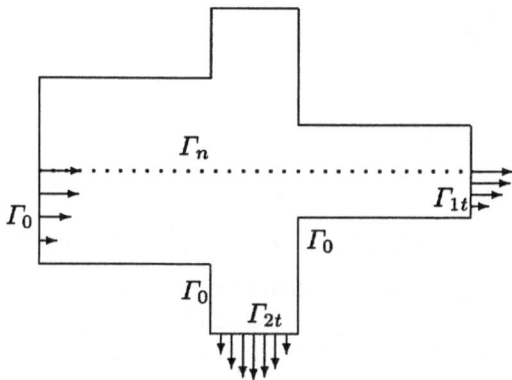

Fig. 1. Different kinds of boundary conditions

In a "nearly developed" flow in a channel or tube the velocity in the cross-stream direction is near to zero, hence the term $(\underline{\underline{\tau}} \cdot \underline{n}) \cdot \underline{n}$ is small and (37) is reduced to a condition for the pressure.

The weak formulation reads as follows:

find $\underline{u} \in \tilde{V}_g = \underline{v} \in (H^1(\Omega))^2 : \Gamma_0 : \underline{v} = \underline{g}, \ \Gamma_n : \underline{v} \cdot \underline{n} = 0, \ \Gamma_t = \Gamma_{1t} \cup \Gamma_{2t} : \underline{v} \cdot \underline{t} = 0$

and $p \in L^2(\Omega)$ with

$$\int \left[\frac{1}{2} \eta (u_{i,j} + u_{j,i})(v_{i,j} + v_{j,i}) + v_{i,i} p \right] \mathrm{d}x = \int \rho f_i v_i \, \mathrm{d}x + \int_\Gamma (-p^g) v_i n_i \, \mathrm{d}\Gamma ,$$

$$\underline{v} \in \tilde{V}_0 , \qquad -\int q u_{i,i} \, \mathrm{d}x = 0, \quad \forall q \in L^2.$$

$$\tag{39}$$

3   Discretization

3.1   General Formulation

We obtain the FEM discretization by replacing the infinite-dimensional spaces of the continuous functions by finite-dimensional subspaces of "simple" functions. In the Galerkin method for the problem (16) the space V_0 is replaced by the N-dimensional space $V_{0h} = \{\underline{v}_h^1, \ldots, \underline{v}_h^N\}$, spanned by the basis $\underline{v}_h^k$, $k = 1, \ldots, N$:

$$\text{find} \quad \underline{u}_h \in V_{0h} \quad \text{with} \quad a(\underline{u}_h, \underline{v}_h) = l(\underline{v}_h) \quad \forall \underline{v}_h \in V_{0h}. \tag{40}$$

The coefficients of the unknown function

$$\underline{u}_h = \sum_{j=1}^{N} u^j \underline{v}_h^j \tag{41}$$

can be found as a solution of the matrix problem:

$$\sum_{j=1}^{N} u^j \, a(\underline{v}_h^j, \underline{v}_h^i) = l(\underline{v}_h^i), \quad i, j = 1, \ldots, N, \tag{42}$$

or

$$\underline{\underline{A}} \, \underline{U} = \underline{G}, \quad A_{ij} = a(\underline{v}_h^j, \underline{v}_h^i), \quad G_i = l(\underline{v}_h^i), \quad U_j = u^j. \tag{43}$$

The functions used to approximate $\underline{u}$ are called trial or form functions. To obtain the weak formulation the equations are multiplied by test functions.

The Galerkin method uses trial and test functions that are identical. If the trial and test spaces are approximated by different function sets, the algorithm is called the Petrov–Galerkin method or the method of weighted residuals.

The matrix $\underline{\underline{A}}$ is called the system matrix or stiffness matrix in the mechanics of solids.

Another approach is the substitution of (41) into the variational problem (17) and the forming of the Euler equations:

$$\frac{\partial}{\partial u^j} \, J\left(\sum_{j=1}^{N} u^j \underline{v}_h^j\right) = 0. \tag{44}$$

This algorithm is called the Ritz method. As a result we obtain the system (43) again.

If we want to use this formalism in the variational formulation (17) or (21) of the Stokes equation we have to have functions that are divergence free in Ω in the strong sense, $\underline{v}_h \in V_0$, or in the weak sense, $\underline{v}_h \in \tilde{V}_0$. Moreover these functions have to form a basis. It is not easy to construct such functions and even if we succeed the functions often do not have a simple form and are not very suitable for computations.

Simpler functions can be employed if we start with the saddle-point formulation (26):

$$\underline{v}_h \in X_0 = \left\{ \underline{v}_h \in (H^1(\Omega))^n : \quad \underline{v}_h|_\Gamma = 0 \right\} , q_h \in L^2(\Omega) .$$

But these functions have to satisfy the inf-sup condition.

Let X_{0h}, Q_h be an approximation for X_0, L_0^2. The saddle point problem reads

$$\begin{aligned} \text{find} \quad & u_h \in X_{0h}, \quad p_h \in Q_h \\ \text{with} \quad & a(\underline{u}_h, \underline{v}_h) + b(\underline{v}_h, p_h) = l(v_h), \quad \forall \underline{v}_h \in X_{0h} , \\ & b(\underline{u}_h, q_h) = 0, \quad \forall q_h \in Q_h . \end{aligned} \quad (45)$$

With the basis $\{\underline{v}_h^1, \ldots, \underline{v}_h^N\}$ of X_{0h} and $\{q_h^1, \ldots, q_h^M\}$ of Q_h we have

$$\underline{u}_h = \sum_{j=1}^N u^j \underline{v}_h^j , \qquad p_h = \sum_{j=1}^M p^j q_h^j$$

and we obtain the following matrix problem, which is characteristic of the approximation of saddle-point problems,

$$\begin{pmatrix} A & B \\ B^T & 0 \end{pmatrix} \begin{pmatrix} U \\ P \end{pmatrix} = \begin{pmatrix} G \\ 0 \end{pmatrix} , \quad (46)$$

with $A_{ij} = a(\underline{v}_h^j, \underline{v}_h^i)$, $B_{ij} = b(\underline{v}_h^i, q_h^j)$, $U_j = u^j$, $P_j = p^j$, $i = 1, \ldots, N$, $j = 1, \ldots, M$. In the FEM this formalism is called the mixed method. The system matrix $\mathcal{A} = \begin{pmatrix} A & B \\ B^t & 0 \end{pmatrix}$ is sparse, symmetrical, but not positive definite. In small and medium-sized problems the solution of (46) can be realized by a direct Gauss algorithm taking advantage of the sparsity of $\mathcal{A}$. In large sized problems, e.g., 3-dimensional one, we have to use specific iteration procedures. Suitable algorithms are penalty methods, preconditioned conjugate gradient (cg) methods and multigrid methods [1, 2].

We emphasize that the usually used iteration procedures are not appropriate for this indefinite problem.

3.2   Finite Elements for Saddle-Point Problems

The key requirement is the satisfaction of the inf-sup stability condition, which involves both velocity and pressure spaces. This condition is one of the most celebrated results in the mathematical theory of finite elements.

Equal-order interpolations of velocity $\underline{u}_h$ and pressure p_h fail to satisfy the LBB condition. Different pathological constellations can occur, e.g., if there are too many constraints due to incompressibility only the trivial solution $\underline{u}_h = 0$ exists (locking effect), or the constraints are not independent and the pressure is not uniquely defined (checkerboard instability). By adding suitable functions for $\underline{v}_h$ (bubble functions) stable elements can be designed. In this way we obtain the mini-element.

Mini-element (P_1 + bubble – P_1).
Let T_k be a triangle (see Fig. 2) and P_1 denote the set of polynomials of first degree on T_k, $P_1 = \{1, x_1, x_2\}$. In the unit triangle the barycentric coordinates are given by

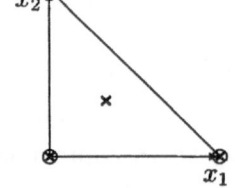

$$\begin{aligned}
\lambda_1 &= x_1 , \\
\lambda_2 &= x_2 , \\
\lambda_3 &= 1 - x_1 - x_2 .
\end{aligned} \qquad (47)$$

Fig. 2. Unit triangle and the position of the degrees of freedom of $\underline{u}_h, p_h$

The "bubble function" $B(x) = \lambda_1 \lambda_2 \lambda_3$ vanishes on the sides of the triangle. The trial functions for $\underline{u}_h$ are the bilinear functions from P_1 combined with bubbles $B(x)$:

$$\begin{aligned}
X_h &= \left\{ \underline{v}_h \in C^0(\Omega)^2, \ \underline{v}_h \in (P_1 \oplus \text{span}\{B(\underline{x})\})^2 \quad \text{on } T_k \right\}, \\
Q_h &= \left\{ q_h \in C^0(\Omega)^2, \ q_h \in P_1 \quad \text{on } T_k \right\} .
\end{aligned} \qquad (48)$$

This approximation fulfills the inf-sup condition.

Taylor–Hood Element (P_2 – P_1). On a triangle (see Fig. 3) we take a quadratic approximation of $\underline{u}_h \in (P_2)^2 = \{1, x_1, x_2, x_1 x_2, x_1^2, x_2^2\}^2$ and a linear one of $p_h \in P_1$.

$$\begin{aligned}
X_h &= \left\{ \underline{v}_h \in C^0(\Omega)^2, \ \underline{v}_h \in (P_2)^2 \quad \text{on } T_k \right\}, \\
Q_h &= \left\{ q_h \in C^0(\Omega)^2, \ q_h \in P_1 \quad \text{on } T_k \right\} .
\end{aligned} \qquad (49)$$

The inf-sup condition is satisfied and the element is approved. We obtain a variant of this approximation by subdividing the triangle into four subtriangles and interpolating the velocity by linear functions on each subelement. This element can also be used with rectangles if the polynomials P_1, P_2 are replaced by $Q_1 = \{1, x_1, x_2, x_1 x_2\}$, $Q_2 = Q_1 \cup \{x_1^2, x_2^2, x_1^2 x_2, x_1 x_2^2, x_1^2 x_2^2\}$. Such an approximation works also on tetrahedrons and bricks in R^3.

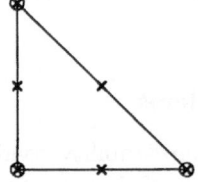
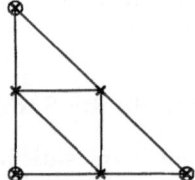

Fig. 3. Nodes for the Taylor–Hood element

P_{k+1} + bubble – P_k discontinuous. In these elements (see Fig. 4) the pressure is discontinuous on the sides and the nodes are located at points that are used for the Gauss integration.

$$\begin{aligned}
X_h &= \left\{ \underline{v}_h \in C^0(\Omega)^2, \ \underline{v}_h \in (P_{k+1} \oplus \text{span}\{B(x)\})^2 \quad \text{on } T_j \right\}, \\
Q_h &= \left\{ q_h \in L^2(\Omega)^2, \ q_h \in P_k \quad \text{on } T_j \right\} .
\end{aligned} \qquad k = 0, 1 \quad (50)$$

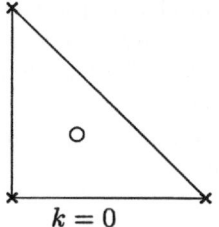 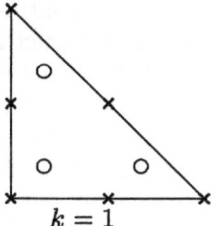

$$k = 0 \qquad\qquad\qquad k = 1$$

Fig. 4. Nodes with discontinuous pressure

$Q_1 - P_0$ Element. In the Stokes equation $\Delta \underline{u}$ and $\underline{\nabla} p$ are in balance. Therefore, one could expect an interpolation for p one degree less than for $\underline{u}$ will work well. Considering the element $Q_1 - P_0$, we see that this assumption does not give automatically stable approximations. Let T_k be a rectangle. We approximate the velocity $\underline{u}_h$ using functions in $Q_1 = \{1, x_1, x_2, x_1 x_2\}$, $\underline{u}_h \in Q_1$ and the pressure p_h using piecewise constant functions, $p \in P_0$, on T_k:

$$
\begin{aligned}
X_h &= \left\{ \underline{v}_h \in C^0(\Omega)^2,\ \underline{v}_h \in (Q_1)^2 \quad \text{on } T_k \right\}, \\
Q_h &= \left\{ q_h \in L^2(\Omega),\ q_h \in P_0 \quad \text{on } T_k \right\}.
\end{aligned}
\tag{51}
$$

The inf-sup condition does not hold and the pressure approximations are unstable (checkerboard instability). The element can be stabilized by adding suitable functions for $\underline{u}_h$ defined on macroelements.

The elements considered above are conforming elements because the trial functions also belong to the spaces of the continuous functions,

$$\underline{v}_h \in X_0,\ q_h \in L^2 \quad \text{or} \quad X_{0h} \subset X_0,\ Q_h \subset L^2. \tag{52}$$

Due to the definition (23) of X_0 the continuity of the velocity $\underline{v}_h$ is required.
It can be shown that there are approximations that do not satisfy this condition but work very well. These methods are called nonconforming. The triangle element with piecewise linear velocities $\underline{v}_h$ discontinuous on the sides belongs to this class.

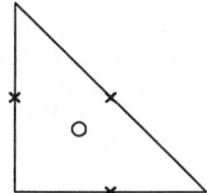

Fig. 5. Nodes of the Crouzeix–Raviart element

Crouzeix–Raviart Element (P_1 discontinuous – P_0 discontinuous). (See Fig. 5)

$$
\begin{aligned}
X_h &= \left\{ \begin{array}{l} \underline{v}_h \in L^2(\Omega)^2,\ \underline{v}_h \in (P_1)^2 \quad \text{on } T_k,\ \underline{v}_h \text{ continuous in the} \\ \hfill \text{middle of the sides} \end{array} \right\}, \\
Q_h &= \left\{ q_h \in L^2(\Omega)^2,\ q_h \in P_0 \quad \text{on } T_k \right\}.
\end{aligned}
\tag{53}
$$

The functions $\underline{v}_h \in X_h$ are continuous only at the nodes located at the middle of the sides. It is easy to construct a basis with weak zero divergence, $b(\underline{v}_h, q_h) = 0$, $\forall q_h \in Q_h$. Hence it is not necessary to use the saddle-point problem (26) but we can directly apply the variational formulation (21).

The precision of the element is not high (order of approximation $O(h)$). A similar construction is possible on tetrahedrons in 3D.

4   Final Remarks

In the FEM, important questions are related to the generation of the finite-element mesh, the assembling of the system matrix including a numerical integration algorithm, and the solution procedure of the sparse but large system of the discrete equations; see [1, 6].

The saddle-point formulation (or mixed method) gives stable solutions only if the combination of the velocity–pressure approximation fulfills the inf-sup condition. In 2D some methods are known to prove this inequality, but in 3D it represents a great challenge.

Therefore a modified weak formulation is sought in order to use elements that do not satisfy the inf-sup condition. This leads to the least-squares method and Petrov–Galerkin method. For this current research refer to [3, 10].

References

[1] D. Braess, *Finite Elemente*, (Springer, Berlin 1992)
[2] J. Cahouet and J.P. Chabart, Int. J. Numer. Methods in Fluids. **8**, 869 (1988)
[3] L.P. Franca and R. Stenberg, SIAM J. Numer. Anal. **26**, 1680 (1991)
[4] R. Glowinski, *Numerical Methods for Nonlinear Variational Problems*, (Springer, Berlin 1985)
[5] V. Girault and P.-A. Raviart, *Finite Element Methods for Navier–Stokes Equations*, (Springer, Berlin 1986)
[6] H. Goering, H.-G. Roos, and L. Tobiska, *Finite-Element-Methode*, (Akademie-Verlag, Berlin 1993)
[7] M. Gunzburger, in *Finite Elements, Theory and Application*, eds. D. Dwoyer, M. Hussaini, R. Voigts, Springer ICASE/NASA LaRC series, (Springer, Berlin 1985), p. 124
[8] O. Pironneau, *Finite Element Methods for Fluids*, (Wiley, Chichester 1989)
[9] A. Sommerfeld, *Vorlesungen über theoretische Physik, Band II Mechanik der deformierbaren Medien*, (Akademische Verlagsgesellschaft Geest & Portig, Leipzig 1957)
[10] T.E. Tezduyar, S. Mittal, S.E. Ray, and R. Shih, Comput. Methods Appl. Mech. Engrg. **95**, 221 (1992)

Principles of Parallel Computers and Some Impacts on Their Programming Models

Wolfgang Rehm and Thomas Radke

Fakultät für Informatik, Technische Universität, D–09107 Chemnitz, Germany
e-mail: `rehm@informatik.tu-chemnitz.de` and
`tomsoft@informatik.tu-chemnitz.de`

Abstract. In this paper we briefly outline some principles of parallel architectures and discuss several impacts on their programming models. First, parallel computers are generally classified. A description of the most important classes – multiprocessors and massively parallel systems – follows, with some details about chosen machines. The corresponding programming models for shared-memory and distributed-memory architectures are introduced. The special relationship between machine architecture and efficient parallel programming is emphasized here. The paper concludes with some hints for the software developer about where to use which parallel programming model.

1 Introduction

The last ten years have seen the employment of parallel computers for the solution of complex scientific, mathematical, and technical problems, with their developing into a key technology. The paradigm shift towards parallelism has led to changes on all levels, from machine hardware to application programs. A broad spectrum of parallel architectures has been developed.

In general, a parallel algorithm can be efficiently implemented only if it is designed for the specific needs of the architecture. Thus the knowledge of primary computer design principles is of course relevant for software developers as well as numerical analysts in the field of computational physics. This fact is often underestimated by software developers.

For this reason, in the following we present a brief introduction to basic architectures of parallel computers.

2 Overview on Architecture Principles

Before the development of "vector computers" in the 1970s, so-called "mainframes" were also used for scientific computing although they have typically been the workhorses of data-processing departments.

The first supercomputer architectures involved the use of one – or, at most, a few – of the fastest processors that could be obtained by increasing the packing density, minimizing switching time, heavily pipelining the system, and employing vector processing techniques, which apply a small set of program instructions repeatedly to multiple data elements.

Table 1. (Vector) supercomputer performance [5]

Computer type	MFLOPS/Processor	Clock Rate [μs]
Cray Y–MP	300	6.0
Cray C90	952	4.2
NEC SX 3/14R	6400	2.5
Fujitsu VP 2600/10	5000	3.2
Hitachi S–3800/180	8000	2.0

Vector processing has proven to be highly effective for certain numerically intensive applications, but much less so for more commercial uses such as online transaction processing or databases.

In fact the sheer computational speed (Tab. 1)was achieved at substantial costs, namely by sophisticated highly specialized architectural hardware design and the renunciation of such techniques as virtual memory (to facilitate the programmability). In particular, the last fact has led to the development of a considerable body of specialized program code.

Another way that respects conventional programmability has led to the design of so-called multiprocessor systems (MPS). Only small changes to earlier uniprocessor systems had to be made by adding a number of processor elements (PEs) of the same type to multiply the performance of a single processor machine. Although there were effects on the programming model, at least the essential fact of a unified global memory could be maintained.

Further developments discarded the demands on a unified global memory because of the impossibility of its physical realization where hundreds and thousands of processors are used. The total memory is distributed over the total number of processors; each one having a fraction in the form of a local memory.

In the 1980s the first massively parallel processors (MPP) began to appear, with the single goal of achieving far greater computational power than with vector computers at greatly improved price/performance ratios by using low-cost standard processors.

A still essentially unsolved problem for the use of such systems is the development of appropriate programming models. No standard programming model that satisfies the needs of all applications has yet been found although a variety of competing models have been developed, including message passing, data-parallel programming, and the virtual shared-memory concept. However, the efficient use of parallel computers with distributed memory requires the exploitation of data locality, which can indeed be found in most important numerical applications.

Because it is easier to bring activities onto established architectures than to do so on parallel machines, high-performance workstations are often still prefered for program implementations. If the performance needs increase then a cluster of interconnected workstations (WSC) can also be considered as a parallel machine.

But typically the interconnection network of such clusters is characterized by relatively small bandwidths (some MBytes/s for 1 KByte messages) and high latency[1] (in the range of milliseconds for 1 KByte messages). Thus suitable applications are of a competitive rather than cooperative type (with naturally high communication requirements).

Nowadays we realize that all the mentioned types – MPS, MPP, and WSC – as well as advanced types of vector computers (multivector computers) are integrated in a network environment and can be combined to form a heterogeneous supercomputer. The recent development of a message-passing interface (MPI) is a landmark achievement in making such systems programmable.

Summarizing we note that each architecture has its strong and weak points and it will take continuous improvement to overcome its drawbacks. Parallel computer development is currently heavily influenced by the technological capabilities. As a consequence we notice a trend to massive parallel arrangements of symmetric multiprocessor systems, which we call MPP/SMP.

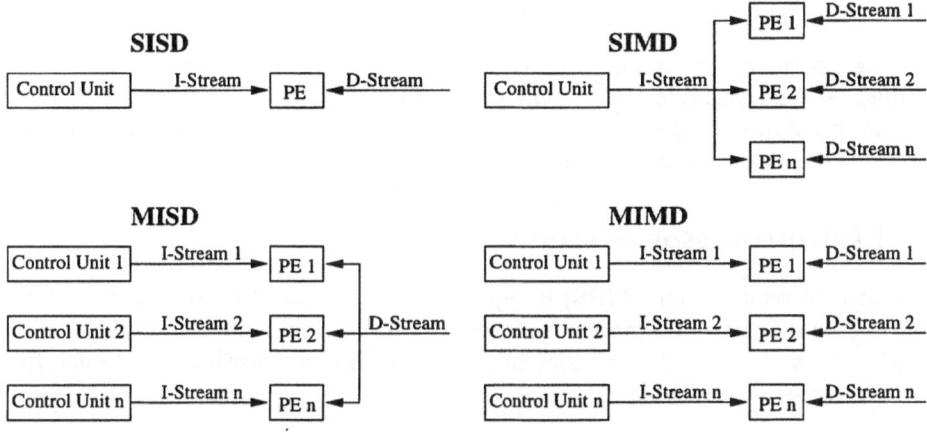

Fig. 1. Flynn's classification of computer architectures

3   General Classification

Michael Flynn [6] introduced a classification of various computer architectures based on notions of instructions (I streams) and data streams (D streams) (Fig. 1). Conventional sequential machines with one processing element (PE) are called

---

[1] Latency is the total amount of time it takes for the sender to pack the message and send it to the receiver, and for the receiver to receive the message and copy (unpack) it into its own buffer.

single instruction single data (SISD) computers. Multiple instruction multiple data (MIMD) machines cover the most popular models of parallel computers.

There are two major classes of parallel computers, namely shared-memory multiprocessors and message-passing multicomputers (Fig. 2).

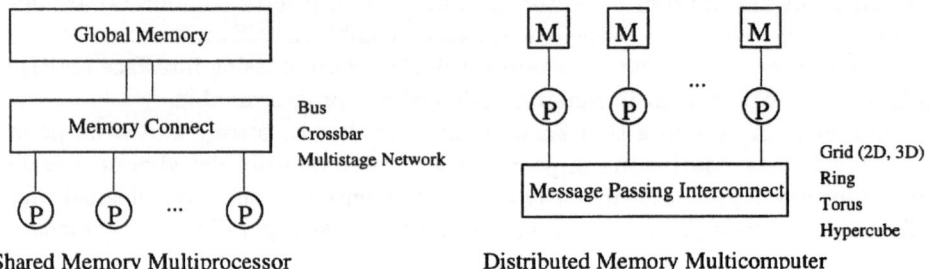

Fig. 2. Parallel computer classes

The processors in a multiprocessor system communicate with each other through shared variables in a common memory, whereas each computer node in a multicomputer system has a local memory, not shared with other nodes. Interprocessor communication is done here through message passing.

4   Multiprocessor Systems

A multiprocessor system (MPS) is typically a RISC-based shared-memory multiprocessor machine designed to provide a moderate amount of parallelism (up to 30 processors) to achieve more power than high-end workstations offer (for RISC processors see Table 2).

Table 2. Performance of some RISC CPUs

CPU Type	Clock Rate [MHz]	Perf. [MIPS]	CPU Type	Clock Rate [MHz]	Perf. [MIPS]
Alpha 21164	300	1200	MPC601	80	240
MPC604	100	400	MPC620	133	532
SuperSparc	60	180	UltraSparc	167	668
PA7200	140	280	R4400SC	150	150
R10000	200	800	MC68060	50	100
Pentium 100	100	200	Pentium Pro	133	399

Table 3. Multiprocessor systems

Company	Model	Scalability [processors]	I/O Type	Bus Bandwidth
Sequent	Symmetry 5000	1...30	symm.	240 MByte/s
Silicon Graphics	PowerChallenge	1...30	symm.	1.2 GByte/s
Sun	SPARCstat. 20 HS14	1...4	symm.	–
Compaq	ProLiant 4000	1...4	symm.	267 MByte/s

Most computer manufacturers have multiprocessor (MP) extensions to their uniprocessor product line (Table 3). All additional processors are attached to the same global bus. Dedicated bus lines are reserved for coordinating the arbitration process between several requestors. The scalability of such systems is restricted to some dozens of processors due to the limited bandwidth of the common bus, which must be shared by all processors. The processors have equal access times to all memory nodes, which is why it is called a uniform memory-access (UMA) multiprocessor model.

On the contrary, in nonuniform memory-access (NUMA) models the access time varies with the location of the memory word. This is because the memory is actually distributed but there are hardware means that the collection of all local memories forms a global address space accessible by all processors. A processor's local memory can be accessed faster than a remote one. Such a logically shared memory based on physically distributed memory is called a virtual shared memory (VSM), especially if there is essential hardware support to realize this (Fig. 3). One special version of a VSM architecture is a cache-only memory architecture (COMA) such as in the KSR-1 machine (Fig. 4). Caches copy data from other caches if necessary. There is a continuous process of data migration. A cache attracts the needed data, and in the ideal case the user is completely freed from predefining the data layout. The drawbacks of such wonderful archi-

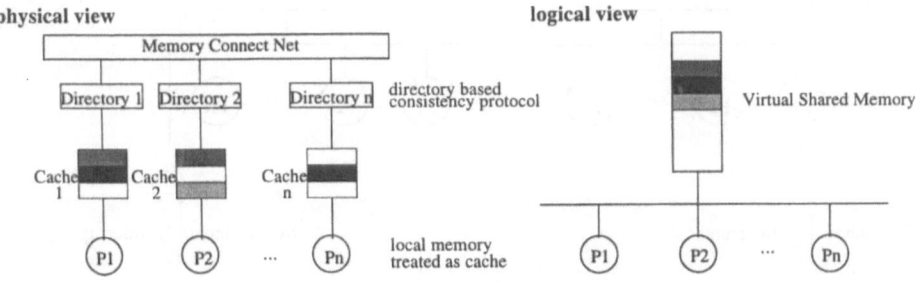

Fig. 3. Virtual shared memory

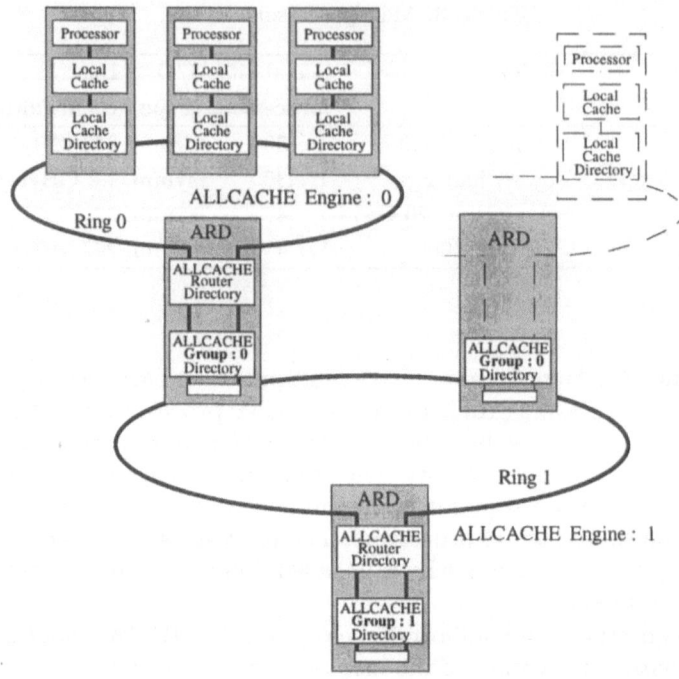

Fig. 4. KSR ALLCACHE architecture [2]

tectures lie in the synchronization costs for maintaining the cache coherency as well as the global synchronization (via semaphores). For further modifications of COMA models see [8].

Another distinction can be made between asymmetric and symmetric multiprocessor systems (Fig. 5). When all processors have equal access to all peripheral

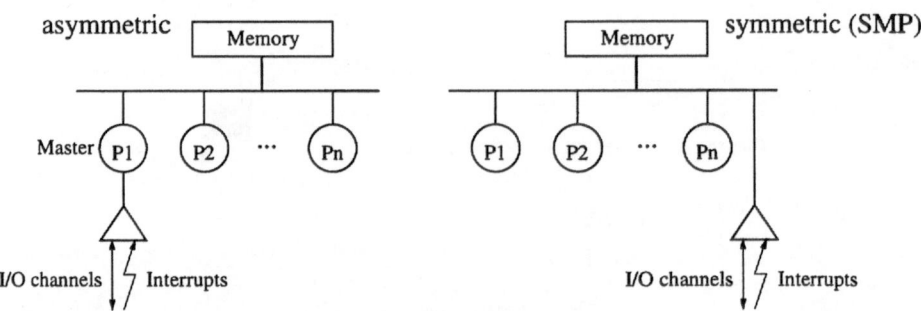

Fig. 5. I/O types of multiprocessor systems

devices the system is called a symmetric multiprocessor (SMP). All processors are equally capable of running the executive programs, such as the operating system kernel and I/O service routines. In asymmetric systems only a master processor can execute the OS and handle I/O. Thus I/O becomes a bottleneck. Today we often find such systems in the form of two-processor stations whereas symmetric solutions signify four-processor board-based workstations or servers (Table 3).

To overcome the drawbacks of the limited speed of a unified common global bus, connection schemes with crossbar technology have recently been developed [3]. The advantage is that more than one connection can be actived (dark points in Fig. 6) at the same time. The achievable transfer rates can be about 600 MByte/s per CPU. The global bus is still in use but only as a broadcast medium for the snooping-bus cache-coherence mechanism [8].

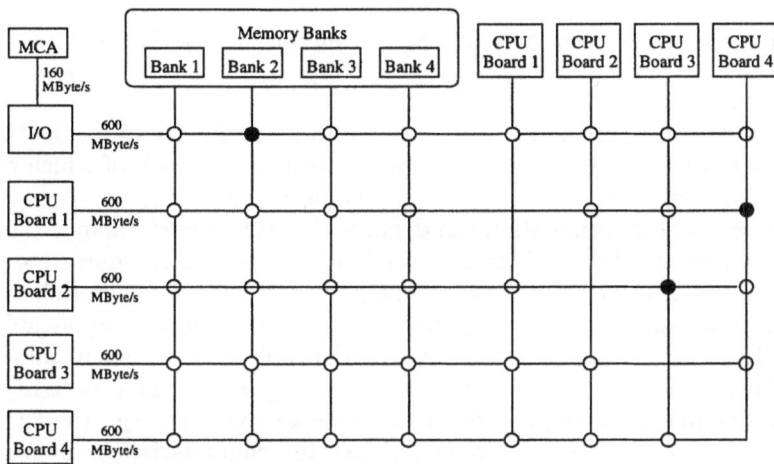

Fig. 6. Crossbar switch

5   Massively Parallel Processor Systems

Massively parallel processor systems (MPP) usually consist of from hundreds to several thousands of identical processors, each of which has its own memory (distributed memory). The processors communicate with each other by message passing. There is no common global memory, although there are some approaches supporting a virtual shared memory by combinations of hardware and software. In this sense the KSR virtual-shared memory computer can be classified as an MPP system.

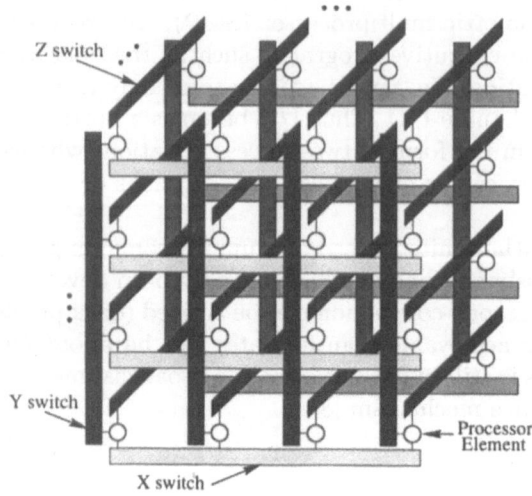

- Message transmission network for linking PEs
- A crossbar switch consists of 3 crossbars, one for each axis, to create 2D and 3D structures.
- At each level, a crossbar switch is capable of switching up to 8×8 connections.
- Data transfer rate: 300 MByte/s in each direction of the bidirectional ports

Fig. 7. Three-dimensional crossbar network in the Hitachi SR2201 [1]

Distributed-memory multicomputers are most useful for problems that can be broken down into many relatively independent parts, each of which requires extensive computation. The interactions should be small because the overhead of interprocessor communication can degrade the system performance. The main limiting factors are the bandwidth and latency. Modern communication system techniques use special latency-reduction protocols such as wormhole routing. Moreover, different latency-hiding methods in software may be applicable.

A fully connected network (clique) is applicable only for small numbers of nodes. To provide high-speed connections among individual processing nodes most parallel machines employ 2D or 3D crossbar switches, e.g., the Cray T3D and Hitachi SR2201 (Fig. 7). Table 4 shows the characteristics for prominent networks.

6   Multiple Shared-Memory Multiprocessors

One approach – a technology-driven one – for building a massively parallel system involves multiple shared-memory multiprocessors connected by a very high-bandwidth interconnect, such as HiPPI, in an optimized topology.

One such interconnection of high-performance shared-memory multiprocessors (of MPP/SMP type), the PowerChallenge array from SGI Corp., has been demonstrated to solve so-called "grand challenge" problems.

A node in a message-passing interconnect is represented by a full SMP. A great advantage of such arrangements is that the computation-to-communication ratio (a measure for the proportion of the maximum computational power and communication peak performance) can be very high. That is, the amount of

Table 4. Properties of interconnection networks

Type	Degree	Connections	Diameter	Bisectional Width	Symm.
Clique	$N-1$	$N(N-1)/2$	1	$(N/2)^2$	yes
Linear chain	2	$N-1$	$N-1$	1	no
Ring	2	N	$[N/2]$	2	yes
Binary tree	3	$N-1$	$2((\log_2 N)-1)$	1	no
2D grid	4	$2N-2\sqrt{N}$	$2(\sqrt{N}-1)$	$2\sqrt{N}$	no
2D torus	4	$2N$	$2[\sqrt{N}/2]$	$2\sqrt{N}$	yes
Hypercube	$\log_2 N$	$N\log_2 N$	$\log_2 N$	$N/2$	yes

message passing is low compared with the amount of work to be done in each SMP node for each message sent.

7   Multithreading Programming Model

With the evolution of MPS originating from conventional uniprocessor machines, the programming of such systems was historically formed by features of UNIX. This classical operating system allowed the quasi-concurrent execution of several tasks (multitasking) and provided some mechanisms for inter-process communication (e.g., pipes, sockets, shared-memory segments). These kernel services were quite expensive in their implementation (they are based on underlying standard network protocols) and caused high overheads. So they seemed to be unsuitable for efficient parallel programming.

For this reason the traditional task concept of UNIX was extended in a manner such that a process can have more than one single execution flow and may be divided into several threads of control that are independent of each other and thus can be executed in parallel. In this programming model a thread can be thought of as a light-weight process with much less state information than a normal UNIX task – it just owns a stack, a register set, and a program counter. All threads see the same address space. Communication between threads is performed through shared-memory variables. Access to these variables is managed by synchronization primitives (e.g., mutexes, semaphores, monitors, and so on).

In general, the application programmer does not have to worry at all about the mapping of his threads onto the processor set of the MPS. This functionality can be fully implemented in the operating system kernel (pure kernel-level threads) or is part of a thread library with kernel support (mixed user/kernel level threads).

The exclusive use of kernel-level threads is reasonable if the number of threads does not exceed the number of processors in the system. Each thread is fixed

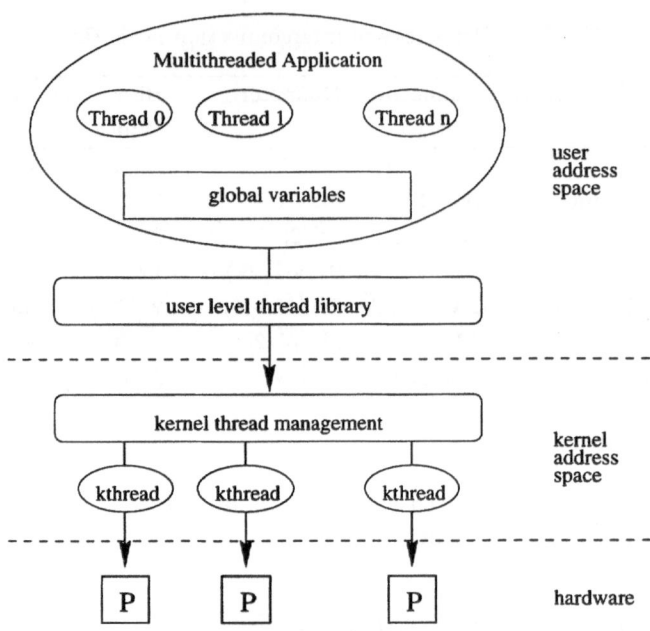

Fig. 8. Multithreading with user- and kernel-level threads

bound to its own processor and can run fully parallel to others. Synchronization can be implemented as busy waiting (the associated processor is not released but spins on a condition become true).

If there are more threads than available processors (and this is the most frequent case) busy waiting between threads is no longer applicable because of possible deadlocks. In this case synchronization can cause a thread switch on a processor. The switching of kernel-level threads can only be done in the kernel, i.e., special system calls are needed. The resulting overhead (kernel-thread context switch plus entering/leaving the kernel) may drastically decrease the efficiency of a pure kernel-level thread management. That is why a mixed thread management is more favorable in this case: the programmer uses user-level threads, which are managed by a thread library; these user-level threads are internally mapped onto some kernel-level threads with their number corresponding to the number of available processors (see Fig. 8). Thread context switches can now completely be done in user mode, and time-expensive system calls are unnecessary.

It is true that the problem of optimal load balancing in MPS is not as difficult as in MPP. Because of the global shared memory every thread can principially be scheduled on any processor without explicit migration. But in NUMA architectures, thread locality must be taken into consideration for achieving efficient multithreading. For instance, if there are still some thread data in a processor's cache, this thread should be scheduled with precedence on that processor again. This technique, called memory-conscious scheduling, is used in particular in systems with multi-level memory hierarchies [4].

The efficiency of I/O intensive multithreaded applications strongly depends on the I/O architecture. In asymmetric systems every I/O operation forces a thread switch onto the master processor (which is capable of serving the request, see Fig. 5). So the master may become a bottleneck. Only SMP systems guarantee a scalable I/O performance because each I/O request can be served on the processor where the thread resides. This circumstance is less decisive for multithreaded programs with a high ratio of computation to communication.

At present, there exist a number of modern commercial and noncommercial standard operating systems that support multithreading and symmetric multiprocessing: Solaris 2.x from SUN, Mach, Linux-SMP as public-domain software, Windows-NT from Microsoft. Research is aimed at the development of a unique multithreading programming interface (currently proposed as "POSIX 1003.4a Threads Extension Draft"). With this, the application programmer should be able to easily port his or her programs on any MPS architecture.

8   Message-Passing Programming Model

Message passing is the natural programming model for distributed-memory architectures. It is based on Hoare's CSP concept (communicating sequential processes [7]), where an application consists of several sequential tasks that communicate with each other by exchanging data over communication channels. These tasks are distributed among the nodes of an MPP and thus are executed in parallel. The communication channels are mapped onto the communication network. The communication hardware in modern MPP systems is capable of operating independently of its assigned compute node so that communication and computation can be done concurrently.

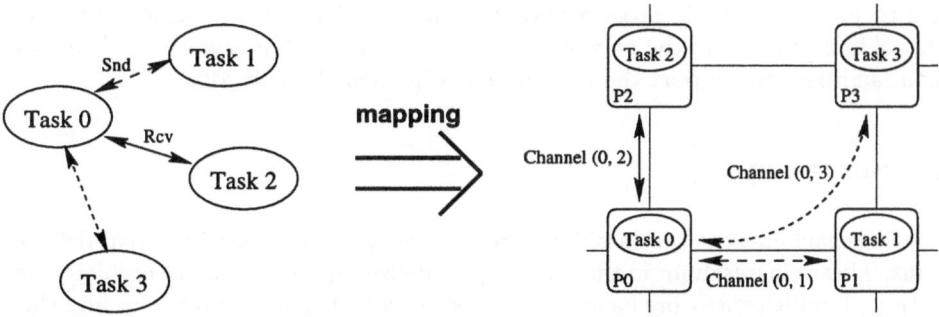

Fig. 9. Mapping of the process graph onto the MPP

The efficiency of the parallel application is essentially determined by the quality of mapping the process graph with its communication edges onto the underlying distributed memory architecture (see Fig. 9). In the ideal case each task gets its own processor, and every communication channel corresponds with a direct physical link between both communication nodes.

This can be realized in most cases from the view of available processors in massively parallel systems. However, scalability requires a relatively simple communication network (2D, 3D grid, ring, torus) so at this point compromises are unavoidable. For instance, a logical communication channel is routed when it passes one or more grid points. This transfer of data takes time, especially if there is no hardware support and the routing must be done by software emulation.

On the one hand, communication paths with different delays arise by nonoptimal mapping of communication channels onto the network. On the other hand, several logical channels are multiplexed on one physical link. From the application programmer's point of view, the usable communication bandwidth is decreased.

Since the beginning of the development of MPP, algorithms for various application classes with static problem sizes emerged that find the optimal mapping scheme for a given machine topology and thus allow the best exploitation of hardware performance. The identical transformation on other topologies is often combined with a loss of efficiency. That is why porting a parallel application requires at least some basic knowledge from the programmer about the target architecture.

In recent years research activities were extended to the field of adaptive parallel algorithms development, i.e., of those application classes for which the process graph is adapted to the problem size dynamically. The decision of how to inbed the actual process graph into the processor graph cannot be made statically at the compile time but only at the runtime. Newly created tasks should be placed on processors with less workload to ensure a load balance. In addition, the communication paths to other tasks should be kept as short as possible and not be overloaded by existing channels.

Those highly complex decisions can no longer be made by the application programmer alone. At this point the operating and runtime system of the MPP has to provide suitable process management functionalities (e.g., for obtaining status information on the current system workload, task placement, and migration facilities) to support the programmers in their difficult job.

9   Summary

Parallel machines can be classified as multiprocessors and massively parallel systems. These classes differ in the scale of parallelism and the memory architecture. Although multiprocessors have equal access to a global shared memory and thus are limited in processor number, they are scalable to hundreds up to thousands of processors with each node having its own local memory.

The hardware architecture determines the way in which the parallel computer is programmed. For multiprocessors the multithreading programming model is prefered. Parallelity in application programs is expressed here as cooperation of several threads of control that share some global data and synchronize with each other. Message passing is used in distributed-memory systems. Processes are executed in parallel on different processor nodes and communicate over channels.

The ratio of communication to computation in a parallel program is decisive for its efficiency. Massively parallel computers provide a high computational power but typically have a lower communication bandwidth so that I/O intensive applications probably achieve poor performance. For this class of applications multiprocessors with their extremely low communication costs would be better suited. Application programmers have to keep these facts in mind when implementing their algorithms on a target architecture.

References

[1] Hitachi Product Sheet SR2201, Hitachi Corp.

[2] KSR *Parallel Programming–Manual*, (Kendall Square Research Corporation, Waltham Massachusetts 1993)

[3] *Cross-Bar und Snooping-Bus*, iX no. 7, (1995)

[4] F. Bellosa, *Memory Concious Scheduling and Processor Allocation on NUMA Architectures*, TR–I4–5–95, (Computer Science Dep., Univ. Erlangen 1995)

[5] J. Dongarra, *Performance of Various Computers using Standard Linear Equation Software*, Tech. Report January 7, 1995, (Computer Science Department, Univ. of Tennessee, Knoxville 1995)

[6] M. Flynn, IEEE Trans. Computers **21**, 948 (1972)

[7] C.A.R. Hoare, *Communicating Sequential Processes*, (Prentice Hall, New York 1989)

[8] K. Hwang, *Advanced Computer Architecture*, (McGraw-Hill, New York 1993)

Parallel Programming Styles: A Brief Overview

Andreas Munke, Jörg Werner, and Wolfgang Rehm

Fakultät für Informatik, Technische Universität, D-09107 Chemnitz, Germany
e-mail: ambu@informatik.tu-chemnitz.de

Abstract. There are several ways to write a parallel program. The parallel programming style depends on the architecture of the parallel machine, including the programming environment, the operating system, and the computer hardware. In this chapter we briefly introduce the three frequently used programming models for multiprocessor systems, shown on three examples of programming environments. The program examples should develop a feeling for the kinds of parallel programming.

1 Introduction

The present-day status of the field of parallel programming is suiteably characterized by A.Reuter, leader of the Institute for Parallel and Distributed High Performance Computing at the University of Stuttgart (Germany):

> *"Parallel Programming is difficult – actually it is not difficult, but when it should be effective it is really difficult."*

Parallel computers open a new dimension and a new world of applications. These new facilities are expensive, both in scientific-technical and commercial terms, because the hardware costs are high and the new algorithms and software tools are still in development.

The education in parallel computing is only now beginning. Today we offer at the Technical University of Chemnitz a specialized education profile "Parallel and Distributed Systems" with practical work on different parallel computers such as Parsytec Multicluster and KSR-1.

2 Programming Models

2.1 Definition

Before translating a problem into a parallel program, the programming model of the computer to be used must be known. This model is dependent on the architecture of the parallel machine. It influences the choice of the algorithm to be implemented. For an effective solution a well-known algorithm may be parallelized. In many cases the development of a new algorithm is needed. The programming model is characterized by Burkhardt [1]:

A *programming model* or *programming style* is the set of the features which describe a task for the solution on a computer. The features are:

- active elements (commands, processes, threads, objects, ...),
- types of communication and synchronization (message passing, global variables, tuple space, logical shared variables, ...), buffers, object inheritance,
- guarded commands (entry conditions for certain program parts, depend on various instruction streams),
- matching of active elements to processors (mapping, load balance),
- behavior by errors (exceptions, redundance).

2.2 Classification

The classification of parallel programming models may follow many points of view. Various classification patterns overlap. This computer-model oriented classification is derived from the well-known classification by Flynn:

- SIMD (Single Instruction stream, Multiple Data stream) for pipeline- and array-computers,
- MIMD-SM (Multiple instruction stream, Multiple data stream with Shared Memory),
- MIMD-DM (Multiple Instruction stream, Multiple Data stream with Distributed Memory).

SIMD model means, that the same operation works on various data. In SIMD-Computers specialized compilers (e.g., for FORTRAN) are normally available. The user may specify array operations, which will be executed in parallel. The members of this class are vector-supercomputers, such as VPP by Fujitsu and various Cray computers.

The *MIMD model with shared memory* has an unique address space over all processor nodes. Each active element may take advantage of all data. No explicit programming of communication between the active elements is needed. The synchronization of parallel processes or threads is the programmers only task. However, the shared address space complicates the maintenance of data consistency. This problem will be solved with compiler systems or special hardware components.

The shared memory is either a global shared memory or a distributed shared memory. Disributed shared memory means a realization through hardware as Virtual Shared Memory (VSM) or through software as Logical Shared Memory (LSM). Important examples are the VSM computers of KSR (Kendall Square Research) and the T3D (Cray).

Although the programmer works with an unique address space, a knowledge of the memory structure is necessary to produce efficient code.

MIMD models with distributed memory use only local data. The Programming of such systems is message passing, based on the CSP model (communicating sequential processes) by Hoare [4]. This model determines the kind of communication and syncronization, but not a special hardware base. Support of applications on heterogenous hardware bases for example could be realized with an overlayed software system such as PVM (parallel virtual machine).

As extension to the CSP model, the SPMD model (single program, multiple data stream), works in a similar way to the SIMD model. An identical program code is loaded on all processor nodes and works with various data. In opposite to SIMD in the SPMD model several program paths get through. A typical example is the GC-Power-Plus (Parsytec).

A special kind of parallel computer is the *homogenous or heterogenous workstation cluster*. The programming of such systems is similar to programming in the MIMD-DM model. A commonly accepted approach is the Message Passing Interface MPI [7], based on well-known communication libraries such as PVM [3] or PARMACS [2].

3   Programming a Shared Memory Computer

3.1   The KSR Programming Model

The KSR is controlled by its own operating system, the KSR-OS, based on the MACH-Kernel and similar to UNIX system V.

The KSR shared memory is based on a physically distributed memory, composed of a collection of local caches. A special hardware engine, the ALL-CACHE™, interconnects the local caches, provides routing and directory services, supports unified address space, and guarantees the maintenance of data consistency.

The carriers of parallelism are pthreads (POSIX threads), started by an active instance (main program or other pthread). The operating system spreads pthreads on processors automatically. The distribution of pthreads by the user is allowed too. Many pthreads on one processor run in time-sharing mode.

Data exchange among processor nodes is realized with shared variables. Each active element knows the same shared variable under the same name. When an active element accesses to a shared data which is not in the local memory, the ALLCACHE searches other local memories with a hierarchical strategy. That is why the users task is only to synchronize some actions. However, working with local data is a necessary condition for producing efficient code because each data exchange decreases the system performance.

A key issue in writing parallal programs in the MIMD-SM style is controlled data sharing in parallel domains. Outside a parallel domain, only the program master pthread executes, and it always uses it's own, private copy of data. Within a parallel domain a team of pthreads that execute the domain use either private or shared copies of data:

1. Private data: Each pthread accesses its own copy of data. A variable reference to private data is interpreted as a reference to a unique memory location, different for each team member.

2. Shared data: Pthreads share access to one copy of the data entity. All variable references are interpreted as references to the same memory location.

All shared data is defined at or before link time and allocated by the operating system once, at process start-up.

3.2 Levels of Parallelism

The KSR allows parallelism on various levels:

- Several users may log in simultanously.
- Each user may open several sessions.
- Each session allows many applications.
- An application can contain many processes (with own address space).
- A process may start various pthreads with unique address spaces.

3.3 Program Implementation

The KSR provides the languages Fortran and C with extensions to describe parallel instances. Parallelizing is supported on various levels (Table 1).

Table 1. Parallelizing tools on the KSR

	Fortran77	C
fully automatic parallelizing	KAP	-
semi automatic parallelizing	PRESTO	SPC
manual parallelizing	library functions	library functions

Fully automatic parallelizing makes it possible for the user to gain the advantages of parallel execution of a program, without needing to know how to parallelize the program. KAP (KSR automatic parallelizer) is a precompiler, which identifies parallel parts in loops and produces parallel source code by tiling the iteration space. The KAP inserts appropriate tiling directives, complete with all tiling parameters.

PRESTO is a programming methodology and run-time system for highlevel parallel programming on the KSR. Most PRESTO constructs apply to Fortran programs. C users are supported by SPC, a simple C interface to PRESTO.

PRESTO provides four major constructs:

- Parallel regions – execute multiple instances of a code segment in parallel.
- Parallel sections – execute various code segments in parallel.
- Tile families – execute loops in parallel.
- Affinity regions – coordinate tiling decisions made for a group of tile families referencing the same data.

For manual parallelizing the explicit use of pthread functions and synchronization utilities is needed. The pthread libraries are part of the operating system and provide statements for creating, terminating, and managing pthreads and teams of pthreads.

3.4   Examples

The first example shows the multiplication of two matrices. Because the calculation of target matrix elements is data independent, no synchronization is needed. The source program will be translated by the KAP. KAP includes control statetements as Fortran comments (C*ksr) into the source code. As strategy the tiling is used. The iteration space is shared and parts of the work are assigned to pthreads. PRESTO finds an efficient partition of work, dependent on number and load of processors. Because the shared memory is based on a physically distributed memory, the tile size is specified as multiples of 128-byte subpages. Figure 1 shows a tiling of 64 rows and 16 columns in dimensions i and j. Each task calculates $64 \times 16 = 1024$ iterations. The variables i and j are shared. Because k is not tiled, KAP declares it as private. Therefore each pthread works with a private copy of k (Fig. 1).

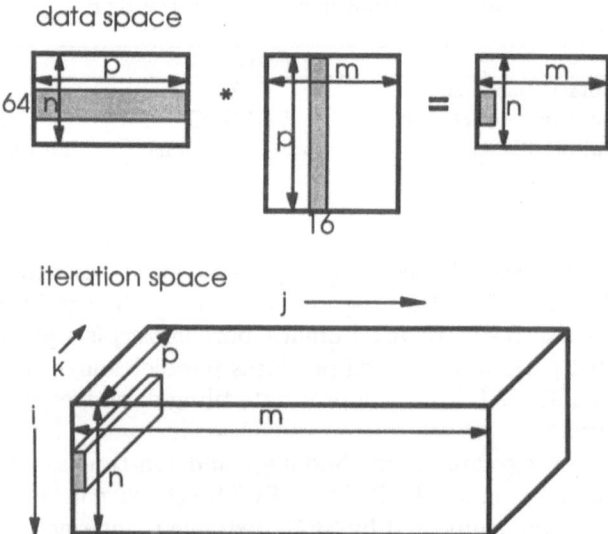

Fig. 1. Two-dimensional tiled iteration space

Program matmul

```
C shared variables
        real a(:,:), b(:,:), c(:,:), t1, t2
        integer n, np
```

```
        print*,'Enter linear dimension of matrices:'
        read(*,*) n
        print *,'Enter number of pthreads:'
        Read(*,*) np
C allocate is a KSR specific statement
        allocate( a(n,n), b(n,n), c(n,n) )
C create a team of pthreads with an user specified number of
C team members np, id is specified by PRESTO
        call ipr_create_team (np, id)
C simple initalization of matrices
        do 10 i=1,n
        do 10 j=1,n
            a(i,j)=1.0
            b(i,j)=1.0
10      continue
        t1 = all_seconds()
C the following directive is included by KAP
C*ksr* tile(i,j,private=k,teamid=id)
C k is private for the pthread
        do 20 i=1,n
        do 20 j=1,n
            c(i,j)=0.0
            do 20 k=1,n
                c(i,j) = c(i,j) + a(i,k) * b(k,j)
20      continue
C*ksr* endtile
        t2= all_seconds()
        print*,'Calculation Time ',t2-t1
        end
        '
```

Result:

```
Enter linear dimension of matrices: 256
Enter number of pthreads: 1
Calculation Time 8.8698763999855146
Enter linear dimension of matrices: 256
Enter number of pthreads: 8
Calculation Time 1.1960556000121869
```

The evaluation compares the run on one processor (a sequential program) with a run on eight processors. The result is a nearly linear speedup. That means the parallelization is effective (only a small overhead is incurred by creating pthreads). Furthermore, the problem is easy to parallelize because no communication or synchronization is needed.

The second example shows the use of PRESTO constructs. The program computes two vectors from a two-dimensional matrix (Fig. 2). The sparse matrix is represented by the array sm. The calculation of the sum of elements in rows (dvr) and columns (dvc) is computed by several code sequences (parallel sections). The rows and columns are computed by multiple instances of the same code (parallel regions). The program solves the problem with parallel regions (each starts a team of four pthreads) nested in two parallel sections. The variables i and j are private because both sections encounter i and j. To assign pthreads to indexes of rows or columns a logical pthread-number (mynum) is used.

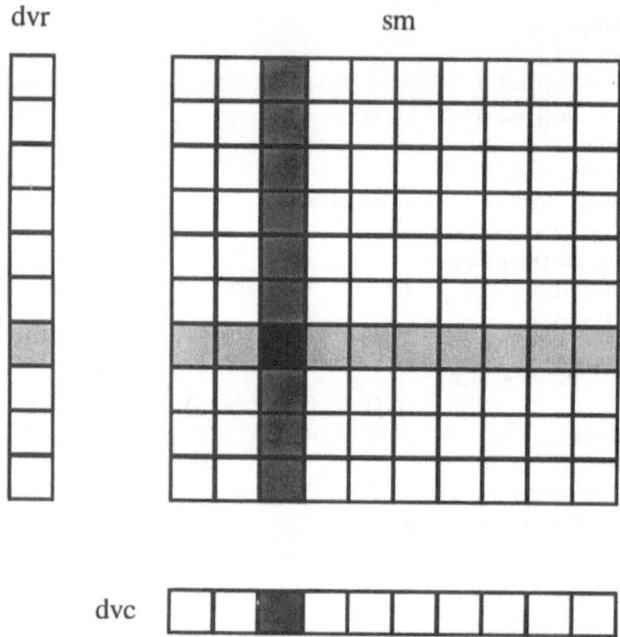

Fig. 2. Compute sums of rows and columns

Program ParSecReg

```
c shared data
        common sm(10,10), dvc(10), dvr(10)
        integer procid, numproc, mynum, tid
        do 10 i = 1,10
                dvr(i)=0
                dvc(i)=0
10      continue
        do 20 i = 1,10
        do 20 j = 1,10
                sm(i,j) = 10 * (i-1) + j
20      continue
        call ipsm_init(numproc)
```

```
                print*,'number of available processors:',numproc
c*ksr* parallel sections (private=(i,j,procid))
c*ksr* section
c*ksr* parallel region (numthreads = 4, private= (i,j, procid, mynum))
                procid = ipsm_mypset()
                tid = ipr_tid()
                mynum = ipr_mid()
                print*,'dvr is running on processor',procid,'tid=',tid,'team_member',mynum
                do 40 i=1,10
                do 40 j=1,10
                        if (mod(i,ipr_tsize()) .eq. mynum) then
                                dvr(i) = dvr(i) + sm(i,j)
                        end if
40      continue
c*ksr* end parallel region
c*ksr* section
c*ksr* parallel region (numthreads = 4, private= (i,j, procid, mynum))
                procid = ipsm_mypset()
                tid = ipr_tid()
                mynum = ipr_mid()
                print*,'dvc is running on processor',procid,'tid=',tid,'team_member',mynum
                do 50 i=1,10
                do 50 j=1,10
                        if (mod(i,ipr_tsize()) .eq. mynum) then
                                dvc(j) = dvc(j) + sm(i,j)
                        end if
50      continue
c*ksr* end parallel region
c*ksr* end parallel sections
                do 50 i=1,10
                        print*,'i:',i,'dvr:', dvr(i), 'dvc:',dvc(i)
50      continue
                end
```

Result:

```
number of available processors: 8
dvr is running on processor 7 tid= 2 team_member 0
dvr is running on processor 2 tid= 2 team_member 2
dvr is running on processor 1 tid= 2 team_member 1
dvr is running on processor 3 tid= 2 team_member 3
dvc is running on processor 0 tid= 3 team_member 0
dvc is running on processor 4 tid= 3 team_member 1
dvc is running on processor 6 tid= 3 team_member 3
dvc is running on processor 5 tid= 3 team_member 2
i: 1 dvr: 55.000000000000000 dvc: 460.00000000000000
i: 2 dvr: 155.00000000000000 dvc: 470.00000000000000
i: 3 dvr: 255.00000000000000 dvc: 480.00000000000000
i: 4 dvr: 355.00000000000000 dvc: 490.00000000000000
```

```
i: 5 dvr: 455.00000000000000 dvc: 500.00000000000000
i: 6 dvr: 555.00000000000000 dvc: 510.00000000000000
i: 7 dvr: 655.00000000000000 dvc: 520.00000000000000
i: 8 dvr: 755.00000000000000 dvc: 530.00000000000000
i: 9 dvr: 855.00000000000000 dvc: 540.00000000000000
i: 10 dvr: 955.00000000000000 dvc: 550.00000000000000
```

4   Programming a Distributed Memory Computer Using PARIX

4.1   What is PARIX

PARIX is a parallel operating environment consisting of software components for machine adminstration, application development, and execution by the user. Some components of PARIX reside on a front-end machine (host), while others run on the parallel target system itself (Fig.3). The name PARIX stands for parallel extensions to UNIX and already explains some basic concepts which are:

- PARIX is based on UNIX,
- the user writes the parallel application by using a high-level language (C, FORTRAN) complemented with parallel extensions.

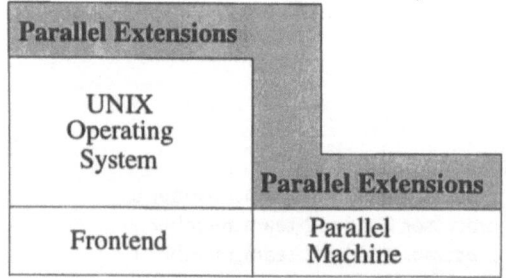

Fig. 3. PARIX software layer model

The parallel extensions are provided by message-passing libraries. PARIX has been designed for distributed memory multiprocessor machines with a range from two up to (theoretically) thousands of processors (Parsytec GC, Xplorer, Multicluster architecture) [10].

4.2   PARIX Hardware Environment

The front-end computer (host), usually a SUN compatible workstation, supplies the user with a development environment and gives parallel applications access to external facilities such as file server, networks, and graphics. The parallel machine consists of computing units and communication networks only. There is no direct support for the facilities mentioned above. Access to this is provided by integration of remote procedure calls (RPC's). All services not supported by the parallel machine itself are shifted to the host.

In order to permit many users to access the parallel machine at the same time, each user usually gets a part of the whole machine. In PARIX terminolgy this part or tile of the parallel machine is called a partition. It is one of the jobs of the system administrator to define numerous partitions with different sizes and locations during installation. The processors of a partition are arranged as a grid. Naturally it is possible to define a partition that contains all processors.

4.3   Communication and Process Model Under PARIX

The PARIX programming model is in most part CSP like. The concept of communicating sequential processes (CSP) is defined by the following basics: In a message-passing system, a process always remains within its own address space. Communication between two processes is realized as message transfer via a communication channel. The data transfer has to be explicitly initiated. This model requires two different mechanisms for memory access.

1. The entire local memory can be accessed using normal operations (individual or local variables)
2. "Remote" memory access requires interprocessor communication.

Under PARIX, a communication channel is a bidirectional, synchronizing, point-to-point communication line between two processes and is called a link. Before a link can be used it has to be built up explicitly by calling a connection management function on both related sides. During the link construction the processes specify the communication partner by a process(or) number and give the link a name. Because address space as well as name space are local, each process can name the link independently. Therefore the given names can differ on both sides of a connection. In a latter data transfer the processes do not specify the communication partner by the destination process but by the named link (Fig. 4).

The server process is responsible for processing remote procedure calls (RPC), e.g. example a simple printf() in C. The RPC channel is the only one that is predefined by the PARIX software.

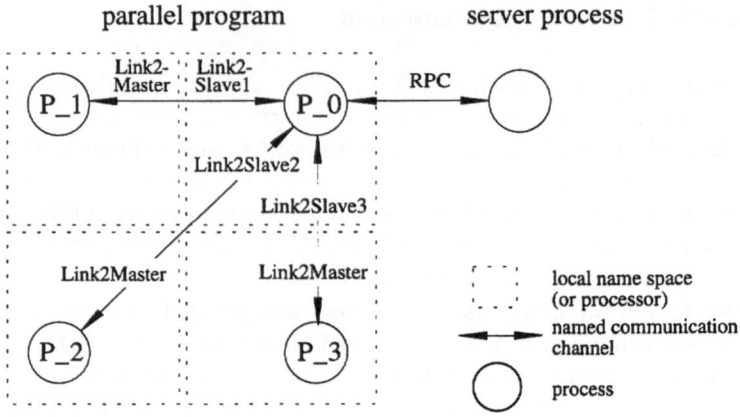

Fig. 4. Communication model of PARIX

4.4   Programming Model

Local Memory. Each processor has its own local memory which is not shared by other processors in the system. Access to remote memory requires data transfer that has to be explicitly initiated.

Main Program Loaded onto all Nodes. Independent of the number of requested processors it is sufficient to write only one program. PARIX includes an initial booting and loading mechanism which distributes an identical main program to all nodes of an allocated partition. Therefore there is no need for explicit mapping instructions that describe how to assign processes to processors. As explained later the user may influence this process indirectly by writing his program in such a manner that the software communication network is optimally assigned to the hardware network.

Identification of the Individual Position Within the Network. The PARIX run-time system initializes a set of global data (called root structure in PARIX terminology) kept on each node. The root structure provides several data about the node position within the 2-d grid, the size of whole partition, etc. The most important information stored in the root structure is a unique processor number, called *ProcID* in the range 0 to (number of processors-1). Based on the *ProcID* the user may assign different instructions to certain processors.

In this way it is possible to branch in different program sections (multiple instruction multiple data) or even load new executables (multiple program multiple data).

...
```
        /* Determine the number of this processor */
int myprocid = GETROOT()->ProcRoot->MyProcID;

if (myprocid == 0)
        do_this(); /* Processor zero executes this part */
else
        do_that(); /* all other processors execute this part */
```
...

Virtual Links. A virtual link is a bidirectional virtual channel between two processes for communication. The links are called virtual because there is no need for a direct hardware connection between the communicating processors. This concept overcomes the restrictions given by the presence of a limited number of physical links per processor. There is no difference between processes which are located on the same processor or on two arbitrary processors.

With the knowledge (obtained from the root structure) about the positions of processors in the physical 2D grid of the requested partition, the user can build up the optimal virtual links to achieve the highest communication performance (e.g., Fig. 5). But this is a question of optimization and should interest only the "power user". Usually the user defines his communication network to be hardware independent.

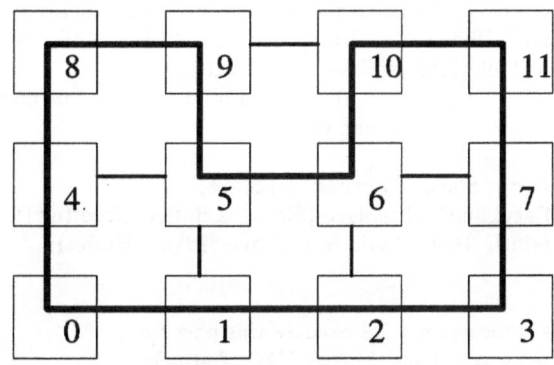

Fig. 5. Optimized mapping of a logical ring on a partition

4.5   An Example, PARIX says "Hello World"

```
/* further explanations to this program follow later */
#include <sys/root.h>
#include <sys/link.h>
```

```c
#include <stdio.h>
#include < stdlib.h>
#include <string.h>
#define TAG 17

int main (void)
{
int nProcs;
int myID;
int Slave, Buffer, error;

        /* LinkCB_t (Link Control Block) is a PARIX specific data */
        /* structure to hold information about a virtual link */
LinkCB_t *Link2Master;
LinkCB_t *Links2Slaves[4];

myID = GET_ROOT()->ProcRoot->MyProcID; /* Who am I */
nProcs = GET_ROOT()->ProcRoot->nProcs; /* How many procs */

if (myID > 3) return; /* all processors with myID > 3 say good bye */

if (myID == 0) { /* Master , processor 0 executes this part */
        printf("Hello World from Master\n ");
        if (nProcs > 4)
                nProcs = 4; /* adaption of nProcs, if there are released processors */
        for (Slave=1; Slave < nProcs; ++Slave) {
                        /* Build up the links to the slave processors */
                Links2Slaves[Slave] = ConnectLink(Slave, TAG, &error);
                if (Links2Slaves[Slave] == NULL)
                        exit(1); /* You should print out an error message before you
                                doing this */
        }
        for (Slave=1; Slave < nProcs; ++Slave) {
                RecvLink(Links2Slaves[Slave], &Buffer, sizeof(int));
                printf("Hello World from Slave %d\n ", Buffer);
        }
}
else { /* Slaves, processors 1,2,3 execute this part */
        Link2Master = ConnectLink(0, TAG, &error);
        if (Link2Master == NULL)
                exit(1); /* You should print out an error message before
                                you doing this */
        SendLink(Link2Master, &myID, sizeof(int));
}
return;
}
```

First, all processors find out which number they have and how many processors were requested. Suppose we are using only four processors. For that reason

all processors with *myID* greater than 3 will be released from work (0,1,2,3 are still in the race). The further execution depends on *myID*. Processor zero, called the master, branches into the true section (the range between *if(myID == 0)* and *else)*. The remaining processors, called slaves, execute the instructions between *else* and *return*.

In order to comunicate with all slaves the master has to establish three links, one each to processor 1,2, and 3. This is done by calling *ConnectLink()* which is executed in a loop. *ConnectLink()* uses as arguments the destination processor (1..3), an identifier (*TAG*), and an error variable. The identifier *TAG* is used to distinguish different links to the same processor. On the master side the links are placed in an array of link control blocks (*LinkCB_t*) and labeled with *Link2Slave[1]*, *Link2Slave[2]*, *Link2Slave[3]*. The entry *Link2Slave[0]* remains unused. Each slave builds up one link to the master. They all name the connection *Link2Master*. If no errors occur the establishment of all links is complete at this point of time. Now the processes are ready to communicate.

After that all slaves send their identifier (*myID*) to the master by calling *SendLink()*. The arguments of *SendLink()* are the named communication channel *Link2Master*, the address of a buffer containing the data to be sent, and the number of bytes of the message. By calling *RecvLink()* the master receives the incoming messages. The slaves are served in order of their processor number. The communication happens synchronously. Both partners have to be ready for transfer data. If one of the processes is busy the other waits until the first is ready.

Finally the master prints a message that obtains the received processor number.

The output should be something like that :

```
Hello World from Master
Hello World from Slave 1
Hello World from Slave 2
Hello World from Slave 3
```

5   Programming Heterogenous Workstation Clusters Using MPI

5.1   Introduction

Message passing is a widely-used paradigm for writing parallel applications. But meanwhile there exist nearly as many variants to this paradigm as different hardware platforms, because each vendor offers his own communication primitives. As a result it is very difficult to port an application written for a particular system to another manufacturer's hardware. In order to solve this problem numerous attempts have been made to propose a standard (Express, PVM, NX/2). For different reasons none of these interfaces has achieved the breakthrough of being acknowledged as a widely accepted standard. But even the availability of a general standard is an important key to growing acceptance of parallel computers in industry and research.

The required standardization process started with a workshop in 1992. Most of the major vendors of concurrent computers and researchers from universities, government laboratories, and industry were involved in the effort of developing a message passing interface standard, called MPI. In November 1993 the working group presented the draft MPI standard at the Supercomputing 93 conference.

The main goal stated by the MPI forum is [7]:

> *"to develop a widely used standard for writing message passing programs. As such the interface should establish a practical, portable, efficient, and flexible standard for message passing".*

Other goals are:

- to allow efficient communication (memory-to-memory copying, overlap of computation und communication),
- to allow for implementations that can be used in heterogenous environments,
- to design an interface that is not too different from current practice, such as PVM, Express.

Based on the message passing paradigm the MPI standard is suitable for developing programs for distributed memory machines, shared memory machines, networks of workstations, and combinations of these. Because the MPI forum only defines the interfaces and the contents of message passing routines, everyone may develop his own implementation. All further explanations are related to the specific MPI implementation of Argonne National Laboratory/Mississippi State University MPICH.

5.2   Basic Structure of MPICH

Each MPI application can be seen as a collection of concurrent processes. In order to use MPI functions the application code is linked with a static library provided by the MPI software package. The library consists of two layers. The upper layer comprises all MPI functions that have been written hardware independent. The lower layer is the native communication subsystem on parallel machines or another message passing system, such as p4 or PVM. p4 offers less functionality than MPI but supports a wide varity of parallel computer systems. The MPI layer accesses the p4 layer through an abstract device interface. In this way all hardware dependencies will be kept out of the MPI layer and the user code.

Processes with identical codes running on the same machine are called clusters in p4 terminology. p4 clusters are not visible to an MPI application. In order to achive peak performance, p4 uses shared memory (if available) for all processes in the same cluster. Special message passing interfaces are used for processes connected by such an interface. All processes have access to the socket interface which is a de facto standard for all UNIX machines. Because the design for data transmission in global networks lacks the performance required for high-performance computers, those provide an additional high-speed message passing interface (see Fig. 6). Using this structure MPI covers a wide range of platforms, from real parallel machines up to networks of workstations [6] (Fig. 6).

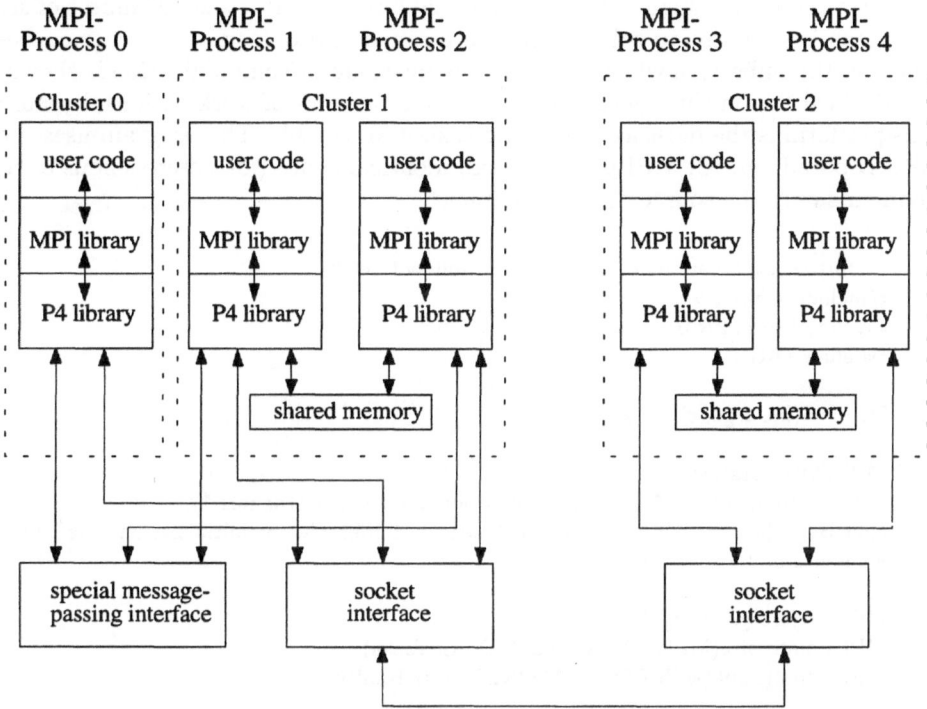

Fig. 6. Basic structure of the MPI implementation MPICH

5.3   What Is Included in MPI?

- Point-to-point communication
- Collective operations
- Process groups
- Communication contexts
- Process topologies
- Bindings for Fortran77 and C
- Environmental management and inquiry
- Profiling interface

5.4   What Does the Standard Exclude?

- Explicit shared-memory operations
- Support for task management
- Parallel I/O functions

5.5   MPI Says "Hello World"

MPI is a complex system that comprises 129 functions. But a small subset of six functions is sufficient to solve a moderate range of problems. The program below uses this subset, in which only basic point-to-point communication is shown. We decided to run this short application on a network of workstations, because this platform is the most available and easiest accessable. The program uses the SPMD paradigm. All MPI processes run identical codes (for heterogenous environments only the source code is common).

```c
/* further explanations to this program follow later */
#include <mpi.h> ...
#define MASTER 0
#define TAG 1

int main(int argc, char *argv[])
{
MPI_Status status;
char Hostname[81]; /* This string contains later the hostnames */
char Buffer[81] = "Me"; /* a communication buffer with preinitialization "Me" */
int myRank, nTasks, Slave;

MPI_Init(&argc, &argv);
MPI_Comm_size(MPI_COMM_WORLD, &nTasks);
MPI_Comm_rank(MPI_COMM_WORLD, &myRank);

gethostname(Hostname, 80);

if (myRank == MASTER)
        for (Slave = 1; Slave < nTasks; Slave++)
                MPI_Send(Hostname,    80,    MPI_CHAR,    Slave,    TAG,
MPI_COMM_WORLD);
else
        MPI_Recv(Buffer, 80, MPI_CHAR, MASTER,TAG, MPI_COMM_WORLD,
                        &status);

printf ("Hello World from Host %s\t rank %d : Master is %s\n",
                        Hostname, myRank, Buffer);

MPI_Finalize();
return 0; }
```

The details of compiling this program depend on the systems you have. MPI does not include a standard for how to start the MPI processes. However all current MPI implementations do provide a programming environment for configuring and starting MPI applications. Under MPICH the best way to describe ones own parallel virtual machine is given by using a configuration file, called a process group file. On a heterogenous network, which requires different executables, it is the only possible way. The process group file contains the machines (first entry), the number of processes to start (second entry) and the full path of the executable programs.

Example process group file *hello.pg* (Explanations below)

```
sun_a    0    /home/jennifer/sun4/hello
sun_b    1    /home/jennifer/sun4/hello
ksr1     3    /home/jennifer/ksr/ksrhello
```

Supposing we call the application *hello*, the process group file should be named *hello.pg*. To run the whole application it suffices to call *hello* on workstation *sun_a*, which serves as a console. A start-up procedure interprets the process group file and starts the specified processes.

```
sun_a > hello
```

The file above specifies five processes, one on both SUN workstations and three on a KSR1 virtual shared memory multiprocessor machine. By calling *hello* on the console (in this case *sun_a*) one process is started by the user directly. For this reason the first line of the process group file contains as number of (additional) processes the entry zero to start on every workstation just one process.

The output should be :

```
Hello World from Host sun_a      rank 0 : Master is Me
Hello World from Host sun_b      rank 1 : Master is sun_a
Hello World from Host ksr1       rank 2 : Master is sun_a
Hello World from Host ksr1       rank 3 : Master is sun_a
Hello World from Host ksr1       rank 4 : Master is sun_a
```

This program demonstrates the most common method for writing MIMD programs. Different processes, running on different processors, can execute different program parts by branching within the program based on an identifier. In MPI this identifier is called rank.

MPI Framework. The functions *MPI_Init()* and *MPI_Finalize()* build the framework around each MPI application. *MPI_Init()* must be called before any other MPI function may be used. After a program has finished its MPI specific part, the call of *MPI_Finalize()* takes care for a tidy clean up. All pending MPI activities will be canceled.

Who Am I, How Many Are We? As earlier mentioned MPI processes are represented by a rank. The function *MPI_Comm_rank()* returns this unique identifier, which simply is a nonnegative integer in range 0 ... (number of processes-1). To find out the total number of processes, MPI provides the function *MPI_Comm_size()*. Both *MPI_Comm_rank()* and *MPI_Comm_size()* use the parameter *MPI_COMM_WORLD*, which marks a determined process scope, called a communicator.

The communicator concept is one of the most important of MPI and distinguishes this standard from other existing message passing interfaces. Communicators provide a local name space for processes and a mechanism for encapsulating communication operations to build up various separate communication "universes". That means a pending communication in one communicator never influences a data transfer in another communicator. The inital communicator *MPI_COMM_WORLD* contains all MPI processes started by the application. Based on *MPI_COMM_WORLD* other communicators may be derivated.

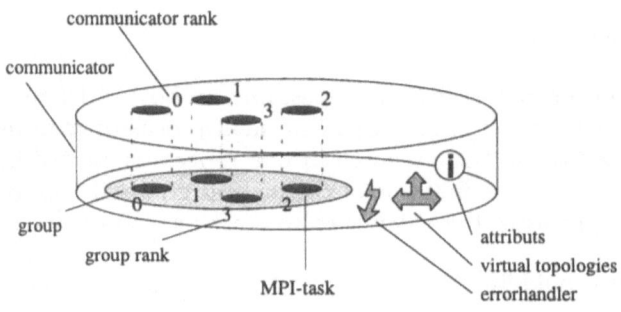

Fig. 7. Basic communicator structure

In a transfered sense it would be possible to consider a communicator as a cover around a group of processes (Fig. 7). A communication operation always specifies a communicator. All processes involved in a communication operation have to be described by their representation on the top side of the cover (communicator rank). There are some other MPI concepts such as virtual topologies and user-defined attributes which may be coupled to a communicator. MPI does not support a dynamic process concept. After start up MPI provides no mechanism to spawn new processes and integrate them into a running application.

Sending/Receiving Messages. A MPI message consists of a data part and a message envelope. The data part is specified by the first three parameters of

MPI_Send()/MPI_Recv() which describe the location, size and datatype of the send buffer. The entire data transfer is based on MPI datatypes which correspond to the basic datatypes of the supported languages. In this example *MPI_CHAR* is used which matches with Char in C. The message envelope describes destination, tag, and communicator of the message. The tag argument can be used to distinguish different types of messages. By using tags the receiver can select particular messages. In this example the master, which is process zero, sends his host name to all other processes, called slaves. The slaves receive this string by using *MPI_Recv()*. After communication is finished all processes print their "Hello World" that appear on the MPI console (Host sun_a). An easier way of obtaining the same result is given by using a broadcast operation.

5.6   Current Available Implementations of MPI

- MPICH implementation from Argonne National Laboratory/Mississippi State University
- LAM implementation from the Ohio Supercomputing Center
- CHIMP implementation from Edinburgh Parallel Computing Center
- UNIFY implementation from Mississippi State University (subset of MPI)

6   Summary

This chapter presented some elementary facts about parallel programming of three widely used parallel architectures. The discussed concepts and models reflect the main trends in the development and application of parallel machines. The reader experiences the ways which are used to explain parallelism on specific architectures and how to write the first parallel programs for shared memory and distributed memory machines and for heterogenous environments. Some characteristic examples help in giving a first view for parallel programming.

References

[1] S. Burkhardt et al., *Parallele Rechnersysteme–Programmierung und Anwendung*, (Verlag Technik, Berlin 1993)
[2] R. Calkin et al., *Portable Programming with the PARMACS Message-Passing Library*, (Parallel Computing **20**, 1994)
[3] A. Geist et al., *PVM 3.0 User's Guide and Reference Manual*, (Oak Ridge National Laboratory, Tennessee Feb. 93)
[4] C.A.R. Hoare, *Communicating Sequential Processes*, (Prentice Hall, Englewood Cliffs 1985)
[5] Kendall Square Research, *KSR Parallel Programming*, User Manual, (Kendall Square Research Corporation, Waltham Massachusetts 1993)
[6] J. Meyer, *Message-Passing Interface for Microsoft Windows 3.1*, (University of Nebraska, Omaha 1994)

[7] Message Passing Interface Forum, *A Message-Passing Interface Standard* , Technical Report CS-94-230, Computer Science Dept., (University of Tennessee, Knoxville 1993)

[8] B. Mukherje, K. Schwan, P. Gopinath, *A Survey of Multiprocessor Operating System Kernels*, (College of Computing Georgia Institute of Technology, Atlanta 1993)

[9] P.S. Pacheco, *Programming Parallel Processors Using MPI*, (University of San Francisco, San Francisco 1995)

[10] Parsytec, *Software Documentation PARIX 1.2 for PowerPC*, (Parsytec, Aachen 1994)

[11] W. Rehm, *Parallelrechner und Parallelprogrammierung*, (Technische Universität Chemnitz-Zwickau 1994)

[12] A.J. van der Steen, *An Overview of (Almost) Available Parallel Systems*, (Stiching Nationale Computer Faciliteiten, Gravenhagen 1993)

Index

Springer-Verlag
and the Environment

We at Springer-Verlag firmly believe that an international science publisher has a special obligation to the environment, and our corporate policies consistently reflect this conviction.

We also expect our business partners – paper mills, printers, packaging manufacturers, etc. – to commit themselves to using environmentally friendly materials and production processes.

The paper in this book is made from low- or no-chlorine pulp and is acid free, in conformance with international standards for paper permanency.